Mathematical Theory of
ELASTICITY

The quality of the materials used in the manufacture of this book is governed by continued postwar shortages.

Mathematical Theory of
ELASTICITY

BY

I. S. SOKOLNIKOFF

*Professor of Mathematics, University
of California at Los Angeles*

WITH THE COLLABORATION OF

R. D. SPECHT

*Assistant Professor of Mathematics
University of Wisconsin*

First Edition
Third Impression

New York   *London*
McGRAW-HILL BOOK COMPANY, INC.
1946

MATHEMATICAL THEORY OF ELASTICITY

Copyright, 1946, by the
McGraw-Hill Book Company, Inc.

PRINTED IN THE UNITED STATES OF AMERICA

*All rights reserved. This book, or
parts thereof, may not be reproduced
in any form without permission of
the publishers.*

PREFACE

The second half of the nineteenth century and the first two decades of the twentieth are likely to be characterized by a mathematical historian of the future as the age of specialists. It is to be expected that a specialist, conditioned in microscopic habits of thought, should set up barriers and overlook the vitalizing influence on mathematics of the real world about him. The present age might well turn out to be the period of transition that will dissipate the notion that the mathematics tinged with reality is an inferior type of intellectual endeavor. (Must one call to mind Euler, James and Daniel Bernoulli, Lagrange, Cauchy, or Poincaré?) Indeed, signs that the period of transition has already set in can be seen in the upsurge of interest in physical mathematics in America and, even more strikingly, in Russia.

This is a book on the mathematical theory of elasticity, and as such it is not concerned with the world of abstraction. The theory of elasticity, in its broad aspects, deals with a study of the behavior of those substances which possess the property of recovering their size and shape when the forces producing deformations are removed. This elastic property is possessed to some extent by all solid bodies, and the prime concern of the mathematical theory of elasticity is to reduce to calculations the stresses and strains in an elastic body subjected to the action of a system of prescribed external forces.

The first three chapters of this book, despite their brevity, contain a comprehensive treatment of the underlying theory of the mechanics of deformable media. Compactness and, it is hoped, clarity are achieved in part by the use of tensor notation. This is a departure from the practice commonly followed in books on applied mathematics (with the exception of treatises on relativity), but no apology need be made for it. The increasing use of tensor calculus in research literature in elasticity, especially in the theory of shells, makes it mandatory to master this elegant and powerful shorthand. The insistence on tensor notation, however, is not carried too far. In specific applications of the theory (such as those contained in Chaps. IV and V), where the economy of thought achieved by tensor symbolism is in some doubt, it is dropped in favor of the customary scalar notation. Thus, a reader familiar with the bases of the theory may proceed directly to Chap. IV and not be hampered by the lack of familiarity with the abridged notation.

Chapter IV gives an up-to-date treatment of extension, torsion, and flexure of homogeneous beams. Certain general methods of approach to the problem of Saint Venant, which make use of conformal representation and were developed mainly by Russian elasticians, are given in detail and are illustrated by numerous applications to problems of technical importance. Bibliographical references are extensive and are believed to contain all significant recent results in the theory of homogeneous and isotropic beams.

Chapter V contains a development of variational methods necessary for the treatment of problems of elasticity. Several procedures for deducing approximate solutions of the boundary-value problems of mathematical physics are outlined and illustrated by their application to the torsion and flexure problems. This chapter includes a derivation of several energy and reciprocity theorems, and a discussion of the Rayleigh-Ritz, Galerkin, and Trefftz methods of obtaining approximate solutions of the boundary-value problems. A brief discussion of the method of finite differences is given.

It is hoped to supplement this book by a companion volume dealing with the two-dimensional problems of elasticity and presenting a systematic treatment of the theory of plates and shells, based on the fundamental differential equations of the theory rather than (as is customary) on a set of special assumptions peculiar to the class of specific problems under consideration.

This volume had its origin in a series of lectures that I gave in 1941 and 1942 in the Program of Advanced Instruction and Research in Mechanics, conducted by the Graduate School of Brown University. In these lectures I stressed the contribution to the theory by the Russian school of elasticians and, in particular, the relatively little known work of great elegance and importance by N. I. Muschelišvili. Suitable acknowledgments to sources are made throughout this volume, but my chief debt is to Prof. N. I. Muschelišvili, member of the U.S.S.R. Academy of

Sciences, whose delightful book on "Some Fundamental Problems of the Theory of Elasticity" (in Russian), published in 1933, has left an indelible imprint.

My colleague and former student, Prof. R. D. Specht, has ably assisted me in the preparation of my mimeographed notes for publication and is responsible for organizing the material included in Chap. V. His help was invaluable.

The author would also like to acknowledge a grant-in-aid from the Wisconsin Alumni Research Foundation, without which the publication of this volume would have been delayed.

I. S. SOKOLNIKOFF.

MADISON, WIS.,
   January, 1946.

CONTENTS

PREFACE v

Chapter I—ANALYSIS OF STRAIN

1. Introduction.. 1
2. Affine Transformation... 2
3. Infinitesimal Affine Deformation...................................... 5
4. A Geometrical Interpretation of the Components of Strain..... 8
5. Strain Quadric of Cauchy.. 11
6. Principal Strains. Invariants.. 13
7. General Infinitesimal Deformation.................................... 18
8. Examples of Strain... 21
9. Notation.. 23
10. Equations of Compatibility.. 24
11. Finite Deformations... 28

Chapter II—ANALYSIS OF STRESS

12. Body and Surface Forces... 35
13. Stress Tensor.. 37
14. Note on Notation and Units.. 40
15. Equations of Equilibrium.. 41
16. Transformation of Coordinates... 44
17. Stress Quadric of Cauchy.. 46
18. Maximum Normal and Shear Stresses..................................... 52
19. Examples of Stress.. 54

Chapter III—STRESS-STRAIN RELATIONS

20. Hooke's Law.. 57
21. Generalized Hooke's Law... 59
22. Homogeneous, Isotropic Bodies... 65
23. Elastic Moduli of Isotropic Bodies.................................... 67
24. Equilibrium Equations for an Isotropic Elastic Solid.......... 72

25. Dynamical Equations of an Isotropic Elastic Solid........... 82
26. The Strain-energy Function and Its Connection with Hooke's Law.. 83
27. Uniqueness of Solution of the Boundary-value Problems of Elasticity.. 92
28. Saint-Venant's Principle..................................... 95

Chapter IV—EXTENSION, TORSION, AND FLEXURE OF HOMOGENEOUS BEAMS

29. Statement of Problem.. 97
30. Extension of Beams by Longitudinal Forces.................. 101
31. Beam Stretched by Its Own Weight........................ 104
32. Bending of Beams by Terminal Couples..................... 108
33. Torsion of a Circular Shaft.................................. 119
34. Torsion of Cylindrical Bars................................ 121
35. Stress Function... 127
36. Torsion of Elliptical Cylinder.............................. 134
37. Simple Solutions of the Torsion Problem. Effect of Grooves.. 139
38. Torsion of a Rectangular Beam and of a Triangular Prism.... 143
39. Complex Form of Fourier Series............................ 150
40. Summary of Some Results of the Complex Variable Theory... 154
41. Theorem of Harnack....................................... 160
42. Formulas of Schwarz and Poisson.......................... 163
43. Conformal Mapping.. 165
44. Solution of the Torsion Problem by Means of Conformal Mapping.. 170
45. Applications of Conformal Mapping........................ 176
46. Membrane and Other Analogies............................ 187
47. Torsion of Hollow Beams.................................. 191
48. Curvilinear Coordinates................................... 197
49. Torsion of Shafts of Varying Circular Cross Section........... 205
50. Local Effects... 209
51. Torsion of Nonisotropic Beams............................. 212
52. Flexure of Beams by Terminal Loads....................... 217
53. Bending by a Load along a Principal Axis.................. 228
54. The Displacement in a Bent Beam......................... 230
55. Flexure of Circular Beams................................. 235
56. Bending of a Beam of Elliptical Cross Section............... 237

CONTENTS

57. Bending of Rectangular Beams.................................. 240
58. Conformal Mapping and the General Problem of Flexure; the Cardioid Section.. 242
59. Bending of Circular Pipe................................... 253
60. Stress Function and Analogies; Beams of Equilateral Triangular Section.. 255
61. Technical Theory of the Bending of Beams.................. 263
62. Some Further Developments in Problems on Beams.......... 273

Chapter V—VARIATIONAL METHODS

63. Introduction.. 277
64. Minimum Potential Energy................................. 278
65. Variational Problem and Euler's Equation.................. 287
66. An Application of the Theorem of Minimum Potential Energy 291
67. Theorems of Work and Reciprocity.......................... 297
68. The Rayleigh-Ritz Method.................................. 304
69. Galerkin's Method... 313
70. The Error Function.. 316
71. Estimates of Error in the Minimal Integral and the Stress Function for Torsion...................................... 323
72. Relaxation of the Boundary Conditions..................... 329
73. The Method of Finite Differences.......................... 334

APPENDIX. Summary of Formulas................................. 347

AUTHOR INDEX.. 365

SUBJECT INDEX... 369

MATHEMATICAL THEORY OF ELASTICITY

CHAPTER I

ANALYSIS OF STRAIN

1. Introduction. A continuous medium is said to be *strained* whenever the relative positions of points in the medium are altered. The change in the position of the material points, when accompanied by a change in distance between them, is called *deformation*.

Let a body τ be referred to a system of orthogonal, rectilinear coordinate axes x_1, x_2, x_3, and let $P(x_1, x_2, x_3)$ represent some point of the body τ when the latter is unstrained. After the deformation has taken place, the coordinates of the same material point will be denoted by (x'_1, x'_2, x'_3), and it will be assumed that the variables x'_1, x'_2, x'_3 are continuous functions of the variables x_1, x_2, x_3 throughout the region τ. We shall write

$$(1.1) \qquad x'_i = x'_i(x_1, x_2, x_3) \equiv x'_i(x), \quad (i = 1, 2, 3).$$

Equations (1.1) can be looked upon as the equations of a transformation of space, so that the region τ, occupied by the body in the x_i-space, is mapped into some region τ' in the x'_i-space. It will be more convenient for our purpose to regard Eqs. (1.1) as those of a transformation of points taking place in the same space, and this point of view will be adhered to in the following unless an explicit statement to the contrary is made.

Physical considerations demand that there be a one-to-one correspondence between the points (x_1, x_2, x_3) and (x'_1, x'_2, x'_3), which means that Eqs. (1.1) must possess a single-valued inverse, so that

$$(1.2) \qquad x_i = x_i(x'_1, x'_2, x'_3), \quad (i = 1, 2, 3),$$

where the functions involved are continuous throughout the deformed region.

Part of the transformation defined by Eqs. (1.1) may represent rigid body motions (that is, translations and rotations) of the body as a whole. This part of the deformation leaves unchanged the length of every vector joining a pair of points within the body and is of no interest in the analysis of strain. The remaining part of the transformation (1.1) will be called *pure deformation*. It will be important to learn how to distinguish between pure deformations and rigid body motions when the latter are present in the equation of transformation (1.1). To this end we shall consider first the simplest case of (1.1), that in which the functions appearing therein are linear functions of the coordinates x_1, x_2, x_3. The Eqs. (1.1) in which the functions are linear define what is called an *affine transformation*.

2. Affine Transformations. The properties of the general linear transformation of points,

$$\begin{aligned} x_1' &= \alpha_{10} + (1 + \alpha_{11})x_1 + \alpha_{12}x_2 + \alpha_{13}x_3, \\ x_2' &= \alpha_{20} + \alpha_{21}x_1 + (1 + \alpha_{22})x_2 + \alpha_{23}x_3, \\ x_3' &= \alpha_{30} + \alpha_{31}x_1 + \alpha_{32}x_2 + (1 + \alpha_{33})x_3, \end{aligned}$$

or, written more compactly,[1]

$$(2.1) \qquad x_i' = \alpha_{i0} + (\delta_{ij} + \alpha_{ij})x_j, \quad (i, j = 1, 2, 3),$$

where the coefficients α_{ij} are constants, are well known. Since physical considerations demand the existence of an inverse, Eqs. (2.1) must be solvable for the variables x_1, x_2, x_3 as functions of x_1', x_2', x_3'. It follows that the determinant $|\delta_{ij} + \alpha_{ij}|$ of the coefficients of the unknowns entering into the right-hand member of (2.1) must not vanish. It is obvious that the inverse transformation

$$(2.2) \qquad x_i = \beta_{i0} + (\delta_{ij} + \beta_{ij})x_j', \quad (i, j = 1, 2, 3),$$

is likewise linear.

---

[1] A repeated subscript indicates summation as the index that is repeated takes the values 1, 2, 3. Thus

$$\alpha_{ij}x_j = \alpha_{i1}x_1 + \alpha_{i2}x_2 + \alpha_{i3}x_3.$$

The symbol δ_{ij}, the Kronecker delta, is defined to have the value one if i equals j, zero if i differs from j. The reason for writing the coefficients of x_1, x_2, and x_3 in the first, second, and third lines as $1 + \alpha_{11}$, $1 + \alpha_{22}$, $1 + \alpha_{33}$ will appear later.

It is easy to see from (2.1) and (2.2) that an affine transformation carries planes into planes, and hence a rectilinear segment joining the points $P^0(x_1^0, x_2^0, x_3^0)$ and $P(x_1, x_2, x_3)$ is transformed into a rectilinear segment joining the corresponding points $P^{0\prime}(x_1^{0\prime}, x_2^{0\prime}, x_3^{0\prime})$ and $P'(x_1', x_2', x_3')$ (Fig. 1). This follows from the fact that the rectilinear segment $\overline{P^0P}$ can be thought of as joining two points P^0 and P on the intersection of two planes S_1 and S_2; under the transformation (2.1) points P^0 and P go over into points $P^{0\prime}$ and P', which lie on the intersection of the planes S_1' and S_2', into which the planes S_1 and S_2 are carried by the transformation.

We shall denote the unit base vectors, directed along the coordinate axes x_1, x_2, and x_3, by e_1, e_2, and e_3, respectively.

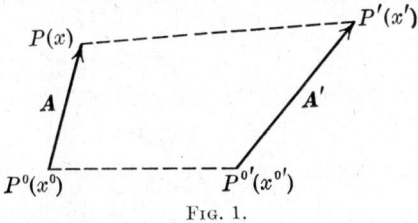

Fig. 1.

Thus, a vector A whose components along the coordinate axes are A_1, A_2, A_3 can be written as

$$A = e_1 A_1 + e_2 A_2 + e_3 A_3 \equiv e_i A_i, \quad (i = 1, 2, 3).$$

Since the vector $A = e_i A_i$ is uniquely determined once its components A_i ($i = 1, 2, 3$) are prescribed, we can represent the vector A by the symbol A_i. Under the transformation (2.1) the vector $A_i = x_i - x_i^0$, joining the points $P^0(x^0)$ and $P(x)$, is carried into another vector $A_i' = x_i' - x_i^{0\prime}$. In general, vectors A_i and A_i' differ in direction and magnitude. From (2.1), which we write in the form

$$x_i' = \alpha_{i0} + x_i + \alpha_{ij} x_j,$$

we have

$$A_i' = x_i' - x_i^{0\prime} = (\alpha_{i0} + x_i + \alpha_{ij} x_j) - (\alpha_{i0} + x_i^0 + \alpha_{ij} x_j^0)$$
$$= (x_i - x_i^0) + \alpha_{ij}(x_j - x_j^0) = A_i + \alpha_{ij} A_j,$$

or

(2.3) $$\delta A_i \equiv A_i' - A_i = \alpha_{ij} A_j, \quad (i, j = 1, 2, 3).$$

It is clear from (2.3) that two vectors A_i and B_i whose components are equal transform into two vectors A'_i and B'_i whose components are again equal. Also two parallel vectors obviously transform into parallel vectors. Hence, two equal and similarly oriented rectilinear polygons located in different parts of the region τ will be transformed into two equal and similarly oriented polygons in the transformed region τ'. Thus, the different parts of the body τ, when the latter is subjected to the transformation (2.1), experience the same deformation independently of the position of the parts of the body. For this reason, the deformation characterized by (2.1) is called a *homogeneous deformation*.

Consider the transformation (2.1) and let the variables x'_i be subjected to another affine transformation,

(2.4) $$x''_k = \gamma_{k0} + (\delta_{ki} + \gamma_{ki})x'_i.$$

Recalling the definition of the Kronecker delta, we can write (2.4) as

$$x''_k = \gamma_{k0} + x'_k + \gamma_{ki}x'_i.$$

Let A''_k be the transform of the vector A'_k; then

$$A''_k \equiv x''_k - x^{0''}_k = (\gamma_{k0} + x'_k + \gamma_{ki}x'_i) - (\gamma_{k0} + x^{0'}_k + \gamma_{ki}x^{0'}_i)$$
$$= (x'_k - x^{0'}_k) + \gamma_{ki}(x'_i - x^{0'}_i) = A'_k + \gamma_{ki}A'_i,$$

or

(2.5) $$\delta A'_k \equiv A''_k - A'_k = \gamma_{ki}A'_i, \quad (i, k = 1, 2, 3).$$

The product of the two successive affine transformations (2.1) and (2.4) is equivalent to the single transformation obtained by substituting in (2.4) the values of x'_i in terms of x_j from (2.1.) Thus one has

$$x''_k = \gamma_{k0} + (\delta_{ki} + \gamma_{ki})[\alpha_{i0} + (\delta_{ij} + \alpha_{ij})x_j]$$
$$= \alpha_{k0} + \gamma_{k0} + (\delta_{kj} + \alpha_{kj} + \gamma_{kj})x_j$$
$$+ \alpha_{i0}\gamma_{ki} + \alpha_{ij}\gamma_{ki}x_j.$$

Now if the coefficients α_{ij} and γ_{ij} are so small that one is justified in neglecting their products in comparison with the coefficients themselves, then

$$x''_k = \alpha_{k0} + \gamma_{k0} + x_k + (\alpha_{kj} + \gamma_{kj})x_j.$$

The product transformation likewise carries the point (x^0_1, x^0_2, x^0_3) to the point $(x^{0''}_1, x^{0''}_2, x^{0''}_3)$ where

$$x^{0''}_k = \alpha_{k0} + \gamma_{k0} + x^0_k + (\alpha_{kj} + \gamma_{kj})x^0_j.$$

The vector $A_k = x_k - x_k^0$ is thus transformed into the vector

$$A_k'' = x_k'' - x_k^{0''} = (x_k - x_k^0) + (\alpha_{kj} + \gamma_{kj})(x_j - x_j^0)$$
$$= A_k + (\alpha_{kj} + \gamma_{kj})A_j,$$

or

(2.6) $\qquad \delta A_k \equiv A_k'' - A_k = (\alpha_{kj} + \gamma_{kj})A_j, \quad (j, k = 1, 2, 3).$

Thus, if one neglects products of the α_{ij} and γ_{ij}, then the coefficients in the resultant transformation (2.6) are obtained by adding the corresponding coefficients α_{ij} and γ_{ij} in the component transformations (2.3) and (2.5). In this event, it is said that the product transformation (2.6) is obtained by superposition of the original transformations. It is clear from the structure of the formulas (2.6) that the resultant transformation is independent of the order in which the transformations are performed. One of the chief sources of the difficulty that confronts one in the study of finite as distinguished from infinitesimal deformations arises from the fact that the principle of superposition of effects and the independence of the order of transformations are no longer valid.

A transformation of the type (2.1), in which the coefficients are so small that their products can be neglected in comparison with the linear terms, is called an *infinitesimal affine transformation*.

3. Infinitesimal Affine Deformations. In this section we shall be concerned with the problem of separating the infinitesimal affine transformation defined by Eq. (2.3),

(3.1) $\qquad \delta A_i \equiv A_i' - A_i = \alpha_{ij} A_j,$

into two component transformations: one of these corresponds to a rigid body motion; the other, which we have termed pure deformation, will be investigated in detail in the next section. We seek first the conditions on the coefficients α_{ij} if the deformation is to be one of rigid body motion (that is, one consisting of rotation and translation) alone.

A rigid body motion may be characterized by saying that the length

$$A = |A| = \sqrt{A_i A_i}$$

of any vector A is unchanged by the transformation. If we replace the A_i in this formula by $A_i + \delta A_i$ and denote the change in length A by δA, we get

(3.2) $\qquad A\, \delta A = A_i\, \delta A_i.$

plus terms of higher order in δA_i, which are neglected, since we are concerned with the infinitesimal affine transformation. When the expressions for δA_i given by (3.1) are inserted in (3.2), one finds that

$$A \, \delta A = \alpha_{ij} A_i A_j,$$

or when written out in full,

$$A \, \delta A = \alpha_{11} A_1^2 + \alpha_{22} A_2^2 + \alpha_{33} A_3^2 \\ + (\alpha_{12} + \alpha_{21}) A_1 A_2 + (\alpha_{23} + \alpha_{32}) A_2 A_3 + (\alpha_{31} + \alpha_{13}) A_3 A_1.$$

Since for a rigid body transformation δA vanishes for all values of A_1, A_2, A_3, we must have

$$\alpha_{11} = \alpha_{22} = \alpha_{33} = 0,$$
$$\alpha_{12} + \alpha_{21} = \alpha_{23} + \alpha_{32} = \alpha_{31} + \alpha_{13} = 0.$$

Hence the necessary and sufficient condition that the infinitesimal transformation (3.1) represent a rigid body motion is

(3.3) $\qquad\qquad\qquad \alpha_{ij} = -\alpha_{ji}, \qquad (i, j = 1, 2, 3).$

In this case, the set of quantities α_{ij} is said to be *skew-symmetric*. When the coefficients α_{ij} are skew-symmetric, the transformation (3.1) takes the form

$$\delta A_1 = \qquad\quad - \alpha_{21} A_2 \;\; + \alpha_{13} A_3,$$
$$\delta A_2 = \;\;\;\; \alpha_{21} A_1 \qquad\quad\; - \alpha_{32} A_3,$$
$$\delta A_3 = -\alpha_{13} A_1 + \alpha_{32} A_2$$

This transformation can be written as the vector product of the infinitesimal rotation vector $\boldsymbol{\omega} = \boldsymbol{e}_i \omega_i$ and the vector $\boldsymbol{A}$, namely

$$\delta \boldsymbol{A} = \boldsymbol{\omega} \times \boldsymbol{A} = \begin{vmatrix} \boldsymbol{e}_1 & \boldsymbol{e}_2 & \boldsymbol{e}_3 \\ \omega_1 & \omega_2 & \omega_3 \\ x_1 - x_1^0 & x_2 - x_2^0 & x_3 - x_3^0 \end{vmatrix},$$

if we take

(3.4) $\qquad \begin{cases} \omega_1 \equiv \alpha_{32} = -\alpha_{23} = \tfrac{1}{2}(\alpha_{32} - \alpha_{23}), \\ \omega_2 \equiv \alpha_{13} = -\alpha_{31} = \tfrac{1}{2}(\alpha_{13} - \alpha_{31}), \\ \omega_3 \equiv \alpha_{21} = -\alpha_{12} = \tfrac{1}{2}(\alpha_{21} - \alpha_{12}). \end{cases}$

The equations representing the rigid body motion can be obtained by observing that $A_i = x_i - x_i^0$ and that

$$\delta A_i = A_i' - A_i = (x_i' - x_i^{0'}) - (x_i - x_i^0)$$
$$= (x_i' - x_i) - (x_i^{0'} - x_i^0) = \delta x_i - \delta x_i^0$$

or $\quad\quad \delta x_i = \delta x_i^0 + \delta A_i = \delta x_i^0 + (\boldsymbol{\omega} \times \boldsymbol{A})_i.$

Then the rigid body portion of the infinitesimal affine transformation (2.1) can be written as

(3.5) $\quad \begin{cases} \delta x_1 = \delta x_1^0 \quad\quad\quad\quad\, - \omega_3(x_2 - x_2^0) + \omega_2(x_3 - x_3^0), \\ \delta x_2 = \delta x_2^0 + \omega_3(x_1 - x_1^0) \quad\quad\quad\, - \omega_1(x_3 - x_3^0), \\ \delta x_3 = \delta x_3^0 - \omega_2(x_1 - x_1^0) + \omega_1(x_2 - x_2^0). \end{cases}$

The quantities $\delta x_i^0 \equiv x_i^{0'} - x_i^0$ are the components of the displacement vector representing the translation of the point $P^0(x^0)$ (see Fig. 1), while the remaining terms of (3.5) represent rotation about the point P^0.

At the beginning of this section, we proposed the problem of separating the infinitesimal affine transformation $\delta A_i = \alpha_{ij} A_j$ into two component transformations, one of which is to represent rigid body motion alone; we have seen that this rigid body motion corresponds to a transformation in which the coefficients are skew-symmetric; that is, $\alpha_{jj} = -\alpha_{jj}$. Now any set of quantities α_{ij} may be decomposed into a symmetric and a skew-symmetric set in one, and only one, way.[1] We can thus write

$$\alpha_{ij} = \tfrac{1}{2}(\alpha_{ij} + \alpha_{ji}) + \tfrac{1}{2}(\alpha_{ij} - \alpha_{ji}).$$

Then Eq. (3.1) can be written as

$$\delta A_i = \alpha_{ij} A_j = [\tfrac{1}{2}(\alpha_{ij} + \alpha_{ji}) + \tfrac{1}{2}(\alpha_{ij} - \alpha_{ji})]A_j,$$

or

(3.6) $\quad\quad\quad \delta A_i = e_{ij} A_j + \omega_{ij} A_j,$

where $\quad\quad\quad\quad e_{ij} = \quad e_{ji} \equiv \tfrac{1}{2}(\alpha_{ij} + \alpha_{ji}),$
$\quad\quad\quad\quad\quad\quad \omega_{ij} = -\omega_{ji} \equiv \tfrac{1}{2}(\alpha_{ij} - \alpha_{ji}).$

The skew-symmetric coefficients ω_{ij} correspond to a rigid body motion, and from (3.4) it can be seen that they are connected with the components of rotation, $\omega_1, \omega_2, \omega_3$, by the relations

$$\omega_{32} = \omega_1, \quad\quad \omega_{13} = \omega_2, \quad\quad \omega_{21} = \omega_3.$$

---

[1] See Prob. 1 at the end of this chapter.

It is clear from Eqs. (3.6) for the transformation of the components of a vector that an infinitesimal affine transformation of the vector A_i can be decomposed into transformation $\delta A_i = \omega_{ij} A_j$, representing rigid body motion, and into transformation

(3.7) $$\delta A_i = e_{ij} A_j,$$

representing pure deformation.

The symmetric coefficients e_{ij} are called *components of the strain tensor*, and they characterize pure deformation. We shall investigate the properties of the strain tensor in the next section.

4. A Geometrical Interpretation of the Components of Strain. The geometrical significance of the components of strain e_{ij} entering into (3.7) can be readily determined by inserting the expressions (3.7) in the formula (3.2), which then takes the form $A \, \delta A = A_i \, \delta A_i = e_{ij} A_i A_j$, or

(4.1) $$\frac{\delta A}{A} = \frac{e_{ij} A_i A_j}{A^2}.$$

If initially the vector A is parallel to the x_1-axis, so that $A = A_1$ and $A_2 = A_3 = 0$, then it follows from (4.1) that

(4.2) $$\frac{\delta A}{A} = e_{11}.$$

Thus, the component e_{11} of the strain tensor represents the *extension* or *change in length per unit length* of a vector originally parallel to the x_1-axis.

Hence, if all components of the strain tensor with the exception of e_{11} vanish, then all unit vectors parallel to the x_1-axis will be extended by an amount e_{11} if this strain component is positive and contracted by the same amount if e_{11} is negative. In this event, one has a homogeneous deformation of material in the direction of the x_1-axis. A cube of material whose edges before deformation are l units long will become a rectangular parallelepiped whose dimensions in the x_1-direction are $l(1 + e_{11})$ units and whose dimensions in the directions of the x_2- and x_3-axes are unaltered. A similar significance can be ascribed to the components e_{22} and e_{33}.

In order to interpret geometrically such strain components as e_{23}, consider two vectors $\boldsymbol{A} = \boldsymbol{e}_2 A_2$ and $\boldsymbol{B} = \boldsymbol{e}_3 B_3$ (Fig. 2), initially directed along the x_2- and x_3-axes respectively. Upon deformation, these vectors become

$$A' = e_1\,\delta A_1 + e_2(A_2 + \delta A_2) + e_3\,\delta A_3,$$
$$B' = e_1\,\delta B_1 + e_2\,\delta B_2 \qquad\quad + e_3(B_3 + \delta B_3).$$

We denote the angle between A' and B' by θ and consider the change $\alpha_{23} = \dfrac{\pi}{2} - \theta$ in the right angle between A and B. From the definition of the scalar product of A' and B', we have

$$A'B'\cos\theta \equiv A'\cdot B' = \delta A_1\,\delta B_1 + (A_2+\delta A_2)\,\delta B_2 + (B_3+\delta B_3)\,\delta A_3$$
$$\doteq A_2\,\delta B_2 + B_3\,\delta A_3,$$

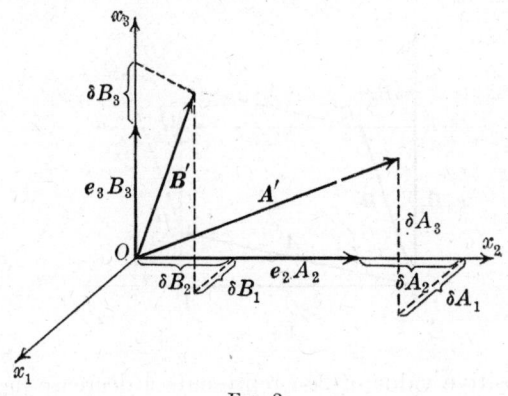

Fig. 2.

if we neglect the products of the changes in the components of the vectors A and B. To the same approximation, we have

$$(4.3)\quad \cos\theta = \frac{A'\cdot B'}{A'B'}$$
$$= \frac{A_2\,\delta B_2 + B_3\,\delta A_3}{\sqrt{(\delta A_1)^2+(A_2+\delta A_2)^2+(\delta A_3)^2}\,\sqrt{(\delta B_1)^2+(\delta B_2)^2+(B_3+\delta B_3)^2}}$$
$$\doteq (A_2\,\delta B_2 + B_3\,\delta A_3)\cdot(A_2+\delta A_2)^{-1}\cdot(B_3+\delta B_3)^{-1}$$
$$\doteq \frac{A_2\,\delta B_2 + B_3\,\delta A_3}{A_2 B_3} = \frac{\delta B_2}{B_3} + \frac{\delta A_3}{A_2}.$$

Since all increments in the components of A and B have been neglected except δA_3 and δB_2, the deformation can be represented

as shown in Fig. 3. If we remember that
$$A_1 = A_3 = B_1 = B_2 = 0,$$
then Eqs. (3.7) yield

(4.4) $\qquad \delta B_2 = e_{23} B_3, \qquad \delta A_3 = e_{23} A_2.$

From (4.3) we have
$$\cos \theta = \cos \left(\frac{\pi}{2} - \alpha_{23} \right) = \sin \alpha_{23}$$
$$\doteq \alpha_{23} = \frac{\delta B_2}{B_3} + \frac{\delta A_3}{A_2} = 2e_{23},$$
or
$$\alpha_{23} = 2e_{23}.$$

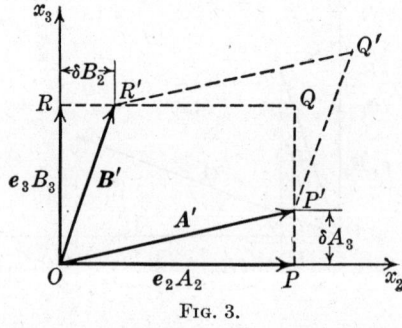

Fig. 3.

Hence a positive value of $2e_{23}$ represents a decrease in the right angle between the vectors A and B, which were initially directed along the positive x_2- and x_3-axes.

Again, from (4.4) and Fig. 3 we see that
$$\angle POP' \doteq \tan POP' = \frac{\delta A_3}{A_2} = e_{23},$$
$$\angle ROR' \doteq \tan ROR' = \frac{\delta B_2}{B_3} = e_{23}.$$

Since the angles POP' and ROR' are equal, it follows that by rotating the parallelogram $R'OP'Q'$ through an angle e_{23} about the origin, one can obtain the configuration shown in Fig. 4. Obviously it represents a slide or a shear of the elements parallel to the $x_1 x_2$-plane, where the amount of slide is proportional to the distance x_3 of the element from the $x_1 x_2$-plane.

A similar interpretation can obviously be made in regard to the components e_{12} and e_{31}.

It is clear that the areas of the rectangle and the parallelogram in Fig. 4 are equal. Likewise a body originally cubical is deformed into a parallelepiped, and the volumes of the cube and parallelepiped are equal if one disregards the products of the changes in the linear elements. Such deformation is called *pure shear*.

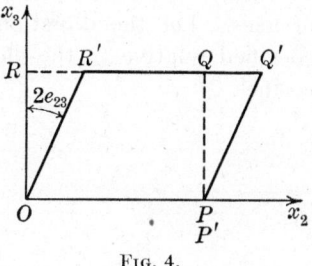

Fig. 4.

5. Strain Quadric of Cauchy. With each point $P^0(x_i^0)$ of a continuous medium, we shall associate a quadric surface, the *quadric of deformation*, which enables one to determine the elongation of any vector $\boldsymbol{A} = \boldsymbol{e}_i(x_i - x_i^0)$ that runs from the point $P^0(x^0)$ to some point $P(x)$.

Now if a local system of axes x_i is introduced, with origin at the initial point $P^0(x^0)$ of the vector $\boldsymbol{A}$ and with axes parallel to the space-fixed axes, then formula (4.1) characterizing the extension $e = \dfrac{\delta A}{A}$ of $\boldsymbol{A}$ can be written as

$$(5.1) \qquad eA^2 = e_{ij}x_ix_j.$$

We consider the quadratic function

$$(5.2) \qquad 2G(x_1, x_2, x_3) \equiv e_{ij}x_ix_j$$

and constrain the end-point $P(x)$ of the vector $\boldsymbol{A}$, as yet unspecified, to lie on the quadric surface

$$(5.3) \qquad 2G(x_1, x_2, x_3) = \pm k^2,$$

where k is any real constant and the sign is chosen so as to make the surface real. Comparison of (5.3) with (5.1) leads to the relation

$$(5.4) \qquad e = \pm \frac{k^2}{A^2},$$

and the strain quadric takes the form

$$(5.5) \qquad e_{ij}x_ix_j = \pm k^2.$$

From (5.4) we see that the extension of any line through $P^0(x^0)$ is inversely proportional to the square of the radius vector that runs along the line from the point (x^0), at which the strain is being studied, to a point (x) on the quadric surface.

We refer the *quadric surface of deformation* (5.5) to a new coordinate system x'_1, x'_2, x'_3, obtained from the old by a rotation of axes. Let the directions of the new coordinate axes x'_i be specified relative to the old system x_i by the table of direction cosines

	x_1	x_2	x_3
x'_1	l_{11}	l_{12}	l_{13}
x'_2	l_{21}	l_{22}	l_{23}
x'_3	l_{31}	l_{32}	l_{33}

in which l_{ij} is the cosine of the angle between the x'_i- and the x_j-axes. The old and the new coordinates are related by the equations

$$x_1 = l_{11}x'_1 + l_{21}x'_2 + l_{31}x'_3,$$
$$x_2 = l_{12}x'_1 + l_{22}x'_2 + l_{32}x'_3,$$
$$x_3 = l_{13}x'_1 + l_{23}x'_2 + l_{33}x'_3,$$

or, more compactly,

(5.6) $$x_i = l_{\alpha i}x'_\alpha.$$

It is readily shown that the inverse transformation is of the form[1]

$$x'_i = l_{i\alpha}x_\alpha.$$

The well-known orthogonality relations between the direction cosines can be written in the form

(5.7) $$l_{i\alpha}l_{j\alpha} = \delta_{ij}, \qquad l_{\alpha i}l_{\alpha j} = \delta_{ij}.$$

When the quadratic surface (5.5) is referred to the x'_i coordinate system, a new set of strains e'_{ij} is determined and (5.5) is replaced by the new equation of the surface, namely,

$$e'_{ij}x'_i x'_j = \pm k^2.$$

The right-hand member of (5.5), however, has a geometrical meaning that is independent of the choice of coordinate system ($\pm k^2 = eA^2$); consequently

(5.8) $$e_{ij}x_i x_j = e'_{ij}x'_i x'_j.$$

[1] See Prob. 5 at the end of this chapter.

In other words, the quadratic form $e_{ij}x_ix_j$ is invariant with respect to an orthogonal transformation of coordinates.

Equations (5.6) and (5.8) together yield

$$e_{ij}l_{\alpha i}l_{\beta j}x'_\alpha x'_\beta = e'_{\alpha\beta}x'_\alpha x'_\beta,$$

and, since the x'_α are arbitrary,

(5.9) $$e'_{\alpha\beta} = l_{\alpha i}l_{\beta j}e_{ij}.$$

Similarly it can be shown that

(5.10) $$e_{\alpha\beta} = l_{i\alpha}l_{j\beta}e'_{ij}.$$

A set of quantities e_{ij} transforming according to the law (5.9) is said to represent a *cartesian tensor of rank two*. We shall meet several such tensors in the subsequent discussion.

Differentiating $2G(x_1, x_2, x_3) = e_{ij}x_ix_j$ and noting from (3.7) that for a pure deformation $\delta A_i = e_{ij}A_j = e_{ij}x_j$, we find that

(5.11) $$\frac{\partial G}{\partial x_i} = e_{ij}x_j = \delta A_i.$$

But $\dfrac{\partial G}{\partial x_i}$ are the direction ratios of the normal ν to the quadric surface (5.5) at the point (x_i), and it follows that the vector $\delta \boldsymbol{A}$ is directed along the normal to the plane tangent to the surface $e_{ij}x_ix_j = \pm k^2$ (see Fig. 5). This property of the strain quadric will prove useful in the next section, where we discuss the principal axes of the quadric surface and their significance for the deformation.

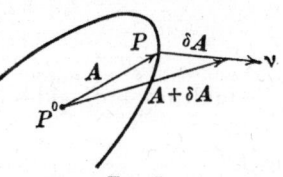

Fig. 5.

6. Principal Strains. Invariants. We seek now the direction ratios of the lines through (x^0) whose orientation is left unchanged by the deformation $\delta A_i = e_{ij}A_j$. If the direction of the vector $\boldsymbol{A}$ is not altered by the strain, then $\delta \boldsymbol{A}$ and $\boldsymbol{A}$ are parallel and their components are proportional. Therefore

$$\delta A_i = eA_i.$$

It should be noted that $e = \dfrac{\delta A_i}{A_i}$ is the extension of each component of $\boldsymbol{A}$ and is thus the extension of $\boldsymbol{A}$ itself, or $e = \dfrac{\delta A}{A}$.

Equation (5.1) then shows that the extension e is given by the expression $e = e_{ij}x_ix_j/A^2$. We return now to $\delta A_i = e_{ij}A_j$, from which it is seen that

(6.1) $$e_{ij}A_j = eA_i = e\,\delta_{ij}A_j$$

or

(6.2) $$(e_{ij} - e\,\delta_{ij})A_j = 0.$$

This set (6.2) of three homogeneous equations in the unknowns A_j possesses a nonvanishing solution if, and only if, the determinant of the coefficients of the A_j is equal to zero; that is,

(6.3) $$|e_{ij} - e\,\delta_{ij}| = 0,$$

or

$$\begin{vmatrix} e_{11} - e & e_{12} & e_{13} \\ e_{21} & e_{22} - e & e_{23} \\ e_{31} & e_{32} & e_{33} - e \end{vmatrix} = 0.$$

We prove next that the three roots e_1, e_2, e_3 of this cubic equation in the elongation e are all real.

Let the three directions determined by the three numbers e_j be given by the vectors[1] $\overset{j}{A}$. In this notation, formula (6.1) becomes for any root $e = e_1$,

$$e_1 \overset{1}{A}_j = e_{jk} \overset{1}{A}_k.$$

We multiply both sides by $\overset{2}{A}_j$ and sum over j, getting

(6.4) $$e_1 \overset{1}{A}_j \overset{2}{A}_j = e_{jk} \overset{1}{A}_k \overset{2}{A}_j.$$

Similarly, from $e_2 \overset{2}{A}_j = e_{jk} \overset{2}{A}_k$
we have

(6.5) $$e_2 \overset{1}{A}_j \overset{2}{A}_j = e_{jk} \overset{1}{A}_j \overset{2}{A}_k = e_{kj} \overset{1}{A}_k \overset{2}{A}_j = e_{jk} \overset{1}{A}_k \overset{2}{A}_j,$$

where j and k have been interchanged and the symmetry of e_{jk} exploited. Comparison of (6.4) and (6.5) shows that

(6.6) $$(e_1 - e_2)\overset{1}{A}_j \overset{2}{A}_j = 0.$$

[1] The index j over A indicates not the jth component but rather the jth vector and its dependence upon the root e_j of the determinantal equation (6.3).

§6 ANALYSIS OF STRAIN

Now if we assume tentatively that (6.3) has complex roots, then these can be written

$$e_1 = E_1 + iE_2, \quad e_2 = E_1 - iE_2, \quad e_3,$$

where E_1, E_2, e_3 are real. If $e_2 = E_1 - iE_2$ is substituted for e in (6.2), it will be found that the resulting solutions $\overset{2}{A}_j \equiv a_j - ib_j$ are the complex conjugates of $\overset{1}{A}_j \equiv a_j + ib_j$, where the latter are obtained by putting $e = e_1 = E_1 + iE_2$.

Therefore
$$\overset{1}{A}_j\overset{2}{A}_j = (a_j + ib_j)(a_j - ib_j)$$
$$= a_1^2 + a_2^2 + a_3^2 + b_1^2 + b_2^2 + b_3^2 \neq 0.$$

Hence it follows from (6.6) that $e_1 - e_2 \equiv 2iE_2 = 0$, or $E_2 = 0$, and the roots e_i are all real.

From (6.6) it follows that if the roots e_1 and e_2 are distinct, then

$$\overset{1}{A}_i\overset{2}{A}_i = \overset{1}{\mathbf{A}} \cdot \overset{2}{\mathbf{A}} = 0,$$

so that the corresponding directions are orthogonal. These directions $\overset{i}{\mathbf{A}}$ are called the *principal directions of strain*, and the strains e_i, which are the extensions of the vectors $\overset{i}{\mathbf{A}}$ in the principal directions, are termed the *principal strains*.

We have seen that at any point (x^0) there are three mutually perpendicular directions $\overset{i}{\mathbf{A}}$ (assuming, for the moment, that the e_i are distinct) that are left unaltered by the deformation; consequently the vectors $\overset{i}{\mathbf{A}}$, the deformed vectors $\overset{i}{\mathbf{A}} + \delta\overset{i}{\mathbf{A}}$, and $\delta\overset{i}{\mathbf{A}}$ are collinear. But (5.11) shows that $\delta\mathbf{A}$ is always normal to the quadric surface (5.5), and therefore the principal directions of strain are also normal to the surface and must be the three principal axes of the quadric $e_{ij}x_ix_j = eA^2$. If some of the principal strains e_i are equal, then the associated directions become indeterminate, but one can always select three directions that are mutually orthogonal. If the quadric surface is a surface of revolution, then one direction $\overset{1}{\mathbf{A}}$, say, will be directed along the axis of revolution, and any two mutually perpendicular vectors lying in the plane normal to $\overset{1}{\mathbf{A}}$ may be taken as the other two principal axes. If $e_1 = e_2 = e_3$, the quadric is a sphere, and any three orthogonal lines may be chosen as the principal axes.

We recall that e_1, e_2, e_3 are the extensions of vectors along the principal axes, while e_{11}, e_{22}, e_{33} are the extensions of vectors along the coordinate axes. If the coordinate axes x_i are taken along the principal axes of the quadric, then the shear strains e_{12}, e_{23}, e_{31} disappear from the equation of the quadric surface, and the latter takes the form

$$e_1 x_1^2 + e_2 x_2^2 + e_3 x_3^2 = \pm k^2.$$

The cubic equation (6.3) can be written in the form

(6.7) $\qquad |e_{ij} - e\,\delta_{ij}| = -e^3 + \vartheta_1 e^2 - \vartheta_2 e + \vartheta_3 = 0,$

where ϑ_1, ϑ_2, ϑ_3 are the sums of the products of the roots taken one, two, and three at a time:

(6.8) $\qquad \begin{cases} \vartheta_1 = e_1 + e_2 + e_3 \equiv \vartheta, \\ \vartheta_2 = e_2 e_3 + e_3 e_1 + e_1 e_2, \\ \vartheta_3 = e_1 e_2 e_3. \end{cases}$

By expanding the determinant (6.7), we see that these expressions can also be written as

(6.9) $\qquad \begin{cases} \vartheta = e_{11} + e_{22} + e_{33}, \\ \vartheta_2 = e_{22}e_{33} + e_{33}e_{11} + e_{11}e_{22} - e_{31}^2 - e_{12}^2 - e_{23}^2 \\ \quad = \begin{vmatrix} e_{22} & e_{23} \\ e_{23} & e_{33} \end{vmatrix} + \begin{vmatrix} e_{11} & e_{31} \\ e_{31} & e_{33} \end{vmatrix} + \begin{vmatrix} e_{11} & e_{12} \\ e_{12} & e_{22} \end{vmatrix}, \\ \vartheta_3 = e_{11}e_{22}e_{33} + 2e_{12}e_{23}e_{31} - e_{11}e_{23}^2 - e_{22}e_{31}^2 - e_{33}e_{12}^2, \\ \quad = \begin{vmatrix} e_{11} & e_{12} & e_{31} \\ e_{12} & e_{22} & e_{23} \\ e_{31} & e_{23} & e_{33} \end{vmatrix}. \end{cases}$

The expressions for ϑ_2 and ϑ_3 can be written compactly by introducing the *generalized Kronecker delta*, $\delta_{pqr\cdots}^{ijk\cdots}$, which we now define. If the subscripts p, q, $r \cdots$ are distinct and if the superscripts i, j, $k \cdots$ are the same set of numbers as the subscripts, then the value of $\delta_{pqr\cdots}^{ijk\cdots}$ is defined to be $+1$ or -1 according as the subscripts and superscripts differ by an even or an odd permutation; the value is zero in all other cases. We can now rewrite the formulas (6.9) in the form

$$(6.10) \quad \begin{cases} \vartheta = e_{ii}, & (i = 1, 2, 3), \\ \vartheta_2 = \dfrac{1}{2!}\delta^{ij}_{pq}e_{pi}e_{qj}, & (i, j, p, q = 1, 2, 3), \\ \vartheta_3 = \dfrac{1}{3!}\delta^{ijk}_{pqr}e_{pi}e_{qj}e_{rk}, & (i, j, k, p, q, r = 1, 2, 3). \end{cases}$$

Since the principal strains, that is, the roots e_1, e_2, e_3 of (6.7), have a geometrical meaning that is independent of the choice of coordinate system, it is clear that ϑ, ϑ_2, and ϑ_3 are invariant with respect to an orthogonal transformation of coordinates. [Note that this invariance could have been used to derive expressions (6.8) from (6.9).]

The quantity ϑ has a simple geometrical meaning. Consider as a volume element a rectangular parallelepiped whose edges are parallel to the principal directions of strain, and let the lengths of these edges be l_1, l_2, l_3. Upon deformation, this element becomes again a rectangular parallelepiped but with edges of lengths $l_1(1 + e_1)$, $l_2(1 + e_2)$, $l_3(1 + e_3)$. Hence the change δV in the volume V of the element is

$$\delta V = l_1 l_2 l_3 (1 + e_1)(1 + e_2)(1 + e_3) - l_1 l_2 l_3$$
$$= l_1 l_2 l_3 (e_1 + e_2 + e_3)$$

plus terms of higher order in e_i. Thus

$$e_1 + e_2 + e_3 = \vartheta = \frac{\delta V}{V},$$

and the first strain invariant ϑ represents the expansion of a unit volume due to strain produced in the medium. For this reason ϑ is called the *cubical dilatation* or simply the *dilatation*.

References for Collateral Reading

A. E. H. LOVE: A Treatise on the Mathematical Theory of Elasticity, Cambridge University Press, London, Secs. 11, 12, pp. 42, 43.

S. TIMOSHENKO: Theory of Elasticity, McGraw-Hill Book Company, Inc., New York, Secs. 58, 59, pp. 190-193.

E. TREFFTZ: Handbuch der Physik, Verlag von Julius Springer, Berlin, vol. 6, Secs. 8-10.

A. G. WEBSTER: Dynamics of Particles and Rigid Bodies, Verlag von Julius Springer, Berlin, Secs. 164-168.

7. General Infinitesimal Deformation.

In the preceding sections, we have discussed the infinitesimal affine transformation (3.7), which carries the vector A_i into the vector $A'_i \equiv A_i + \delta A_i$, where

$$(7.1) \qquad \delta A_i = \alpha_{ij} A_j = \left(\frac{\alpha_{ij} + \alpha_{ji}}{2} + \frac{\alpha_{ij} - \alpha_{ji}}{2} \right) A_j$$
$$= (e_{ij} + \omega_{ij}) A_j;$$

the e_{ij} and ω_{ij} were constants and so small that their products could be neglected in comparison with their first powers. Now we consider the general functional transformation and its relation to the affine deformation.

Consider an arbitrary material point $P^0(x_1^0, x_2^0, x_3^0)$ in a continuous medium, and let the same material point assume after deformation the position $P^{0\prime}(x_1^{0\prime}, x_2^{0\prime}, x_3^{0\prime})$ (see Fig. 1). We denote the small displacement of the point P^0 by

$$u_i(x_1^0, x_2^0, x_3^0) = x_i^{0\prime} - x_i^0.$$

The quantities u_1, u_2, u_3 are called the *components of displacement*. It is clear from physical considerations that one must demand that the functions u_i be single-valued and continuous throughout the region occupied by the body. For reasons that will become apparent in Sec. 10, it will be assumed that the functions $u_i(x_1, x_2, x_3)$ are continuous of class $C^{(3)}$ (that is, the u_i together with their first, second, and third derivatives are continuous).

The character of the deformation in the neighborhood of the point P^0 can be determined by analyzing the change in the vector A joining the point $P^0(x_1^0, x_2^0, x_3^0)$ with an arbitrary neighboring point $P(x_1, x_2, x_3)$ of the undeformed medium. If $P'(x_1', x_2', x_3')$ is the deformed position of P, then the displacement u_i at the point P is

$$u_i(x_1, x_2, x_3) = u_i(x_1^0 + A_1, x_2^0 + A_2, x_3^0 + A_3) = x_i' - x_i.$$

The deformed vector A' has components $A'_i = x'_i - x_i^{0\prime}$, and for the components of $\delta A = A' - A$ we have

$$\delta A_i = (x'_i - x_i^{0\prime}) - (x_i - x_i^0)$$
$$= (x'_i - x_i) - (x_i^{0\prime} - x_i^0)$$
$$= u_i(x_1^0 + A_1, x_2^0 + A_2, x_3^0 + A_3) - u_i(x_1^0, x_2^0, x_3^0)$$
$$= \left(\frac{\partial u_i}{\partial x_j} \right)_0 A_j$$

plus the remainder in the Taylor's expansion of the function $u_i(x_1^0 + A_1, x_2^0 + A_2, x_3^0 + A_3)$. The subscript zero indicates that the derivative is to be evaluated at the point P^0. The derivative $\dfrac{\partial u_i}{\partial x_j}$ will be written by introducing the symbol $u_{i,j}$ so that

(7.2) $$\frac{\partial u_i}{\partial x_j} \equiv u_{i,j},$$

and the subscript can be dropped without confusion, since we shall deal only with vectors at P^0. If the region in the vicinity of P^0 is chosen sufficiently small, that is, if A is sufficiently small, then one has the formulas analogous to (7.1),

(7.3) $$\delta A_i = u_{i,j} A_j.$$

Comparison of formulas (7.3) and (7.1) shows that the transformation of the neighborhood of the point P^0 is affine and that

$$\alpha_{ij} = u_{i,j}.$$

Now if we assume further that the displacements u_i, as well as their partial derivatives, are so small that their products can be neglected, then (7.3) defines an infinitesimal affine transformation of the neighborhood of the point in question. Hence the considerations of the earlier sections are immediately applicable; the transformation (7.3) can be decomposed into pure deformation and rigid body motion,

(7.4) $$\delta A_i = u_{i,j} A_j = \left(\frac{u_{i,j} + u_{j,i}}{2} + \frac{u_{i,j} - u_{j,i}}{2} \right) A_j$$
$$= (e_{ij} + \omega_{ij}) A_j,$$
(7.5) $$e_{ij} = \tfrac{1}{2}(u_{i,j} + u_{j,i}), \qquad \omega_{ij} = \tfrac{1}{2}(u_{i,j} - u_{j,i}).$$

It is clear that the transformation defined by (7.5) is in general no longer homogeneous, inasmuch as both the strain components e_{ij} and the components of rotation ω_{ij} are functions of the coordinates of the medium. The dilatation

$$\vartheta = e_{11} + e_{22} + e_{33}$$
$$= \frac{\partial u_1}{\partial x_1} + \frac{\partial u_2}{\partial x_2} + \frac{\partial u_3}{\partial x_3} = u_{i,i}$$

or the *divergence* of the displacement vector u_i will likewise differ, in general, from point to point of the body.[1]

In order to indicate the advantages of notation adopted here over the customary one in use by writers of technical treatises on elasticity and hydrodynamics, we rewrite (7.5) by setting

$$x_1 = x, \quad x_2 = y, \quad x_3 = z, \quad e_{11} = e_{xx}, \quad e_{12} = e_{xy}, \text{ etc.},$$

and denote the components of the displacement vector (u_1, u_2, u_3) by (u, v, w). The components of the strain tensor become:

$$e_{xx} = \frac{\partial u}{\partial x}, \quad e_{yy} = \frac{\partial v}{\partial y}, \quad e_{zz} = \frac{\partial w}{\partial z},$$

$$e_{zy} = \frac{1}{2}\left(\frac{\partial w}{\partial y} + \frac{\partial v}{\partial z}\right), \quad e_{xz} = \frac{1}{2}\left(\frac{\partial u}{\partial z} + \frac{\partial w}{\partial x}\right), \quad e_{yx} = \frac{1}{2}\left(\frac{\partial v}{\partial x} + \frac{\partial u}{\partial y}\right),$$

so that the dilatation ϑ is

$$\vartheta = e_{xx} + e_{yy} + e_{zz} = \frac{\partial u}{\partial x} + \frac{\partial v}{\partial y} + \frac{\partial w}{\partial z}.$$

---

[1] Some of the important relations of vector analysis will now be written in tensor notation. In cartesian coordinates the divergence, gradient, and Laplacian operators can be written as follows:

$$\text{div } \mathbf{A} \equiv \nabla \cdot \mathbf{A} = \frac{\partial A_1}{\partial x_1} + \frac{\partial A_2}{\partial x_2} + \frac{\partial A_3}{\partial x_3} = \frac{\partial A_i}{\partial x_i} = A_{i,i},$$

$$\text{grad } \varphi \equiv \nabla \varphi = \frac{\partial \varphi}{\partial x_i} \equiv \varphi_{,i},$$

$$\nabla^2 \varphi = \text{div}(\nabla \varphi) = \frac{\partial^2 \varphi}{\partial x_1^2} + \frac{\partial^2 \varphi}{\partial x_2^2} + \frac{\partial^2 \varphi}{\partial x_3^2} = \frac{\partial^2 \varphi}{\partial x_i \, \partial x_i} = \varphi_{,ii}.$$

The Green-Gauss Theorem, namely,

$$\int_\tau \text{div } \mathbf{A} \, d\tau = \int_\sigma \mathbf{A} \cdot \mathbf{v} \, d\sigma,$$

takes the form

$$\int_\tau A_{i,i} \, d\tau = \int_\sigma A_i \nu_i \, d\sigma,$$

where $d\sigma$ is an element of area, $d\tau$ is an element of volume, and $\mathbf{v}$ is the exterior normal to the surface σ. If we set $A_i = \dfrac{\partial \varphi}{\partial x_i} \equiv \varphi_{,i}$, then there results the identity

$$\int_\tau \varphi_{,ii} \, d\tau = \int_\sigma \varphi_{,i} \nu_i \, d\sigma,$$

or

$$\int_\tau \nabla^2 \varphi \, d\tau = \int_\sigma \frac{d\varphi}{d\nu} \, d\sigma.$$

The components of rotation ω_{ij} read:

$$\omega_{zy} = \frac{1}{2}\left(\frac{\partial w}{\partial y} - \frac{\partial v}{\partial z}\right), \qquad \omega_{xz} = \frac{1}{2}\left(\frac{\partial u}{\partial z} - \frac{\partial w}{\partial x}\right),$$

$$\omega_{yx} = \frac{1}{2}\left(\frac{\partial v}{\partial x} - \frac{\partial u}{\partial y}\right).$$

While the unabridged notation, just explained, has some advantages in the discussion of specific problems, the compactness of the tensor notation and the economy of thought to which it leads in general developments are unquestionable.

8. Examples of Strain. Several important examples of strain will be considered next, and since there are no great advantages in using tensor notation in specific problems, we make use of the unabridged notation explained in the closing paragraphs of the preceding section.

a. Uniform Dilatation. If the strain quadric is a sphere, then any three orthogonal lines through the point may be used as the principal axes. In this case, the strain quadric has the equation

(8.1) $$e_{xx}x^2 + e_{yy}y^2 + e_{zz}z^2 = \pm k^2,$$

where $e_{xx} = e_{yy} = e_{zz} \equiv e$, and $e_{xy} = e_{yz} = e_{xz} = 0$. The linear extension (or contraction) in any direction is the same and is equal to one-third of the dilatation, since

$$\vartheta = e_{xx} + e_{yy} + e_{zz} = 3e.$$

b. Simple Extension. Consider a simple extension of magnitude e in the direction of the x'-axis, whose direction cosines relative to the system of axes x, y, z are (l_{11}, l_{12}, l_{13}). Referred to the x', y', z' coordinate system, the strain quadric has the equation $ex'^2 = k^2$. By use of the transformation equations (5.10), we obtain in the x, y, z-system

(8.2) $$\begin{cases} e_{xx} = el_{11}^2, & e_{yy} = el_{12}^2, & e_{zz} = el_{13}^2, \\ e_{xy} = el_{11}l_{12}, & e_{yz} = el_{12}l_{13}, & e_{zx} = el_{13}l_{11}. \end{cases}$$

Thus, a simple extension in the direction (l_{11}, l_{12}, l_{13}) may be specified in any x, y, z coordinate system by means of the six strain components given in (8.2).

c. *Shearing Strain.* Let the equation of the strain quadric when referred to the x', y', z'-system of coordinates be given by

$$(8.3) \qquad 2sx'y' = \pm k^2,$$

so that the only strain component is a shearing strain of magnitude s along the directions of the x'- and y'-axes. This is the equation of a hyperbolic cylinder asymptotic to the $x'z'$- and $y'z'$-planes. The equation of the quadric (8.3) assumes the form

$$sx^2 - sy^2 = \pm k^2$$

when the axes are rotated through an angle of 45° about the z'-axis. A comparison of this equation with the general equation of the strain quadric when the latter is referred to the principal axes of strain,

$$e_{xx}x^2 + e_{yy}y^2 + e_{zz}z^2 = \pm k^2,$$

shows that we must have $e_{zz} = 0$, $e_{xx} = -e_{yy} = s$. Thus equal extension and contraction of two orthogonal linear elements are equivalent to a shearing strain of equal magnitude, which is associated with directions bisecting the angles between the elements.

d. *Plane Strain.* Suppose that the principal extension in the direction of the z'-axis is zero. Then for the x, y, z-system (assuming the directions of the z'- and z-axes to be the same), the strain quadric has the form

$$e_{xx}x^2 + e_{yy}y^2 + 2e_{xy}xy = \pm k^2,$$

corresponding to

$$e'_{x'x'}x'^2 + e'_{y'y'}y'^2 = \pm k^2 \quad \text{in the } x', y', z'\text{-system,}$$

$e'_{x'x'}$ and $e'_{y'y'}$ being principal extensions. In the case of simple extension (see part b, above), the quadric consists of two parallel planes; in the case of shearing strain (see part c), it consists of a rectangular hyperbolic cylinder. If the quadric is a circular cylinder, the state of strain is such that there is equal extension (or contraction) in all directions perpendicular to that of the z'-axis.

In the case of plane strain, the relative displacements u and v are functions of x and y alone, and w is a constant.

PROBLEMS

1. Verify the invariance of the functions ϑ, ϑ_2, and ϑ_3 [see Eqs. (6.10)] of the strains in the case of simple extension.

2. Find the dilatation and the principal strains, and describe the strain quadric for the case of simple extension.

3. Show that the examples of strain given in this section can be described by the following displacement components:
  (a) uniform dilatation, $u = ex$, $\quad v = ey$, $\quad w = ez$;
  (b) simple extension, $u' = ex'$, $\quad v' = w' = 0$;
  (c) shearing strain, $u' = 2sy'$, $\quad v' = w' = 0$;
  (d) plane strain, $u = u(x,y)$, $\quad v = v(x,y)$, $\quad w = 0$.

4. Show that in the examples of strain given in this section the rotation components are given by:
  (a) uniform dilatation, $\omega_{xy} = \omega_{yz} = \omega_{zx} = 0$;
  (b) simple extension, $\omega'_{xy} = \omega'_{yz} = \omega'_{zx} = 0$;
  (c) shearing strain, $\omega'_{xy} = S$, $\quad \omega'_{yz} = \omega'_{zx} = 0$;
  (d) plane strain, $\omega_{xy} = \dfrac{1}{2}\left(\dfrac{\partial u}{\partial y} - \dfrac{\partial v}{\partial x}\right)$, $\quad \omega_{yz} = \omega_{zx} = 0$.

9. Notation. The values of the shear components e_{xy}, e_{zx}, e_{yz} of the strain tensor e_{ij}, defined in (7.5), differ from the quantities e_{xy}, e_{zx}, e_{yz} used by Love,[1] who writes

$$e_{yz} = \frac{\partial w}{\partial y} + \frac{\partial v}{\partial z}, \quad e_{zx} = \frac{\partial u}{\partial z} + \frac{\partial w}{\partial x}, \quad e_{xy} = \frac{\partial v}{\partial x} + \frac{\partial u}{\partial y}.$$

The factor ½ was inserted in the formulas (7.5) in order that the set of quantities may transform according to the tensor algorism.

Trefftz[2] writes for the components of his strain tensor

$$\gamma_{xx} = 2\frac{\partial u}{\partial x}, \quad \gamma_{yy} = 2\frac{\partial v}{\partial y}, \quad \gamma_{zz} = 2\frac{\partial w}{\partial z},$$

$$\gamma_{yz} = \frac{\partial w}{\partial y} + \frac{\partial v}{\partial z}, \quad \gamma_{zx} = \frac{\partial u}{\partial z} + \frac{\partial w}{\partial x}, \quad \gamma_{xy} = \frac{\partial v}{\partial x} + \frac{\partial u}{\partial y},$$

while Timoshenko[3] uses

$$\epsilon_x = \frac{\partial u}{\partial x}, \quad \epsilon_y = \frac{\partial v}{\partial y}, \quad \epsilon_z = \frac{\partial w}{\partial z}.$$

---

[1] A. E. H. Love, A Treatise on the Mathematical Theory of Elasticity.

[2] E. Trefftz, Handbuch der Physik, vol. 6, Mechanik der elastischen Körper.

[3] S. Timoshenko, Theory of Elasticity.

for the components of normal strain and agrees in notation with Trefftz for the components of shearing strain.

References for Collateral Reading

S. P. Timoshenko: Theory of Elasticity, McGraw-Hill Book Company, Inc., New York, Sec. 5, pp. 5–7.

A. E. H. Love: A Treatise on the Mathematical Theory of Elasticity, Cambridge University Press, London, Secs. 1–14, pp. 32–46.

R. V. Southwell: Theory of Elasticity for Engineers and Physicists, Oxford University Press, New York, Secs. 292–307, pp. 285–297.

A. G. Webster: Dynamics of Particles and Rigid Bodies, Verlag von Julius Springer, Berlin, Sec. 169.

10. Equations of Compatibility. The defining formulas for the strain components e_{ij}, namely

$$(10.1) \qquad u_{i,j} + u_{j,i} = 2e_{ij},$$

will be looked upon in this section as a system of partial differential equations for the determination of the displacements u_i when the strain components e_{ij} are prescribed functions of the coordinates. We shall discuss first a necessary condition for the uniqueness of the solutions u_i of Eqs. (10.1). Thereupon we shall raise the question:

What restrictions must be placed on the given functions $e_{ij}(x_1, x_2, x_3)$ to ensure the existence of single-valued, continuous solutions $u_i(x_1, x_2, x_3)$ of Eqs. (10.1)?

It is clear first of all that specification of the e_{ij} does not determine the displacements u_i uniquely, for the strain components characterize the pure deformation of the medium in the neighborhood of the point (x), while the functions u_i may involve rigid body motions which do not affect the e_{ij}. In fact, if one obtains some solution

$$(10.2) \qquad u_i = u_i(x_1, x_2, x_3)$$

of the system (10.1), and if $P^0(x_1^0, x_2^0, x_3^0)$ is an arbitrary point of the body, then the addition to the right-hand member of (10.2) of the terms[1]

$$(10.3) \qquad u_j = u_j^0 + \omega_{jk}^0(x_k - x_k^0),$$

[1] Cf. formulas (3.5).

representing the motion of the body as a whole, will not affect the values of the prescribed components of strain entering into (10.1). It thus becomes clear that the solution of the system (10.1) cannot be unique unless one specifies the components of displacement u_i^0 and the components of rotation ω_{ij}^0 of some point P^0 of the medium, and we shall suppose in the following discussion that this has been done.

Inasmuch as there are six conditions imposed on the three functions u_i by Eqs. (10.1), one cannot expect in general that the system (10.1) will possess a solution for an arbitrary choice of the functions e_{ij}. We seek the further conditions that must be imposed on the functions e_{ij} if the system of Eqs. (10.1) is to possess a solution for the triplet of functions u_i.

The fact that the strain components e_{ij} cannot be prescribed arbitrarily can be seen from the following rough geometrical considerations. Imagine that a body τ is subdivided into small volume elements, which in the interior of τ may be assumed to have the form of cubes. The strain components e_{ij} are given on the faces of each cube, and the displacements u_i of those faces are to be calculated. If each individual cube is subjected to a deformation so that it becomes a parallelepiped, then it may happen that it is impossible to arrange the parallelepipeds to form a continuous distorted body τ'. The points that were coincident on the interfaces of the cubes may no longer coincide on the interfaces of the parallelepipeds. In fact, there may even be gaps between the pairs of initially coincident points. The requirements of continuity and single-valuedness imposed on the components of displacement place some restrictions on the choice of the strain components e_{ij} if the differential equations (10.1) are to possess solutions.

Let $P^0(x_1^0, x_2^0, x_3^0)$ be some point of a simply connected region[1] τ, at which the displacements $u_j^0(x_1^0, x_2^0, x_3^0)$ and the components of rotation $\omega_{ij}^0(x_1^0, x_2^0, x_3^0)$ are known. We determine the displacements u_i at any other point $P'(x_1', x_2', x_3')$ in terms of the

---

[1] A region of space is said to be simply connected if every closed curve drawn in the region can be shrunk to a point, by continuous deformation, without passing out of the boundaries of the region. Thus the region between two concentric spheres is simply connected, but the interior of an anchor ring (torus) is not.

known functions[1] e_{ij} by means of a line integral over a simple[2] continuous curve C joining the points P^0 and P':

$$u_j(x_1', x_2', x_3') = u_j^0 + \int_{P^0}^{P'} du_j = u_j^0 + \int_{P^0}^{P'} u_{j,k}\, dx_k$$
$$= u_j^0 + \int_{P^0}^{P'} e_{jk}\, dx_k + \int_{P^0}^{P'} \omega_{jk}\, dx_k,$$

where the last step comes from the definition (7.5). An integration by parts yields

$$\int_{P^0}^{P'} \omega_{jk}\, dx_k \equiv \int_{P^0}^{P'} \omega_{jk}\, d(x_k - x_k')$$
$$= (x_k' - x_k^0)\omega_{jk}^0 + \int_{P^0}^{P'} (x_k' - x_k)\omega_{jk,l}\, dx_l,$$

and hence

(10.4) $\quad u_j(x_1', x_2', x_3') = u_j^0 + (x_k' - x_k^0)\omega_{jk}^0$
$$+ \int_{P^0}^{P'} [e_{jl} + (x_k' - x_k)\omega_{jk,l}]\, dx_l.$$

We express the derivatives of the components of rotation $\omega_{jk,l}$ in terms of the known functions e_{ij} by using the definitions (7.5) and writing

$$\omega_{jk,l} = \frac{\partial}{\partial x_l}\frac{1}{2}(u_{j,k} - u_{k,j})$$
$$= \frac{1}{2}(u_{j,kl} - u_{k,jl}) + \frac{1}{2}(u_{l,jk} - u_{l,jk})$$
$$= \frac{\partial}{\partial x_k}\frac{1}{2}(u_{l,j} + u_{j,l}) - \frac{\partial}{\partial x_j}\frac{1}{2}(u_{k,l} + u_{l,k}),$$

where the continuity of the mixed derivatives has been used. It follows from the preceding equation that

(10.5) $\qquad \omega_{jk,l} = e_{lj,k} - e_{kl,j}.$

When (10.5) is inserted in (10.4), it is seen that the determination of the displacements u_i at any point (x) has now been reduced to a quadrature,

(10.6) $\quad u_j(x_1', x_2', x_3') = u_j^0 + (x_k' - x_k^0)\omega_{jk}^0 + \int_{P^0}^{P'} U_{jl}\, dx_l,$

where the integrand

(10.7) $\qquad U_{jl} = e_{jl} + (x_k' - x_k)(e_{lj,k} - e_{kl,j})$

is a known function.

---

[1] The functions e_{ij} are assumed to be of class $C^{(2)}$ (see Sec. 7).
We use the term *simple* curve in the sense of *rectifiable* curve.

§10   ANALYSIS OF STRAIN   27

Inasmuch as the displacements u_i must be independent of the path of integration, the integrands $U_{jl}\,dx_l$ must be exact differentials. Hence, applying the necessary and sufficient condition that the integrands in (10.6) be exact differentials, namely
$$U_{ji,l} - U_{jl,i} = 0,$$
we have

(10.8) $e_{ji,l} - \delta_{kl}(e_{ij,k} - e_{ki,j}) - e_{jl,i} + \delta_{ki}(e_{lj,k} - e_{kl,j})$
$$+ (x_k' - x_k)(e_{ij,kl} - e_{ki,jl} - e_{lj,ki} + e_{kl,ji}) = 0.$$

The first line of (10.8) vanishes identically, and since this equation must be true for an arbitrary choice of $(x_k' - x_k)$, it follows that

(10.9) $e_{ij,kl} + e_{kl,ij} - e_{ik,jl} - e_{jl,ik} = 0.$

The system (10.9) consists of $3^4 = 81$ equations, but some of these are identically satisfied, and some are repetitions because of the symmetry in indices ij and kl. A little reflection will show that only 6 of the 81 equations (10.9) are essential, and when these are written out in unabridged notation, one has

(10.10)
$$\begin{cases} \dfrac{\partial^2 e_{xx}}{\partial y\,\partial z} = \dfrac{\partial}{\partial x}\left(-\dfrac{\partial e_{yz}}{\partial x} + \dfrac{\partial e_{zx}}{\partial y} + \dfrac{\partial e_{xy}}{\partial z}\right), \\ \dfrac{\partial^2 e_{yy}}{\partial z\,\partial x} = \dfrac{\partial}{\partial y}\left(-\dfrac{\partial e_{zx}}{\partial y} + \dfrac{\partial e_{xy}}{\partial z} + \dfrac{\partial e_{yz}}{\partial x}\right), \\ \dfrac{\partial^2 e_{zz}}{\partial x\,\partial y} = \dfrac{\partial}{\partial z}\left(-\dfrac{\partial e_{xy}}{\partial z} + \dfrac{\partial e_{yz}}{\partial x} + \dfrac{\partial e_{zx}}{\partial y}\right), \\ 2\dfrac{\partial^2 e_{xy}}{\partial x\,\partial y} = \dfrac{\partial^2 e_{xx}}{\partial y^2} + \dfrac{\partial^2 e_{yy}}{\partial x^2}, \\ 2\dfrac{\partial^2 e_{yz}}{\partial y\,\partial z} = \dfrac{\partial^2 e_{yy}}{\partial z^2} + \dfrac{\partial^2 e_{zz}}{\partial y^2}, \\ 2\dfrac{\partial^2 e_{zx}}{\partial z\,\partial x} = \dfrac{\partial^2 e_{zz}}{\partial x^2} + \dfrac{\partial^2 e_{xx}}{\partial z^2}. \end{cases}$$

The six equations (10.10) ensuring the continuity of displacements are known as the *equations of compatibility* and were obtained by Saint-Venant in 1860, in a way different from that outlined above.[1]

[1] The essential features of the method of derivation of the compatibility equations given above are due to E. Cesaro, *Rendiconto dell' accademia delle scienze fisiche e matematiche, Classe della società reale di Napoli* (1906). See also a memoir by V. Volterra, "L'Equilibre des corps élastiques," *Annales de l'école normale supérieure*, vol. 24 (1907). The necessity of conditions (10.10) can be proved easily. See Prob. 6 at the end of this Chapter.

One can verify by direct substitution that the displacements given by (10.6) actually satisfy the differential equations (10.1). We have already seen that the displacements specified by (10.3) contribute nothing to the strain components e_{ij}. Equation (10.6) shows that, conversely, if the e_{ij} vanish identically, then the resulting solutions are given by Eqs. (10.3), and these obviously represent a rigid body motion.

If the region of integration τ is multiply connected, then the functions u_i may turn out to be multiple-valued. As is well known, a multiply connected region may be reduced to a simply connected one, provided suitable barriers or crosscuts are introduced. In this case, the displacements u_i will be single-valued functions of the coordinates when evaluated by means of a line integral taken along any curve C that does not pass through one of these crosscuts. If the curve C does intersect the crosscut, then to ensure that the u_i be single-valued, we must demand in addition to the satisfaction of the compatibility relations that the limiting values of $u_i(x_1, x_2, x_3)$ be the same when the cut is approached from either side.

References for Collateral Reading

A. E. H. Love: A Treatise on the Mathematical Theory of Elasticity, Cambridge University Press, London, Secs. 17–18, pp. 48–51.

S. Timoshenko: Theory of Elasticity, McGraw-Hill Book Company, Inc., New York, Sec. 61, p. 196; Sec. 62, p. 199.

R. V. Southwell: Theory of Elasticity for Engineers and Physicists, Oxford University Press, New York, Sec. 308, p. 297.

11. Finite Deformations. The preceding sections of this chapter contain all the principal results of the classical theory of infinitesimal strain. It is clear from the general discussion of the affine transformation in Sec. 2 that the linearization of the equations appearing there led to a consideration of infinitesimal transformations that permits the application of the principle of superposition of effects. Some recent developments in the theory of elasticity connected with the problems of buckling and stability have indicated the desirability of considering finite deformations; that is, deformations in which the displacements u together with their derivatives are no longer small. This section contains only a brief introduction to the theory of finite strains and provides an admirable illustration of the complications that appear in the development of a theory when the fundamental equations become nonlinear.

There are two modes of description of the deformation of a continuous medium, the *Lagrangian* and the *Eulerian*. The Lagrangian description employs the coordinates a_i of a typical particle in the initial state as the independent variables, while in the case of Eulerian coordinates the independent variables are the coordinates x_i of a material point in the deformed state.

In the preceding sections, we have used the Lagrangian viewpoint as the natural means of describing the deformation of the neighborhood of the point (a_1, a_2, a_3). When we come, in the next chapter, to the discussion of the stresses acting throughout the medium, we shall find that these stresses must satisfy equilibrium conditions in the deformed body, and hence Eulerian coordinates are indicated. In this section, we shall describe the deformation from both points of view, and we shall see that when the deformation is infinitesimal (that is, when products of the derivatives of the displacements can be neglected), these two viewpoints, Lagrangian and Eulerian, coalesce, and we need make no distinction between them.

Consider an aggregate of particles in a continuous medium that lie along a curve C in the undeformed state. Just as in the preceding sections, it will be convenient to use the same reference frame for the location of particles in the deformed and undeformed states. Let the coordinates of a particle lying on a curve C_0 (before deformation) be denoted by (a_1, a_2, a_3), and let the coordinates of the same particle after deformation (now lying on some curve C) be (x_1, x_2, x_3). Then the elements of arc of the curves C_0 and C are given, respectively, by

(11.1) $$ds_0^2 = da_1^2 + da_2^2 + da_3^2 = da_i\, da_i,$$

and

(11.2) $$ds^2 = dx_i\, dx_i.$$

We consider first the Eulerian description of the strain and write $a_i = a_i(x_1, x_2, x_3)$. Substituting[1] $da_i = a_{i,j}\, dx_j = a_{i,k}\, dx_k$ in (11.1) yields

$$ds_0^2 = a_{i,j} a_{i,k}\, dx_j\, dx_k,$$

[1] The notation $a_{i,j} = \dfrac{\partial a_i}{\partial x_j}$ denotes differentiation with respect to the jth independent variable, which in this case is x_j.

while $ds^2 = dx_i\, dx_i = \delta_{jk}\, dx_j\, dx_k$. It is evident that the equality of ds^2 and ds_0^2 for all curves C_0 is the necessary and sufficient condition that the transformation $a_i = a_i(x_1, x_2, x_3)$ be one of rigid body motion; hence we shall take the difference $ds^2 - ds_0^2$ as the measure of the strain and write

$$(11.3) \qquad ds^2 - ds_0^2 = 2\eta_{jk}\, dx_j\, dx_k.$$

From the expressions given above for ds^2 and ds_0^2, we get

$$2\eta_{jk} = \delta_{jk} - a_{i,j} a_{i,k}.$$

We now write the strains η_{jk} in terms of the displacement components $u_i = x_i - a_i$. Since $a_i = x_i - u_i$, we have

$$\begin{aligned} a_{i,j} a_{i,k} &= (\delta_{ij} - u_{i,j})(\delta_{ik} - u_{i,k}) \\ &= \delta_{jk} - u_{j,k} - u_{k,j} + u_{i,j} u_{i,k} \end{aligned}$$

and hence

$$(11.4) \qquad 2\eta_{jk} = u_{j,k} + u_{k,j} - u_{i,j} u_{i,k}.$$

The functions η_{jk} are called the *Eulerian strain components*.

If, on the other hand, Lagrangian coordinates are used, so that the a_i are regarded as the independent variables and the equations of transformation are of the form $x_i = x_i(a_1, a_2, a_3)$, then we can write $dx_i = x_{i,j}\, da_j$ and

$$ds^2 = dx_i\, dx_i = x_{i,j} x_{i,k}\, da_j\, da_k,$$

while $ds_0^2 = da_j\, da_j = \delta_{jk}\, da_j\, da_k$. The Lagrangian components of strain ϵ_{jk} are defined by

$$(11.5) \qquad ds^2 - ds_0^2 = 2\epsilon_{jk}\, da_j\, da_k,$$

and since $x_i = a_i + u_i$, we have

$$x_{i,j} x_{i,k} = (\delta_{ij} + u_{i,j})(\delta_{ik} + u_{i,k})$$
$$= \delta_{jk} + u_{j,k} + u_{k,j} + u_{i,j} u_{i,k},$$

and
$$ds^2 - ds_0^2 = 2\epsilon_{jk}\, da_j\, da_k$$
$$= (u_{j,k} + u_{k,j} + u_{i,j} u_{i,k})\, da_j\, da_k,$$

with

$$(11.6) \qquad 2\epsilon_{jk} = u_{j,k} + u_{k,j} + u_{i,j} u_{i,k}.$$

In order to exhibit the fact that the differentiation in (11.4) is carried out with respect to the variables x_i, while in (11.6) the

§11 ANALYSIS OF STRAIN

u_i are regarded as the independent variables, we write out the typical expressions for η_{ij} and ϵ_{ij} in unabridged notation,

$$\eta_{xx} = \frac{\partial u}{\partial x} - \frac{1}{2}\left[\left(\frac{\partial u}{\partial x}\right)^2 + \left(\frac{\partial v}{\partial x}\right)^2 + \left(\frac{\partial w}{\partial x}\right)^2\right],$$

$$\epsilon_{aa} = \frac{\partial u}{\partial a} + \frac{1}{2}\left[\left(\frac{\partial u}{\partial a}\right)^2 + \left(\frac{\partial v}{\partial a}\right)^2 + \left(\frac{\partial w}{\partial a}\right)^2\right],$$

$$2\eta_{xy} = \frac{\partial u}{\partial y} + \frac{\partial v}{\partial x} - \left(\frac{\partial u}{\partial x}\frac{\partial u}{\partial y} + \frac{\partial v}{\partial x}\frac{\partial v}{\partial y} + \frac{\partial w}{\partial x}\frac{\partial w}{\partial y}\right),$$

$$2\epsilon_{ab} = \frac{\partial u}{\partial b} + \frac{\partial v}{\partial a} + \left(\frac{\partial u}{\partial a}\frac{\partial u}{\partial b} + \frac{\partial v}{\partial a}\frac{\partial v}{\partial b} + \frac{\partial w}{\partial a}\frac{\partial w}{\partial b}\right).$$

It was shown in Sec. 4 that e_{11}, e_{22}, and e_{33} can be interpreted as extensions of vectors originally parallel to the coordinate axes, while e_{12}, e_{23}, and e_{31} represent shears or changes of angle between vectors originally at right angles. When the strain components are large, however, it is no longer possible to give simple geometrical interpretations of the functions ϵ_{ij} and η_{ij}.

Consider a line element with $ds_0 = da_1$, $da_2 = da_3 = 0$, and define the extension E_1 of this element by $E_1 = \dfrac{(ds - ds_0)}{ds_0}$, or

(11.7) $$ds = (1 + E_1)\, ds_0.$$

From (11.5) we have

$$ds^2 - ds_0^2 = 2\epsilon_{jk}\, da_j\, da_k = 2\epsilon_{11}\, da_1^2,$$

and the insertion of (11.7) in this expression yields

$$(1 + E_1)^2 = 1 + 2\epsilon_{11},$$

or

(11.8) $$E_1 = \sqrt{1 + 2\epsilon_{11}} - 1.$$

When ϵ_{11} is small, this reduces to

$$E_1 \doteq \epsilon_{11},$$

as was shown in the discussion of infinitesimal strain in Sec. 4.

Consider next two line elements, $ds_0 = da_2$, $da_1 = da_3 = 0$, and $d\bar{s}_0 = d\bar{a}_3$, $d\bar{a}_1 = d\bar{a}_2 = 0$, that lie initially along the a_2- and a_3-axes. Let θ denote the angle between the corresponding deformed elements dx_i and $d\bar{x}_i$, of lengths ds and $d\bar{s}$ respectively. Then

$$ds\, d\bar{s} \cos\theta = dx_i\, d\bar{x}_i = x_{i,\alpha} x_{i,\beta}\, da_\alpha\, d\bar{a}_\beta$$
$$= x_{i,2} x_{i,3}\, da_2\, d\bar{a}_3 = 2\epsilon_{23}\, da_2\, d\bar{a}_3.$$

If $\alpha_{23} = \pi/2 - \theta$ denotes the change in the right angle between the line elements in the initial state, then we have

$$\sin \alpha_{23} = 2\epsilon_{23} \frac{da_2\, d\bar{a}_3}{ds\, d\bar{s}},$$

and by (11.7) and (11.8)

(11.9) $$\sin \alpha_{23} = \frac{2\epsilon_{23}}{\sqrt{1 + 2\epsilon_{22}}\sqrt{1 + 2\epsilon_{33}}}.$$

Again, if the strains ϵ_{ij} are so small that their products can be neglected, then

$$\alpha_{23} \doteq 2\epsilon_{23},$$

as was seen in Sec. 4.

If the displacements and their derivatives are small, then it is immaterial whether the derivatives of the displacements are calculated at the position of a point before or after deformation. In this case, we may neglect the nonlinear terms in the partial derivatives in (11.4) and (11.6) and reduce both sets of formulas to Eqs. (7.5), which were obtained for an infinitesimal transformation. Unless a statement to the contrary is made, we shall deal with infinitesimal strain and shall write

$$e_{ij} = \tfrac{1}{2}(u_{i,j} + u_{j,i}).$$

The ratio of the volume element in the strained state to the corresponding element of volume in the unstrained state is equal to the functional determinant

(11.10) $$\frac{\partial(x_1, x_2, x_3)}{\partial(a_1, a_2, a_3)} = \left|\frac{\partial x_i}{\partial a_j}\right| = \left|\frac{\partial(a_i + u_i)}{\partial a_j}\right| = |\delta_{ij} + u_{i,j}|.$$

If this is denoted by $1 + \Delta$, then Δ is the change of volume per unit volume at a point and is called the *cubical dilatation*. It is obvious that for small strains

$$\Delta \doteq u_{i,i} = e_{11} + e_{22} + e_{33} = \vartheta$$

As was done in the infinitesimal case, Eqs. (11.4) [or (11.6)] can be looked upon as the differential equations for the determination of the functions u_i, where the components of strain η_{ij} (or ϵ_{ij}) are prescribed functions. Since these equations are nonlinear, the problem of integration is much more involved.

While it is not difficult to formulate the conditions on the functions η_{ij} (or ϵ_{ij}) if the set of Eqs. (11.4) [or (11.6)] is to possess a solution with suitable properties, this will not be pursued here.[1]

Because of the difficulty of the problem, the only case of finite strain that has been treated with any degree of completeness is that of finite homogeneous strain.[2]

Of the recent work in finite strains, the contributions of Murnaghan,[3] Kappus,[4] Biot,[5] and Seth[6] should be mentioned. The theory developed by Murnaghan has been applied by Panov[7] and by Riz and Zvolinsky.[8]

In concluding this brief treatment of finite strains, it should be emphasized that the transformations of finite homogeneous strain are not in general commutative and that the simple superposition of effects is no longer applicable to finite deformations. These facts are responsible, in the main, for the absence of satisfactory solutions for all but the simplest problems, such as

[1] For a brief treatment, see A. E. H. Love, A Treatise on the Mathematical Theory of Elasticity, Sec. 31, pp. 65–66.

[2] A readable account of this case is found in A. E. H. Love, A Treatise on the Mathematical Theory of Elasticity, Secs. 32–40, pp. 66–73, and in Léon Lecornu, Théorie mathématique de l'élasticité, pp. 34–40.

[3] F. D. MURNAGHAN, "Finite Deformations of an Elastic Solid," *American Journal of Mathematics*, vol. 59 (1937), p. 235.

[4] R. KAPPUS, "Zur Elastizitätstheorie endlicher Verschiebungen," *Zeitschrift für angewandte Mathematik und Mechanik*, vol. 19 (1939), pp. 271–285, 344–361.

[5] M. A. BIOT, "Non-linear Theory of Elasticity and the Linearized Case for a Body Under Initial Stress," *Philosophical Magazine*, ser. 7, vol. 27 (1939), pp. 449–452, 468–469.

[6] B. R. SETH, "Finite Strain in Elastic Problems," *Philosophical Transactions of the Royal Society* (London), ser. A, vol. 234 (1935), pp. 231–264.

[7] D. PANOV, "On Secondary Effects Arising at the Torsion of an Elliptical Cylinder," *Comptes rendus (Doklady) de l'académie des sciences de l'U.R.S.S.*, new ser., vol. 22 (1939), pp. 158–160.

[8] See, for example, P. M. Riz and N. V. Zvolinsky, "Torsion of a Prismatic Bar Which Is Simultaneously Subjected to Tension," *Comptes rendus (Doklady) de l'académie des sciences de l'U.R.S.S.*, new ser., vol. 20 (1938), pp. 101–104. See also N. V. Zvolinsky and P. M. Riz, "On Certain Problems of Non-linear Theory of Elasticity," *Applied Mathematics and Mechanics* (Akad. Nauk S.S.S.R., Prikl. Mat. Mech.), vol. 2 (1939), No. 4; and P. M. Riz, "General Solution of Torsion Problem in Non-linear Theory of Elasticity," *Applied Mathematics and Mechanics* (Akad. Nauk S.S.S.R., Prikl. Mat. Mech.), vol. 7 (1943), pp. 149–154. (These are in Russian with English summaries.)

simple tension and the torsion of an elliptical cylinder, which become trivial when the equations are linearized.

References for Collateral Reading

L. Lecornu: Théorie mathématique de l'élasticité, Mémorial des sciences mathématiques, Gauthiers-Villars & Cie, Paris, pp. 34–40.

A. E. H. Love: A Treatise on the Mathematical Theory of Elasticity, Cambridge University Press, London, Secs. 32–40, pp. 66–73.

E. Trefftz: Handbuch der Physik, Verlag von Julius Springer, Berlin, vol. 6, Secs. 8–9.

PROBLEMS

1. Show that a tensor α_{ij} can be decomposed into a symmetric tensor $e_{ij} = e_{ji}$ and a skew-symmetric tensor $\omega_{ij} = -\omega_{ji}$ in one, and only one, way. *Hint:* Assume that the decomposition can be made in two ways:

$$\alpha_{ij} = e_{ij} + \omega_{ij} = \bar{e}_{ij} + \bar{\omega}_{ij}$$

2. From $\delta A_i = e_{ij} A_j$, find δA and δA for a vector lying initially along the x-axis, that is, $A = iA$, and justify the statement of Sec. 4 that in this case $\dfrac{\delta A}{A} = e_{xx}$. Does δA lie along the x-axis?

3. Derive Eq. (5.10) from (5.8) and (5.9).

4. Derive Eq. (8.2) by using the invariance of the strain quadric and the equation of transformation $x'_1 = l_{1j} x_j$.

5. Show that the inverse of the transformation (5.6) is $x'_i = l_{i\alpha} x_\alpha$.

6. Show by differentiation of the strain components

$$e_{ij} = \tfrac{1}{2}(u_{i,j} + u_{j,i})$$

that the equations of compatibility are necessary conditions for the existence of continuous single-valued displacements. *Hint:*

$$e_{ij,kl} = \tfrac{1}{2}(u_{i,jkl} + u_{j,ikl})$$

and, by interchange of i, k and likewise of j, l, $e_{kl,ij} = \tfrac{1}{2}(u_{k,lij} + u_{l,kij})$. Add these, interchange j and k, and show that the compatibility Eqs. (10.9) result.

7. Show that the shear strain e_{23}, for example, can be interpreted as the extension of the diagonal OQ of the rectangle $OPQR$ (Fig. 4), provided the rectangle is a square.

CHAPTER II
ANALYSIS OF STRESS

12. Body and Surface Forces. Consider a continuous medium, the points of which are referred to a rectangular cartesian system of axes, and let τ represent the region occupied by the medium and $\Delta\tau$ an element of volume of τ. In analyzing the forces acting on the volume element $\Delta\tau$, it is necessary to take into account two types of forces:

1. Body (or volume) forces; that is, the forces which are proportional to the mass contained in the volume element $\Delta\tau$;
2. Surface forces, which act on the surface $\Delta\sigma$ of the volume element $\Delta\tau$.

It will be assumed throughout this discussion that the volume forces are continuous functions of class $C^{(1)}$ and the surface forces are piecewise continuous functions of the coordinates (x_1, x_2, x_3) of the points of the medium.

As a typical example of body force, one can take the force of gravity, $\rho g\, \Delta\tau$, acting on the mass contained in the volume element $\Delta\tau$ of the medium whose density is ρ, and where g is the gravitational acceleration. An example of surface force is the tension acting on any horizontal section of a steel rod suspended vertically. Thus, if one imagines that the rod is cut by a horizontal plane into two parts, the upper and the lower, then the action of the weight of the lower part of the rod is transmitted to the upper part across the surface of the cut. A hydrostatic pressure on the surface of a submerged solid body provides another example of surface force.

Let the vector $\boldsymbol{F} = \boldsymbol{e}_i F_i$ represent the body force per unit volume of the medium. The resultant $\boldsymbol{R} = \boldsymbol{e}_i R_i$ of the body forces $\boldsymbol{F}$ can be represented as the volume integral $\boldsymbol{R} = \int_\tau \boldsymbol{F}\, d\tau$, or

(12.1) $$R_i = \int_\tau F_i\, d\tau.$$

The resultant moment $M = e_i M_i$ due to the body force F can be written as the integral over τ of the vector product of the position vector $r = e_i x_i$ and the force vector F; that is, $M = \int_\tau (r \times F) \, d\tau$, or[1]

$$(12.2) \qquad M_i = \int_\tau e_{ijk} x_j F_k \, d\tau.$$

Consider next an element $\Delta\sigma$ of a surface situated either in the interior or on the boundary of the medium, and let the force acting on the element $\Delta\sigma$ be $\overline{T} \Delta\sigma$. Because of the assumed continuity of forces, we have $\lim_{\Delta\sigma \to 0} \dfrac{\overline{T} \Delta\sigma}{\Delta\sigma} = T(x_1, x_2, x_3)$, where the vector T represents the surface force per unit area of the surface acting at the point (x_i) and is called the *stress vector*.

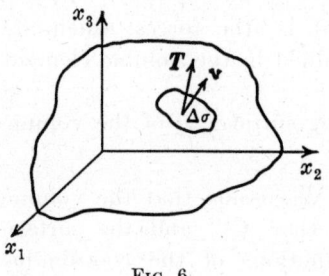

Fig. 6.

If $\Delta\sigma$ is a surface element in the interior of the medium, we agree to call one side of the element $\Delta\sigma$ positive and the other side negative; the force $T \Delta\sigma$ will be thought to represent the action of the part of the body lying on the positive side upon the part on the negative side. Hence, if a unit normal ν is drawn (Fig. 6) to the surface element $\Delta\sigma$ so that it points in the direction of the positive side, then the action of the matter lying on the negative side of the normal upon that on the positive is $-T \Delta\sigma$. This latter statement follows directly from Newton's third law of motion.

It is obvious that the surface forces developed in a solid body are of much more complicated character than those in an ideal fluid at rest, since they need not be normal to the elements of surface. Clearly, the surface forces depend not only on the position of the surface element, but on its orientation as well. In order to bring into explicit evidence the dependence of the

---

[1] The alternating tensor e_{ijk} is defined to be zero if any two subscripts are equal, $+1$ if i, j, k is a cyclic permutation of 1, 2, 3, and -1 if i, j, k is a cyclic permutation of 1, 3, 2. We have, for example,

$$M_1 = \int_\tau (x_2 F_3 - x_3 F_2) \, d\tau.$$

stress T on the orientation of the element of surface, the stress vector will be written as $\overset{\nu}{T}$. It must be noted that in general $\overset{\nu}{T}$ is not in the direction ν.

13. Stress Tensor. It will be shown in this section that the state of stress at any point of the medium is completely characterized by the specification of nine quantities that are called the components of the stress tensor.

Let $P(x)$ be any point of the medium, and draw through P three planar elements parallel to the coordinate planes. The positive direction of the normals to these planar elements will be taken to coincide with the positive directions of the coordinate axes.

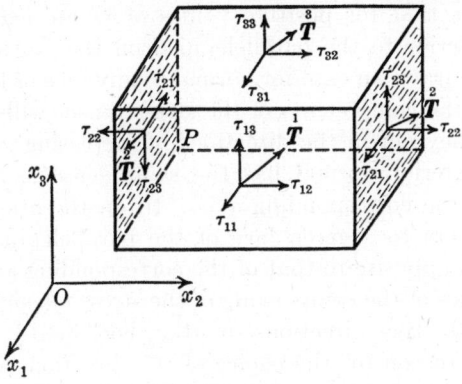

FIG. 7.

The planar elements so drawn will determine a small rectangular parallelepiped, shown in Fig. 7.

Let $\overset{i}{T}$ denote the stress vector acting on a face of the parallelepiped that is perpendicular to the x_i-axis, and resolve the vector $\overset{i}{T}$ into components along the coordinate axes to give $\overset{i}{T} = e_j \overset{i}{T}_j$. It will be convenient to write

(13.1) $$\overset{i}{T}_j = \tau_{ij}$$

so that

$$\overset{i}{T} = e_j \tau_{ij}.$$

We shall show that the nine quantities τ_{ij} are the components of a symmetric tensor, the *stress tensor*, and that the τ_{ij} determine completely the state of stress in the neighborhood of the point P, so that the stress vector $\overset{\nu}{T}$ can be calculated from them for any orientation $\mathbf{v}$. The quantities τ_{ij} are called the *components of the stress tensor*. We note that in τ_{23} (for example)

> the first subscript, 2, indicates the direction of the *normal* to the plane under consideration, while
> the second subscript, 3, indicates the direction of the *component* of the stress vector $\overset{2}{T}$.

The convention in regard to the signs of the scalar quantities $\tau_{11}, \tau_{12}, \cdots, \tau_{33}$ is the following: the arrows in Fig. 7 indicate the vectors $\mathbf{e}_1 \tau_{11}, \mathbf{e}_2 \tau_{12}, \cdots, \mathbf{e}_3 \tau_{33}$, representing the directions of the forces that for positive values of τ_{ij} are exerted by the material exterior to the parallelepiped on the matter within it. Thus, if one draws an exterior normal to any face of the parallelepiped, then the components of the stress tensor will be reckoned positive if they act in the directions of increasing x_1, x_2, and x_3 when the exterior normal has the same sense as the positive direction of the corresponding axis. If, on the other hand, the exterior normal to a given face of the parallelepiped points in the direction opposite to that of the corresponding axis, then the positive values of the components of the stress tensor are directed oppositely to the directions of the coordinate axes. This explains the reason for the choice of the directions of the forces acting on the shaded faces of the parallelepiped in Fig. 7.

It follows from this convention that tensile stresses are regarded as positive, compressive stresses as negative. In other words, if the material to the right of the parallelepiped pulls on the shaded face (Fig. 7), then the positive component τ_{22} indicates tension. The positive direction of the tensile stress τ_{22} acting on the shaded face on the left is opposite to that of the positive sense of the x_2-axis.

We shall call τ_{11}, τ_{22}, τ_{33} the *normal components* of stress; the other components are called *shears*.

In order to establish the connection between the stress tensor τ_{ij} and the stress vector $\overset{\nu}{T}$, which acts on an element of surface passing through the point $P(x)$ and with normal $\mathbf{v}$, draw through P three planar elements parallel to the coordinate planes, and pass a fourth plane ABC normal to $\mathbf{v}$ and at a distance h from P (Fig. 8). Let the area of the face ABC be σ; then the face

§13   ANALYSIS OF STRESS   39

normal to the x_i-axis will have an area $\sigma_i = \sigma \cos(x_i, \nu) = \sigma \nu_i$. The equilibrium of the tetrahedral element $PABC$ requires the vanishing of the resultant force acting on the matter within $PABC$, and we proceed to calculate the x_i-component of this force.

Let $\overset{\nu}{T}_i$, τ_{ij}, and F_i be the values of the stress vector, stress tensor, and body force at the point P; then, on account of the assumed continuity of the stress vector $\overset{\nu}{T}_i$, the x_i-component of the force acting on the face ABC of the tetrahedron is $(\overset{\nu}{T}_i + \epsilon_i)\sigma$,

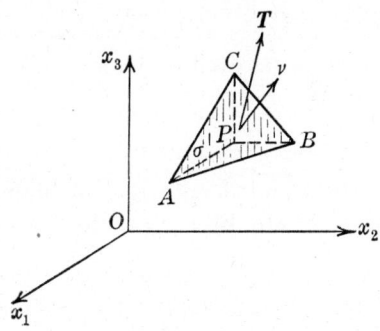

Fig. 8.

where $\lim_{h \to 0} \epsilon_i = 0$. The corresponding component of force due to stresses acting on the faces of areas σ_j is $(-\tau_{ji} + \epsilon_{ji})\sigma_j$, where $\lim_{h \to 0} \epsilon_{ij} = 0$ and the τ_{ij} are taken with the negative sign because the exterior normals to the faces of areas σ_j are directed oppositely to the direction of increasing x_j-coordinate. Finally, the contribution of the body force F_i to the x_i-component of the resultant force is $(F_i + \epsilon'_i)\tfrac{1}{3}h\sigma$, where $\tfrac{1}{3}h\sigma = \Delta\tau$ is the volume of the element $PABC$ and $\lim_{h \to 0} \epsilon'_i = 0$. Thus, for equilibrium of the tetrahedron we must have

(13.2) $(\overset{\nu}{T}_i + \epsilon_i)\sigma + (-\tau_{ji} + \epsilon_{ji})\sigma_j + (F_i + \epsilon'_i)\tfrac{1}{3}h\sigma = 0.$

If in (13.2) we set $\sigma_j = \sigma\nu_j = \sigma \cos(x_j, \nu)$, divide through by σ, and pass to the limit as $h \to 0$, we get

(13.3) $\overset{\nu}{T}_i = \tau_{ji}\nu_j.$

It is clear from (13.3) that having specified the components of the stress tensor τ_{ij} at any point $P(x)$ of the medium, one can calculate the stress vector $\overset{\nu}{T}$ on any element of surface whose orientation is determined by $\mathbf{v}$ and which passes through the point in question.

14. Note on Notation and Units. There is a deplorable lack of uniformity of notation and terminology in use by various writers on the theory of elasticity. Many British writers have adopted the notation for the components of the stress tensor introduced by Kirchhoff and write

$$\tau_{11} \equiv X_x, \qquad \tau_{12} \equiv Y_x, \qquad \tau_{21} \equiv X_y, \cdots, \qquad \tau_{33} \equiv Z_z.$$

Most American writers (as well as many Russian and German authors) write

$$\sigma_x \equiv \tau_{11}, \qquad \sigma_y = \tau_{22}, \qquad \sigma_z = \tau_{33}$$

for the *normal stresses* and denote the remaining six *tangential* or *shear* stresses τ_{12}, τ_{31}, etc. by τ_{xy}, τ_{zx}, etc.

The notation

$$\tau_{11} = \widehat{xx}, \qquad \tau_{12} = \widehat{xy}, \cdots, \qquad \tau_{33} = \widehat{zz}$$

has been suggested by K. Pearson and is quite convenient when one contemplates using orthogonal curvilinear coordinates. When it appears desirable to exhibit the dependence of the components of the stress tensor on the x, y, z-system of coordinates, we shall write $\tau_{11} = \tau_{xx}, \tau_{22} = \tau_{yy}, \tau_{23} = \tau_{yz}, \tau_{21} = \tau_{yx}$, etc. In this notation, formulas (13.3) read:

$$\overset{\nu}{T}_x = \tau_{xx} \cos (x, \nu) + \tau_{yx} \cos (y, \nu) + \tau_{zx} \cos (z, \nu)$$
$$\overset{\nu}{T}_y = \tau_{xy} \cos (x, \nu) + \tau_{yy} \cos (y, \nu) + \tau_{zy} \cos (z, \nu)$$
$$\overset{\nu}{T}_z = \tau_{xz} \cos (x, \nu) + \tau_{yz} \cos (y, \nu) + \tau_{zz} \cos (z, \nu).$$

From the definition of the stress vector, it follows that the stress vector T has the dimensions of

$$\frac{\text{force}}{\text{area}} \quad \text{or} \quad \frac{M}{LT^2}.$$

In the c.g.s. system, the stress is measured in dynes per square centimeter, while in practical units it is measured in pounds per square inch or in tons per square inch.

References for Collateral Reading

A. E. H. Love: A Treatise on the Mathematical Theory of Elasticity, Cambridge University Press, London, Secs. 41–48, pp. 74–80.

E. Trefftz: Handbuch der Physik, Verlag von Julius Springer, Berlin, vol. 6, Secs. 1–3.

R. V. Southwell: Theory of Elasticity for Engineers and Physicists, Oxford University Press, New York, Secs. 258–268, pp. 259–264.

S. Timoshenko: Theory of Elasticity, McGraw-Hill Book Company, Inc., New York, Secs. 1–4, 52.

15. Equations of Equilibrium. Consider a continuous medium every portion of which, contained within the volume τ and bounded by the closed surface σ, is in equilibrium. For equilibrium, the resultant force acting on the matter within τ must vanish, and we calculate now the x_i-component of this force.

Both body forces F and surface forces $\overset{\nu}{T}$ must be considered; the condition of equilibrium of forces requires that

$$\int_\tau F_i \, d\tau + \int_\sigma \overset{\nu}{T}_i \, d\sigma = 0,$$

or, making use of (13.3),

(15.1) $$\int_\tau F_i \, d\tau + \int_\sigma \tau_{ji} \nu_j \, d\sigma = 0.$$

Now if it is assumed that the functions τ_{ji} and their first partial derivatives $\tau_{ji,k} \equiv \dfrac{\partial \tau_{ji}}{\partial x_k}$ are continuous and single-valued in τ, then the Divergence Theorem[1] can be applied to the surface integral in (15.1) to yield

$$\int_\sigma \tau_{ji} \nu_j \, d\sigma = \int_\tau \tau_{ji,j} \, d\tau,$$

and (15.1) takes the form

(15.2) $$\int_\tau (F_i + \tau_{ji,j}) \, d\tau = 0.$$

Since the region of integration τ is arbitrary (every part of the medium is in equilibrium!) and since the integrand of (15.2) is continuous, it follows that the latter must vanish identically. Thus,

(15.3) $$\tau_{ji,j} = -F_i,$$

[1] See p. 20.

or, when written out in full in the notation explained in Sec. 14,

$$\frac{\partial \tau_{xx}}{\partial x} + \frac{\partial \tau_{yx}}{\partial y} + \frac{\partial \tau_{zx}}{\partial z} = -F_x,$$

$$\frac{\partial \tau_{xy}}{\partial x} + \frac{\partial \tau_{yy}}{\partial y} + \frac{\partial \tau_{zy}}{\partial z} = -F_y,$$

$$\frac{\partial \tau_{xz}}{\partial x} + \frac{\partial \tau_{yz}}{\partial y} + \frac{\partial \tau_{zz}}{\partial z} = -F_z.$$

Consider next the consequence of the vanishing of the resultant moment, which is produced by body and surface forces. Recalling the formula (12.2), the condition that the resultant moment due to body and surface forces vanishes can be written as

(15.4) $$M_i = \int_\tau e_{ijk} x_j F_k \, d\tau + \int_\sigma e_{ijk} x_j \overset{\nu}{T}_k \, d\sigma = 0.$$

With the aid of (13.3) and the Divergence Theorem, the surface integral in (15.4) can be transformed as follows:

$$\int_\sigma e_{ijk} x_j \overset{\nu}{T}_k \, d\sigma = \int_\sigma e_{ijk} x_j \tau_{lk} \nu_l \, d\sigma$$
$$= \int_\tau (e_{ijk} x_j \tau_{lk})_{,l} \, d\tau = \int_\tau e_{ijk} (x_j \tau_{lk,l} + \delta_{jl} \tau_{lk}) \, d\tau.$$

But $\delta_{jl} \tau_{lk} = \tau_{jk}$, and from equilibrium equations (15.3),

$$\tau_{lk,l} = -F_k,$$

so that the foregoing expression gives

$$\int_\sigma e_{ijk} x_j \overset{\nu}{T}_k \, d\sigma = \int_\tau e_{ijk} (-x_j F_k + \tau_{jk}) \, d\tau.$$

Accordingly, Eq. (15.4) becomes

$$\int_\tau e_{ijk} \tau_{jk} \, d\tau = 0,$$

and, since the integrand is continuous and the volume τ is arbitrary, we must have

(15.5) $$e_{ijk} \tau_{jk} = 0.$$

Equation (15.5) can be expanded to give, for example,

$$e_{123} \tau_{23} + e_{132} \tau_{32} = 0,$$

or, since $e_{123} = -e_{132} = +1$, $\tau_{23} = \tau_{32}$; one obtains similarly

$$\tau_{12} = \tau_{21} \quad \text{and} \quad \tau_{13} = \tau_{31}.$$

In short,

(15.6) $$\tau_{ij} = \tau_{ji};$$

that is, the stress tensor is symmetric. The symmetry of the components of the stress tensor allows us to write (15.3) as $\tau_{ij,j} = -F_i$ or, recalling the definition (13.1), $\overset{i}{T}_{j,j} = -F_i$; that is,

(15.7) $$\operatorname{div} \overset{i}{T} = -F_i.$$

Since the nine stress components τ_{ij} are bound by the three relations (15.6), we see that the state of stress at any point is completely characterized by the six quantities τ_{11}, τ_{22}, τ_{33}, $\tau_{12} = \tau_{21}$, $\tau_{23} = \tau_{32}$, $\tau_{31} = \tau_{13}$.

It follows from the foregoing that the six components of stress must satisfy the three partial differential equations (15.3),

$$\tau_{ij,j} = -F_i,$$

in the interior of the medium, and that on the surface bounding the medium they must satisfy the three boundary conditions (13.3),

$$\tau_{ij}\nu_j = \overset{\nu}{T}_i,$$

in which the functions F_i and $\overset{\nu}{T}_i$ are prescribed. It is clear that these equations are not sufficient for the complete determination of the state of stress, and one must have further information concerning the constitution of the body in order that the solution of Eqs. (15.3) be unique.

PROBLEM

Consider an elastic solid acted upon by body forces that exert moments M_i per unit volume (as in the case of a polarized dielectric solid under the action of an electric field). Show that in this case, Eq. (15.5) must be replaced by

$$e_{ijk}\tau_{jk} + M_i = 0.$$

What can be said in this case about the symmetry of the stress components? See in this connection Eric Reissner, "Note on the Theorem of the Symmetry of the Stress Tensor," *Journal of Mathematics and Physics*, vol. 23 (1944), pp. 192–194.

16. Transformation of Coordinates.

The symmetry of the shear components of the stress tensor ($\tau_{ij} = \tau_{ji}$) established in Sec. 15 is but a special case of a general theorem that will prove useful in establishing the laws of transformation of the components of the stress tensor under an orthogonal transformation of coordinate axes. We prove the following theorem:

THEOREM: *Let the surface elements $\Delta\sigma$ and $\Delta\sigma'$, with unit normals $\mathbf{v}$ and $\mathbf{v}'$, pass through the point P; then the component of the stress vector $\overset{v}{\mathbf{T}}$ (acting on $\Delta\sigma$) in the direction of $\mathbf{v}'$ is equal to the component of the stress vector $\overset{v'}{\mathbf{T}}$ (acting on $\Delta\sigma'$) in the direction of the normal $\mathbf{v}$.*

In vector notation, the theorem reads:

$$(16.1) \qquad \overset{v'}{\mathbf{T}} \cdot \mathbf{v} = \overset{v}{\mathbf{T}} \cdot \mathbf{v}'.$$

The proof of the theorem employs only Eq. (13.3) and the symmetry of the stress components. For

$$\overset{v'}{\mathbf{T}} \cdot \mathbf{v} = \overset{v'}{T}_i v_i = \tau_{ji} v'_j v_i$$
$$= (\tau_{ij} v_i) v'_j = \overset{v}{T}_j v'_j = \overset{v}{\mathbf{T}} \cdot \mathbf{v}',$$

and the theorem is proved.

The formula

$$(16.2) \qquad \overset{v'}{\mathbf{T}} \cdot \mathbf{v} = \tau_{ij} v'_i v_j,$$

obtained above, enables one to compute the component in any direction $\mathbf{v}$ of the stress vector acting on any given element with normal $\mathbf{v}'$. It will be used now to derive the formulas of transformation of the components of the stress tensor τ_{ij} when the latter is referred to a new coordinate system x'_i obtained from the old by a rotation of axes.

Since the stress component $\tau'_{\alpha\beta}$ (referred to the x'_i-system of coordinates) is the projection on the x'_β-axis of the stress vector acting on a surface element normal to the x'_α-axis, we can write

$$(16.3) \qquad \tau'_{\alpha\beta} = \overset{\alpha}{T}_\beta = \overset{v'}{\mathbf{T}} \cdot \mathbf{v},$$

where $\mathbf{v}'$ is parallel to the x'_α-axis and $\mathbf{v}$ is parallel to the x'_β-axis. Thus, (16.2) and (16.3) give

$$\tau'_{\alpha\beta} = \tau_{ij} v'_i v_j.$$

Then
$$\nu'_i = \cos(x'_\alpha, x_i) \equiv l_{\alpha i},$$
$$\nu_j = \cos(x'_\beta, x_j) \equiv l_{\beta j},$$
and we get
(16.4) $$\tau'_{\alpha\beta} = l_{\alpha i} l_{\beta j} \tau_{ij}.$$

The equations of transformation from the τ'_{ij} to $\tau_{\alpha\beta}$ have the form
(16.5) $$\tau_{\alpha\beta} = l_{i\alpha} l_{j\beta} \tau'_{ij}.$$

The law of transformation (16.4) is identical with that deduced in Sec. 5 for the transformation of the strain tensor and exhibits the tensor character of the quantities τ_{ij}. Indeed, these equations represent the transformation under rotation of axes of any tensor of rank two that is referred to a cartesian coordinate system.

If we set $\beta = \alpha$ in (16.4) and use the orthogonality relations
$$l_{\alpha i} l_{\alpha j} = l_{i\alpha} l_{j\alpha} = \delta_{ij},$$
we see that
$$\tau'_{\alpha\alpha} = l_{\alpha i} l_{\alpha j} \tau_{ij} = \delta_{ij} \tau_{ij} = \tau_{ii}$$
or
$$\tau'_{11} + \tau'_{22} + \tau'_{33} = \tau_{11} + \tau_{22} + \tau_{33}.$$

This result can be stated as a theorem.

THEOREM: *The expression*
$$\Theta = \tau_{11} + \tau_{22} + \tau_{33}$$
is invariant relative to an orthogonal transformation of coordinates.

This theorem states, in effect, that whatever be the orientation of three mutually orthogonal planes passing through a given point, the sum of the normal stresses is independent of the orientation of the planes.

References for Collateral Reading

A. E. H. LOVE: A Treatise on the Mathematical Theory of Elasticity, Cambridge University Press, London, Sec. 48, p. 80; Secs. 54–55, 56a, 56b, pp. 84–87.

R. V. SOUTHWELL: Theory of Elasticity for Engineers and Physicists, Oxford University Press, New York, Secs. 270–275, pp. 265–268.

E. TREFFTZ: Handbuch der Physik, Verlag von Julius Springer, Berlin, vol. 6, Secs. 4, 5, 7.

PROBLEMS

1. Show from (16.1), $\overset{\nu'}{T} \cdot \nu = \overset{\nu}{T} \cdot \nu'$, that if $\overset{P}{T}$ is the stress vector across a plane P, then the stress vector on any plane Q that contains $\overset{P}{T}$ lies in the plane P.

2. Show that the symmetry of the stress components $\tau_{ij} = \tau_{ji}$ follows from (16.1), $\overset{\nu'}{T} \cdot \mathbf{v} = \overset{\nu}{T} \cdot \mathbf{v}'$.

3. If $\overset{}{\underset{P}{T}}$ and $\overset{}{\underset{Q}{T}}$ are the stress vectors at a point and acting across planes P and Q, find the direction of the stress vector $\overset{}{\underset{R}{T}}$ on a plane R containing both $\overset{}{\underset{P}{T}}$ and $\overset{}{\underset{Q}{T}}$.

4. Show with the help of (16.1) that the normal stress has a stationary value (maximum or minimum) when the shear stress is zero. *Hint:*

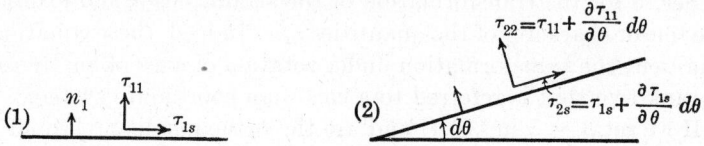

Let τ_{11} be the normal and τ_{1s} the shear stress across plane (1). Then by (16.1),

$$\left(\tau_{11} + \frac{\partial \tau_{11}}{\partial \theta} d\theta\right) \cos d\theta + \left(\tau_{1s} + \frac{\partial \tau_{1s}}{\partial \theta} d\theta\right) \sin d\theta = \tau_{11} \cos d\theta - \tau_{1s} \sin d\theta$$

or

$$\frac{\partial \tau_{11}}{\partial \theta} = -2\tau_{1s}.$$

That is, the normal stress across a surface element varies as the element is rotated and at a rate which is twice the shear component (with sign changed) perpendicular to the axis of rotation.

17. Stress Quadric of Cauchy. For the purpose of studying the nature of the distribution of stresses throughout a continuous medium, we define at each point $P(x)$ a quadric surface, the stress quadric of Cauchy. The discussion of this quadric will parallel closely that of the strain quadric in Secs. 5 and 6.

Consider an element of area with normal $\mathbf{v}$ and containing a point $P^0(x^0)$, and let $\overset{\nu}{T}$ be the stress vector acting on this surface element (Fig. 9). We introduce a local system of axes x_i with origin at P^0, and we denote by $\mathbf{A}$ the vector, in the direction of the normal $\mathbf{v}$, from P^0 to some point $P(x)$. The normal component N of $\overset{\nu}{T}$ can be written with the help of (16.2) as

$$N = \overset{\nu}{T} \cdot \mathbf{v} = \overset{\nu}{T}_i \nu_i = \tau_{ij} \nu_i \nu_j.$$

or, since $x_i = A\nu_i$,

(17.1) $$NA^2 = \tau_{ij}x_ix_j.$$

This suggests that we consider the quadratic function

(17.2) $$2F(x_1, x_2, x_3) = \tau_{ij}x_ix_j.$$

The length of the vector $\boldsymbol{A}$ is as yet unspecified; we restrict the coordinates x_i by requiring the end point $P(x)$ of $\boldsymbol{A}$ to lie on the quadric surface

(17.3) $$2F(x_1, x_2, x_3) = \tau_{ij}x_ix_j = \pm k^2,$$

where k is an arbitrary real constant and where the sign is chosen so as to make the surface real. From (17.3) and (17.1) it is seen that

(17.4) $$N = \pm \frac{k^2}{A^2}.$$

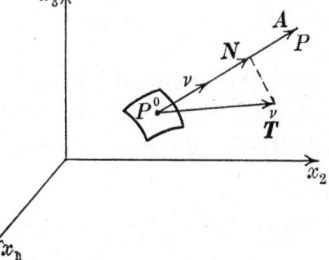

Fig. 9.

Since A^2 is a positive quantity, k^2 will be taken with the positive sign whenever the normal component N of $\overset{\nu}{\boldsymbol{T}}$ represents tension and with the negative sign when it represents compression. (Note the convention adopted in Sec. 13.)

If the coordinate axes are rotated to give a new coordinate system x'_i, then new stress components τ'_{ij} are determined, and the equation of the stress quadric becomes

$$\tau'_{ij}x'_ix'_j = NA^2 = \pm k^2.$$

But both N and A have values that do not depend on the particular coordinate system used, and hence

(17.5) $$\tau_{ij}x_ix_j = \tau'_{ij}x'_ix'_j.$$

Thus, the quadratic form $\tau_{ij}x_ix_j$ has a value that is independent of the choice of coordinate system. In other words, it is *invariant* with respect to an orthogonal transformation of coordinates.

The invariance of the form $\tau_{ij}x_ix_j$, shown by Eq. (17.5), affords an easy means of calculating the equations of transformation

(16.4). For (cf. Sec. 5) if one substitutes in the right-hand member of (17.5) the expressions for x_i' in terms of the x_i, namely,

$$x_i' = l_{i\alpha} x_\alpha,$$

then the resulting expression

$$\tau_{\alpha\beta} x_\alpha x_\beta = l_{i\alpha} l_{j\beta} \tau_{ij}' x_\alpha x_\beta$$

is an identity in the variables x_i. From this we get the equations of transformation (16.5),

$$\tau_{\alpha\beta} = l_{i\alpha} l_{j\beta} \tau_{ij}'.$$

From $2F(x_1, x_2, x_3) = \tau_{ij} x_i x_j$ and Eq. (13.3), it is seen that

(17.6) $$\frac{\partial F}{\partial x_i} = \tau_{ij} x_j = \tau_{ij} \nu_j A = A \overset{\nu}{T}_i.$$

Thus, the quadratic form $F(x_1, x_2, x_3)$ has some attributes of a potential function, since its derivatives with respect to the variables x_i are proportional to the corresponding components of force.

Since the $\dfrac{\partial F}{\partial x_i}$ are the direction ratios of the normal n to the

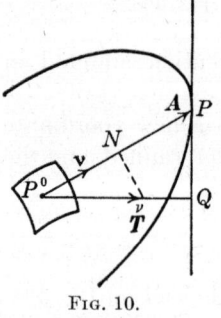

Fig. 10.

plane tangent to the quadric surface (17.3) at the point $P(x)$, we see from (17.6) that the stress vector $\overset{\nu}{T}$ is also normal to this tangent plane. This gives an easy means of constructing the stress vector $\overset{\nu}{T}$ from the knowledge of its normal component N. All that is necessary is to draw the quadric surface (17.3) and construct the tangent plane to the quadric through the terminus $P(x)$ of the vector A (Fig. 10). Then the vector $\overset{\nu}{T}$ is directed along the perpendicular $P^0 Q$ to the tangent plane. If the magnitude of N is known, one can readily determine the length of the vector $\overset{\nu}{T}$.

If the direction ν is taken along one of the principal axes of the quadric, then ν (and A) will be normal to the plane tangent to the surface at (x_i). But $\overset{\nu}{T}$ is perpendicular to the tangent plane so that, in this case, $\overset{\nu}{T}$ and ν coincide in direction; hence their components must be proportional. Thus,

(17.7) $$\overset{\nu}{T}_i = \tau \nu_i = \tau\, \delta_{ij} \nu_j$$

when ν lies along a principal axis of the stress quadric. Since ν is a unit vector and $\overset{\nu}{T} = \tau\nu$, the constant τ denotes the magnitude of the stress vector $\overset{\nu}{T}$ that acts on an element normal to a principal axis of the surface. For any direction ν we have $\overset{\nu}{T}_i = \tau_{ij}\nu_j$, and therefore $\tau_{ij}\nu_j = \tau\, \delta_{ij}\nu_j$, or

(17.8) $$(\tau_{ij} - \tau\, \delta_{ij})\nu_j = 0.$$

This set of three homogeneous equations in the unknown directions ν has a nonvanishing solution if, and only if, the determinant of the coefficients of the ν_j is equal to zero; that is,

(17.9) $$|\tau_{ij} - \tau\, \delta_{ij}| = 0.$$

This cubic equation in the stress τ is entirely analogous to Eq. (6.3) for the principal strains. Like the latter equation, it has three real roots τ_1, τ_2, τ_3, which are called the *principal stresses*. If τ in (17.8) is replaced by any one of these roots τ_i, then the resulting set of equations may be solved for the corresponding direction $\overset{i}{\nu}$. The three directions $\overset{i}{\nu}$ are termed the *principal directions of stress*, and the argument of Sec. 6 shows that these directions are orthogonal. The planes normal to the principal directions are called the *principal planes of stress*. If the vector ν is a principal direction $\overset{i}{\nu}$, then the associated stress vector $\overset{\nu}{T} = \tau\overset{i}{\nu}$ lies along the normal $\overset{i}{\nu}$, and the stress is normal. In other words, there is no shearing stress on a surface element tangent to a principal plane.

In general, there are only three mutually orthogonal principal axes of the quadric, so that at each point $P^0(x^0)$ of the medium one can find three mutually orthogonal directions $\overset{i}{\nu}$ such that the surface elements normal to these directions will experience no tangential stress. If the quadric surface is a surface of revolution, there will be infinitely many such directions $\overset{i}{\nu}$; one of them will be directed along the axis of revolution, and any two mutually perpendicular directions lying in the plane normal to the axis of revolution may be taken as the remaining principal axes. If $\tau_1 = \tau_2 = \tau_3$, the quadric is a sphere, and any three orthogonal lines may be chosen as the principal axes. In this case, whatever be the orientation of the surface element at the center of the sphere, the stress experienced by it will be purely normal.

We recall that τ_1, τ_2, τ_3 are the only stresses acting on the surface elements perpendicular to the principal directions $\overset{1}{\mathbf{v}}$, $\overset{2}{\mathbf{v}}$, $\overset{3}{\mathbf{v}}$, while τ_{11}, τ_{22}, τ_{33} are the normal stresses on elements perpendicular to the coordinate axes. If the coordinate axes are taken along the principal axes of the quadric, then the shear stresses τ_{12}, τ_{23}, τ_{31} disappear from the equation of the surface $\tau_{ij}x_ix_j = \pm k^2$, which now takes the form

(17.10) $\qquad \tau_1 x_1^2 + \tau_2 x_2^2 + \tau_3 x_3^2 = \pm k^2 = NA^2.$

The cubic equation (17.9) can be written as

$$|\tau_{ij} - \tau\,\delta_{ij}| = -\tau^3 + \Theta_1\tau^2 - \Theta_2\tau + \Theta_3 = 0,$$

where Θ_1, Θ_2, Θ_3 are the invariants of the stress tensor:[1]

(17.11) $\begin{cases} \Theta_1 = \tau_1 + \tau_2 + \tau_3 = \tau_{11} + \tau_{22} + \tau_{33} \equiv \Theta, \\ \Theta_2 = \tau_1\tau_2 + \tau_2\tau_3 + \tau_3\tau_1 \\ \quad = \begin{vmatrix} \tau_{22} & \tau_{23} \\ \tau_{23} & \tau_{33} \end{vmatrix} + \begin{vmatrix} \tau_{11} & \tau_{31} \\ \tau_{31} & \tau_{33} \end{vmatrix} + \begin{vmatrix} \tau_{11} & \tau_{12} \\ \tau_{12} & \tau_{22} \end{vmatrix}, \\ \Theta_3 = \tau_1\tau_2\tau_3 \\ \quad = \begin{vmatrix} \tau_{11} & \tau_{12} & \tau_{13} \\ \tau_{21} & \tau_{22} & \tau_{23} \\ \tau_{31} & \tau_{32} & \tau_{33} \end{vmatrix}. \end{cases}$

A reference to formulas (16.5) shows that one can write down at once the expressions for the components of the stress tensor τ_{ij} in terms of the principal stresses. Thus, if the direction cosines of the principal axes of stress X_i are given by the table

	X_1	X_2	X_3
x_1	l_{11}	l_{12}	l_{13}
x_2	l_{21}	l_{22}	l_{23}
x_3	l_{31}	l_{32}	l_{33}

[1] Cf. Eq. (6.9).

then one has the simple formula

(17.12) $$\tau_{ij} = \sum_{\alpha=1}^{3} l_{i\alpha} l_{j\alpha} \tau_\alpha.$$

The character of the distribution of stress at the point $P^0(x^0)$ depends on the signs of the principal stresses. (Note the agreement above concerning the choice of the sign of k^2.) If the principal stresses are all positive, then the equation of the stress quadric has the form

$$\tau_1 x_1^2 + \tau_2 x_2^2 + \tau_3 x_3^2 = k^2,$$

and the surface is an ellipsoid. Equation (17.4) now reads $N = k^2/A^2$, from which it follows that the force acting on every surface element passing through the point P^0 is tensile. If, on the other hand, all τ_i are negative, then (17.10) takes the form

$$\tau_1 x_1^2 + \tau_2 x_2^2 + \tau_3 x_3^2 = -k^2.$$

This surface is again an ellipsoid, but the normal component N of the stress vector $\overset{\nu}{T}$ this time is $N = -k^2/A^2$, and the stress is compressive.

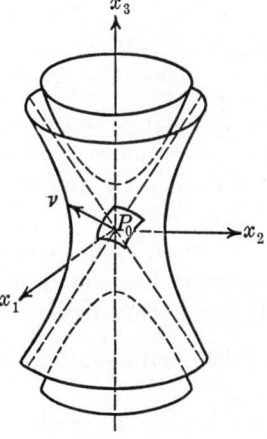

Fig. 11.

Consider next the case when $\tau_1 > 0$, $\tau_2 > 0$, $\tau_3 < 0$; Eq. (17.10) has one of the forms

$$\tau_1 x_1^2 + \tau_2 x_2^2 - |\tau_3| x_3^2 = k^2$$

or

$$\tau_1 x_1^2 + \tau_2 x_2^2 - |\tau_3| x_3^2 = -k^2,$$

depending on the orientation of the surface element at $P^0(x^0)$. The first of these equations represents an unparted hyperboloid and the second a biparted one (Fig. 11). If the normal to the surface element at P^0 cuts the biparted hyperboloid, then $N = -k^2/A^2$, so that the stress is compressive, while if the normal cuts the unparted hyperboloid, then $N = k^2/A^2$, and the stress is tensile. Vectors A that lie on the surface of the asymptotic cone

$$\tau_1 x_1^2 + \tau_2 x_2^2 - |\tau_3| x_3^2 = 0$$

do not cut either of the hyperboloids. In this case, $NA^2 = 0$, and hence $N = 0$. Accordingly, the elements of surface whose normals are directed along the generators of the cone experience only tangential stress.

It is easily shown that the case of $\tau_1 < 0$, $\tau_2 < 0$, $\tau_3 > 0$ does not differ essentially from that just considered. The only difference is in the regions in which the medium experiences compression and tension.

18. Maximum Normal and Shear Stresses. A further property of the principal stresses is worth noting. It is observed from Eq. (17.4) that the component N in the direction $\boldsymbol{\nu}$ of the stress vector $\overset{\nu}{T}$ is inversely proportional to the square of the radius vector A. Inasmuch as the extreme values of the radius vector are directed along the principal axes of the quadric (17.3), it follows that τ_1, τ_2, τ_3 represent the extremal values of the normal components τ_{11}, τ_{22}, τ_{33} of the stress tensor as the element of area is rotated to various orientations. Hence if $|\tau_1| < |\tau_2| < |\tau_3|$, then τ_1 is the least and τ_3 the greatest of the normal stresses acting at a given point $P^0(x^0)$ of the medium for various orientations of the planar elements passing through the point $P^0(x^0)$.

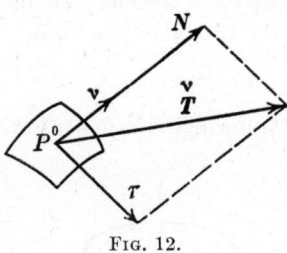

Fig. 12.

In order to investigate the maximum shearing stresses for various orientations $\boldsymbol{\nu}$ of the surface element passing through a given point $P(x)$, we employ a special coordinate system, namely, one that has the coordinate axes directed along the principal directions of stress. Then the shear components τ_{12}, τ_{23}, τ_{31} vanish, and equations $\overset{\nu}{T}_i = \tau_{ij}\nu_j$ [see (13.3)] give

$$\overset{\nu}{T}_1 = \tau_1\nu_1, \quad \overset{\nu}{T}_2 = \tau_2\nu_2, \quad \overset{\nu}{T}_3 = \tau_3\nu_3,$$

where ν_1, ν_2, ν_3 are the direction cosines of the normal $\boldsymbol{\nu}$ to any given plane. Hence

$$|\overset{\nu}{T}|^2 = \overset{\nu}{T}_1^2 + \overset{\nu}{T}_2^2 + \overset{\nu}{T}_3^2 = \tau_1^2\nu_1^2 + \tau_2^2\nu_2^2 + \tau_3^2\nu_3^2.$$

If we put $x_i^2 = A^2\nu_i^2$, Eq. (17.10) reads

$$N = \tau_1\nu_1^2 + \tau_2\nu_2^2 + \tau_3\nu_3^2.$$

§18 ANALYSIS OF STRESS

Therefore the square of the tangential component of $\overset{\nu}{T}$ on the same plane is (see Fig. 12)

$$(18.1) \quad \tau^2 = |\overset{\nu}{T}|^2 - N^2 = \tau_1^2 \nu_1^2 + \tau_2^2 \nu_2^2 + \tau_3^2 \nu_3^2 \\ - (\tau_1 \nu_1^2 + \tau_2 \nu_2^2 + \tau_3 \nu_3^2)^2.$$

In order to find the surface element with a maximum shearing stress τ, it is merely necessary to maximize (18.1) subject to the constraining relation

$$(18.2) \quad \nu_1^2 + \nu_2^2 + \nu_3^2 = 1.$$

When ν_3 is eliminated from (18.1) with the aid of (18.2), the resulting expression differentiated with respect to ν_1 and ν_2, and the derivatives set equal to zero, there result two equations for the extremal values of τ, namely,

$$(18.3) \quad \begin{cases} (\tau_1 - \tau_3)\nu_1[(\tau_1 - \tau_3)\nu_1^2 + (\tau_2 - \tau_3)\nu_2^2 - \tfrac{1}{2}(\tau_1 - \tau_3)] = 0, \\ (\tau_2 - \tau_3)\nu_2[(\tau_1 - \tau_3)\nu_1^2 + (\tau_2 - \tau_3)\nu_2^2 - \tfrac{1}{2}(\tau_2 - \tau_3)] = 0. \end{cases}$$

These equations are obviously satisfied if $\nu_1 = \nu_2 = 0$, $\nu_3 = \pm 1$, and this corresponds to a direction $\boldsymbol{\nu}$ along the x_3-axis and to the shearing stress $\tau = 0$. From symmetry, or by eliminating first ν_2 and then ν_1 from (18.1), we get the other two directions for which the shearing stress vanishes. In all, we have

$$\nu_1 = \nu_2 = 0, \quad \nu_3 = \pm 1, \quad \boldsymbol{\nu} \text{ along } x_3\text{-axis},$$
$$\nu_1 = \nu_3 = 0, \quad \nu_2 = \pm 1, \quad \boldsymbol{\nu} \text{ along } x_2\text{-axis},$$
$$\nu_2 = \nu_3 = 0, \quad \nu_1 = \pm 1, \quad \boldsymbol{\nu} \text{ along } x_1\text{-axis}.$$

For these three orientations of the normal the shear stresses vanish, as they should, since the directions of the coordinate axes were assumed to be the principal directions.

If in Eqs. (18.3) ν_1 is set equal to zero,[1] one obtains $\nu_1 = 0$, $\nu_2 = \pm 1/\sqrt{2}$, $\nu_3 = \pm 1/\sqrt{2}$, and with these values Eq. (18.1) yields $\tau = \pm \tfrac{1}{2}(\tau_2 - \tau_3)$. Again by symmetry, or by eliminating ν_2 and ν_1 in turn from (18.1), one finds two other directions and

---

[1] Note that if the roots are distinct, Eqs. (18.3) are inconsistent unless ν_1 or ν_2 vanishes. If $\tau_1 = \tau_2$, then $\nu_1^2 + \nu_2^2 = \tfrac{1}{2}$, $\nu_3^2 = \tfrac{1}{2}$, and

$$\tau = \pm \tfrac{1}{2}(\tau_1 - \tau_3),$$

and the directions ν_1 and ν_2 are not uniquely determined. This is to be expected, since in this case the stress quadric is a surface of revolution.

the corresponding shearing stresses. The three values of the shear stress are

(18.4) $\quad \tau = \pm \tfrac{1}{2}(\tau_2 - \tau_3), \ \tau = \pm \tfrac{1}{2}(\tau_1 - \tau_3), \ \tau = \pm \tfrac{1}{2}(\tau_1 - \tau_2),$

and these act, respectively, on planes whose normals have the direction cosines

(18.5) $\quad \begin{cases} \nu_1 = 0, & \nu_2 = \pm \dfrac{1}{\sqrt{2}}, & \nu_3 = \pm \dfrac{1}{\sqrt{2}}, \\ \nu_2 = 0, & \nu_1 = \pm \dfrac{1}{\sqrt{2}}, & \nu_3 = \pm \dfrac{1}{\sqrt{2}}, \\ \nu_3 = 0, & \nu_1 = \pm \dfrac{1}{\sqrt{2}}, & \nu_2 = \pm \dfrac{1}{\sqrt{2}}. \end{cases}$

It is to be noted that each of the planes whose normals are given by (18.5) passes through one of the principal axes and bisects the angle between the two remaining principal axes. We can state the following theorem.

THEOREM: *The maximum shearing stress is equal to one-half the difference between the greatest and least principal stresses and acts on the plane that bisects the angle between the directions of the largest and smallest of the principal stresses.*

Obviously one can obtain the same result by maximizing the expressions for the shear components $\tau_{12}, \tau_{23}, \tau_{31}$ given by the formulas (17.12), subject to the customary constraining relations between the direction cosines,

$$l_{\alpha i} l_{\alpha j} = l_{i\alpha} l_{j\alpha} = \delta_{ij}.$$

References for Collateral Reading

A. E. H. LOVE: A Treatise on the Mathematical Theory of Elasticity, Cambridge University Press, London, Secs. 49–53.

R. V. SOUTHWELL: Theory of Elasticity for Engineers and Physicists, Oxford University Press, New York, Secs. 272–279.

E. TREFFTZ: Handbuch der Physik, Verlag von Julius Springer, Berlin, vol. 6, Secs. 4–5.

S. TIMOSHENKO: Theory of Elasticity, McGraw-Hill Book Company, Inc., New York, Secs. 53–56, 60.

19. Examples of Stress. This section contains several examples closely paralleling those in Sec. 8. As in that section, we prefer to use the unabridged notation.

a. *Purely Normal Stress.* If for every plane passing through a point $P^0(x^0)$ the stress vector $\overset{\nu}{T}$ is normal to the plane, that is, if it is directed along the normal ν or opposite to it, then for any choice of rectangular coordinates

$$\tau_{xy} = \tau_{xz} = \tau_{yz} = 0, \quad \text{and} \quad \tau_{xx} = \tau_{yy} = \tau_{zz}.$$

The stress quadric in this case is a sphere whose equation is

$$x^2 + y^2 + z^2 = \frac{\pm k^2}{\tau_{xx}}.$$

Any set of orthogonal axes that pass through the point P^0 may be taken as principal axes of the quadric. This case corresponds to hydrostatic pressure if τ_{xx} is negative.

b. *Simple Tension or Compression.* A state of simple tension or compression is characterized by the fact that the stress vector for one plane through the point is normal to that plane, and the stress vector for any plane perpendicular to this one vanishes. Hence if the x'-, y'-, and z'-axes coincide with the principal axes of stress, then the stress quadric (17.3) has the equation

$$\tau_1 x'^2 = \pm k^2.$$

Transforming to any other orthogonal coordinate system x, y, z with the aid of (17.12), we obtain the following stress components:

$$\tau_{xx} = \tau_1 l_{11}^2, \qquad \tau_{yy} = \tau_1 l_{21}^2, \qquad \tau_{zz} = \tau_1 l_{31}^2,$$
$$\tau_{xy} = \tau_1 l_{11} l_{21}, \qquad \tau_{yz} = \tau_1 l_{21} l_{31}, \qquad \tau_{zx} = \tau_1 l_{31} l_{11},$$

where l_{11}, l_{21}, l_{31} are the direction cosines of the x'-axis relative to the axes x, y, z. A positive value of τ_1 represents tension, and a negative represents compression.

c. *Shearing Stress.* Consider a stress quadric

(19.1) $$2\tau x' y' = \pm k^2,$$

which is a hyperbolic cylinder whose elements are parallel to the z'-axis and which represents a shearing stress of magnitude τ. Equation (19.1) takes the form

$$\tau x^2 - \tau y^2 = \pm k^2,$$

when the axes are rotated through an angle of 45° about the z'-axis. A comparison of this equation with the general equation of the stress quadric when the latter is referred to the principal axes of stress

(19.2) $$\tau_{xx}x^2 + \tau_{yy}y^2 + \tau_{zz}z^2 = \pm k^2$$

shows that we must have

$$\tau_{zz} = 0, \qquad \tau_{xx} = -\tau_{yy} = \tau.$$

Thus, the shearing stress is equivalent to tension across one plane and compression of equal magnitude across a perpendicular plane.

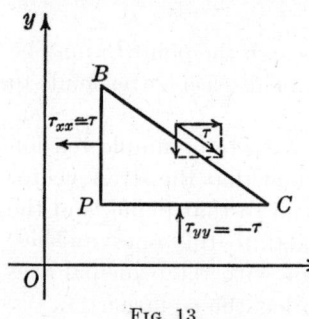

Fig. 13.

This can also be shown geometrically by considering the equilibrium of the element PBC (Fig. 13). Hence the stress on the face BC is a pure shear of magnitude $\tau = -\tau_{yy} = +\tau_{xx}$. This type of shearing stress would tend to slide planes of the material originally perpendicular to the y'-axis in a direction parallel to the x'-axis and planes of the material originally perpendicular to the x'-axis in a direction parallel to the y'-axis.

d. *Plane Stress.* If one of the principal stresses vanishes, then the stress quadric becomes a cylinder whose base is a conic, the stress conic. A state of stress, in this case, is said to be *plane*. The base of the cylinder lies in a plane containing the directions of the nonvanishing principal stresses. For example, if this plane is perpendicular to the z-axis, the equation of the quadric is

$$\tau_{xx}x^2 + \tau_{yy}y^2 + 2\tau_{xy}xy = \pm k^2.$$

For simple tension in the x-direction, the stress conic reduces to the pair of lines

$$x = \pm \sqrt{\frac{\pm k^2}{\tau_{xx}}}.$$

For the case of shear, the stress conic is a rectangular hyperbola

$$xy = \pm \frac{k^2}{2\tau_{xy}}.$$

If the stress conic is a circle, there is equal tension or compression in all directions in the plane of the circle.

CHAPTER III

STRESS-STRAIN RELATIONS

20. Hooke's Law. It has already been noted that the treatment contained in Chaps. I and II is applicable to all material bodies that can be represented with sufficient accuracy as continuous bodies; this chapter will be concerned with the characterization of elastic solids.

The first attempt at a scientific description of the strength of solids was made by Galileo. He treated bodies as inextensible, however, since at that time there existed neither experimental data nor physical hypotheses that would yield a relation between the deformation of a solid body and the forces responsible for the deformation. It was Robert Hooke who, some forty years after the appearance of Galileo's Discourses (1638), gave the first rough law of proportionality between the forces and displacements. Hooke published his law first in the form of an anagram "ceiiinosssttuu" in 1676, and two years later gave the solution of the anagram: "*ut tensio sic vis,*" which can be translated freely as "the extension is proportional to the force." To study this statement further, we discuss the deformation of a thin rod subjected to a tensile stress.

Consider a thin rod (of a low-carbon steel, for example), of initial cross-sectional area a_0, which is subjected to a variable tensile force F. If the stress is assumed to be distributed uniformly over the area of the cross section, then the *nominal stress* $T = F/a_0$ can be calculated for any applied load F. The *actual stress* is obtained, under the assumption of a uniform stress distribution, by dividing the load at any stage of the test by the actual area of the cross section of the rod at that stage. The difference between the nominal and the actual stress is negligible, however, throughout the elastic range of the material.

If the nominal stress T is plotted as a function of the extension e (change in length per unit length of the specimen), then for some ductile metals a graph like that in Fig. 14 is secured. The

graph is very nearly a straight line with the equation
(20.1) $$T = Ee$$
until the stress reaches the *proportional limit* (point P in Fig. 14). The position of this point, however, depends to a considerable extent upon the sensitivity of the testing apparatus. The constant of proportionality is known as Young's modulus.

In most metals, especially in soft and ductile materials, careful observation will reveal very small permanent elongations resulting from very small additional tensile forces. In many metals, however (steel and wrought iron, for example), if these very small permanent elongations are neglected (less than 1/100,000 of the length of a bar under tension), then the graph of stress against extension is a straight line, as noted above, and practically all the deformation disappears after the force has been removed. The greatest stress that can be applied without producing a permanent deformation is called the *elastic limit* of the material. When the applied force is increased beyond this fairly sharply defined limit, the material exhibits both elastic and plastic properties. The determination of this limit requires successive loading and unloading by ever larger forces until a permanent set is recorded. For many materials the proportional limit is very nearly equal to the elastic limit, and the distinction between the two is sometimes dropped, particularly since the former is more easily obtained.

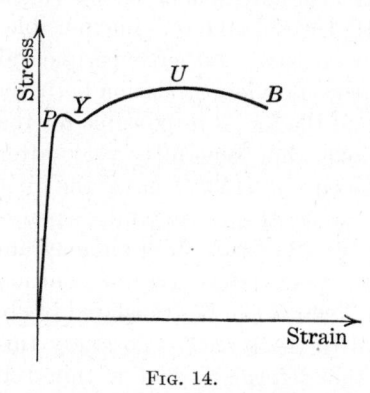

Fig. 14.

When the stress increases beyond the elastic limit, a point is reached (Y on the graph) at which the rod suddenly stretches with little or no increase in the load. The stress at point Y is called the *yield-point stress*.

The nominal stress T may be increased beyond the yield point until the *ultimate stress* (point U) is reached. The corresponding force $F = Ta_0$ is the greatest load that the rod will bear. When the ultimate stress is reached, a brittle material (such as a high-carbon steel) breaks suddenly, while a rod of some ductile metal begins to "neck"; that is, its cross-sectional area is greatly reduced over a small portion of the length of the rod. Further elongation is accompanied by an increase in actual stress but by a decrease in total load, in cross-sectional area, and in nominal stress until the rod breaks (point B).

The elastic limit of low-carbon steels is about 35,000 lb. per sq. in.; the ultimate stress is about 60,000 lb. per sq. in. Hard steels may be prepared with an ultimate strength greater than 200,000 lb. per sq. in.

We shall consider only the behavior of elastic materials subjected to stresses below the proportional limit; that is, we shall be concerned only with those materials and situations in which Hooke's law, expressed by Eq. (20.1), or a generalization of it, is valid.[1]

A natural generalization of Hooke's law immediately suggests itself, namely, one can invoke the principle of superposition of effects and assume that at each point of the medium the strain components e_{ij} are linear functions of the stress components τ_{ij}. This generalization was made by Cauchy, and the resulting law bears the name of *generalized Hooke's law*. It should be remembered that the relations between stresses and strains do not admit of direct experimental verification, since strains within the body are not susceptible to direct experimental measurement. Hence the validity of the law can be tested only indirectly by comparing the predictions of the theory, based on the law, with observed displacements of the points of a body when the latter is subjected to tests that were not even contemplated in the original formulation of the law.

Reference for Collateral Reading

A. E. H. LOVE: A Treatise on the Mathematical Theory of Elasticity, Cambridge University Press, London, Secs. 76–84.

21. Generalized Hooke's Law. We shall find it convenient to introduce the following notation:

[1] In order to give the reader some feeling regarding the magnitude of deformations with which the theory of elasticity deals, we note that a rod of iron with proportional limit of 25,000 lb. per sq. in., a yield point of 30,000 lb. per sq. in., and Young's modulus of 30,000,000 lb. per sq. in. will elongate under a load of 13,000 lb. per sq. in. about 0.0004 in. Even if the rod is loaded to the yield point, the determination of the extension will require very refined measurements.

(21.1) $\begin{cases} \tau_{11} = \tau_1,\ \tau_{22} = \tau_2,\ \tau_{33} = \tau_3,\ \tau_{23} = \tau_4,\ \tau_{31} = \tau_5,\ \tau_{12} = \tau_6, \\ e_{11} = e_1,\ e_{22} = e_2,\ e_{33} = e_3,\ 2e_{23} = e_4,\ 2e_{31} = e_5,\ 2e_{12} = e_6. \end{cases}$

The principal stresses will be denoted by τ_I, τ_II, τ_III, and the principal strains will be written as e_I, e_II, e_III.

Let it be supposed that at each point of the medium the components of the strain tensor e_{ij} are linear functions of the components of the stress tensor τ_{kl}, and conversely. We thus write

$$\tau_1 = c_{11}e_1 + c_{12}e_2 + c_{13}e_3 + c_{14}e_4 + c_{15}e_5 + c_{16}e_6,$$
$$\cdots\cdots\cdots\cdots\cdots\cdots\cdots\cdots\cdots\cdots\cdots\cdots$$
$$\tau_6 = c_{61}e_1 + c_{62}e_2 + c_{63}e_3 + c_{64}e_4 + c_{65}e_5 + c_{66}e_6.$$

These equations may be written in the compact form

(21.2) $\qquad\qquad \tau_i = c_{ij}e_j, \quad (i, j = 1, 2, \cdots, 6).$

The coefficients c_{ij} entering into (21.2) are independent of the strain components e_{ij}, but they may vary from point to point of the body. We shall term the coefficients c_{ij} the *elastic coefficients* of the material.

If the elastic properties of a body are the same in all directions about any given point, then the body is said to be *isotropic*. If it happens that the elastic properties of the body are independent of the position of the point, then the body is said to be *homogeneous*.

It should be noted that (21.2) is a very natural generalization of (20.1). Thus, if we assume the body to be isotropic and if we choose the x_i-axes as the axes of principal strain (and stress[1]), then (21.2) yields

$$\tau_\mathrm{I} = c_{11}e_\mathrm{I} + c_{12}e_\mathrm{II} + c_{13}e_\mathrm{III}.$$

Now if the body has the shape of a long, thin rod, then a tensile stress along the axis of the rod will produce an extension e_I in the direction of the axis and contractions e_II and e_III in the linear dimensions of the cross section of the rod. It is reasonable to assume that for small applied loads the stress τ_I is a linear function of e_I, e_II, and e_III.

Most structural materials are formed of crystalline substances, and hence very small portions of such materials cannot be regarded as isotropic. Nevertheless, the assumption of isotropy

---

[1] It will be shown in the next section that for an isotropic medium the principal axes of strain coincide with the principal axes of stress.

and homogeneity, when applied to an entire body, does not lead to serious discrepancies between experimental and theoretical results.[1] The reason for this agreement lies in the fact that the dimensions of most crystals are small in comparison with the dimensions of the entire body and that they are chaotically distributed, so that, in the large, the behavior of the substance is isotropic. It may be remarked that the process of rolling frequently produces a definite orientation of crystals, so that many rolled metals are anisotropic. Such a structural material as wood, for example, is definitely anisotropic, and the elastic properties of wood in the direction of the grain differ greatly from those in the directions perpendicular to the grain.

Symmetry of structure of the medium may introduce relations among the quantities c_{ij}. Because of their practical importance, we shall discuss two particular cases of elastic symmetry. These are: (1) symmetry with respect to a plane (in which case 13 independent elastic coefficients are involved[2]) and (2) symmetry with respect to three mutually perpendicular planes (involving nine independent functions c_{ij}). In the next section we shall prove that for an isotropic body the number of essential elastic coefficients reduces to two.

Consider a substance elastically symmetric with respect to the $x_1 x_2$-plane. This symmetry is expressed by the statement that the elastic functions c_{ij} are invariant under the transformation

$$x_1 = x_1', \qquad x_2 = x_2', \qquad x_3 = -x_3'.$$

The table of direction cosines of this transformation is

	x_1	x_2	x_3
x_1'	1	0	0
x_2'	0	1	0
x_3'	0	0	-1

and from formulas (5.9) and (16.4) it is seen that

$$\tau_i' = \tau_i, \qquad e_i' = e_i, \qquad (i = 1, 2, 3, 6),$$
$$\tau_4' = -\tau_4, \qquad e_4' = -e_4, \qquad \tau_5' = -\tau_5, \qquad e_5' = -e_5.$$

---

[1] Many cast metals are notable exceptions.
[2] See Eq. (21.5).

The first equation of (21.2) becomes

$$\tau_1' = c_{11}e_1' + c_{12}e_2' + c_{13}e_3' + c_{14}e_4' + c_{15}e_5' + c_{16}e_6',$$

or

$$\tau_1 = c_{11}e_1 + c_{12}e_2 + c_{13}e_3 - c_{14}e_4 - c_{15}e_5 + c_{16}e_6.$$

Comparison of this equation with the expression for τ_1 given by (21.2) shows that

$$c_{14} = c_{15} = 0.$$

Similarly, by considering $\tau_2', \cdots, \tau_6'$, we find that

$$c_{24} = c_{25} = c_{34} = c_{35} = c_{64} = c_{65} = 0,$$
$$c_{41} = c_{42} = c_{43} = c_{46} = c_{51} = c_{52} = c_{53} = c_{56} = 0.$$

For a material with one plane of elastic symmetry (which is taken to be the x_1x_2-plane), the matrix of the coefficients of the linear forms in (21.2) can be written as follows.

(21.3)
$$\begin{pmatrix} c_{11} & c_{12} & c_{13} & 0 & 0 & c_{16} \\ c_{21} & c_{22} & c_{23} & 0 & 0 & c_{26} \\ c_{31} & c_{32} & c_{33} & 0 & 0 & c_{36} \\ 0 & 0 & 0 & c_{44} & c_{45} & 0 \\ 0 & 0 & 0 & c_{54} & c_{55} & 0 \\ c_{61} & c_{62} & c_{63} & 0 & 0 & c_{66} \end{pmatrix}$$

Such materials as wood, for example, have three mutually orthogonal planes of elastic symmetry and are said to be *orthotropic*. In considering such materials, we shall choose the axes of coordinates so that the coordinate planes coincide with the planes of elastic symmetry. In this case, some of the coefficients c_{ij} exhibited in the array (21.3) vanish. Besides the symmetry with respect to the x_1x_2-plane, expressed by (21.3), the elastic functions c_{ij} must also be invariant under the transformation of coordinates defined by the following table of direction cosines.

	x_1	x_2	x_3
x_1'	-1	0	0
x_2'	0	1	0
x_3'	0	0	1

This change of coordinates is a reflection in the x_2x_3-plane and leaves the τ_i and e_i unchanged with the following exceptions:

$$\tau_5' = -\tau_5, \quad e_5' = -e_5, \quad \tau_6' = -\tau_6, \quad e_6' = -e_6.$$

§21 STRESS-STRAIN RELATIONS

From (21.3) we have

$$\tau_1 = c_{11}e_1 + c_{12}e_2 + c_{13}e_3 + c_{16}e_6.$$

This becomes

$$\tau_1' = c_{11}e_1' + c_{12}e_2' + c_{13}e_3' + c_{16}e_6',$$

or

$$\tau_1 = c_{11}e_1 + c_{12}e_2 + c_{13}e_3 - c_{16}e_6,$$

from which it follows that $c_{16} = 0$. By considering in a similar way the transformed expressions for $\tau_2, \cdots, \tau_6$, we find that[1] $c_{26} = c_{36} = c_{45} = c_{54} = c_{61} = c_{62} = c_{63} = 0$. Thus, for orthotropic media the matrix of the c_{ij} takes the following form.

(21.4)
$$\begin{pmatrix} c_{11} & c_{12} & c_{13} & 0 & 0 & 0 \\ c_{21} & c_{22} & c_{23} & 0 & 0 & 0 \\ c_{31} & c_{32} & c_{33} & 0 & 0 & 0 \\ 0 & 0 & 0 & c_{44} & 0 & 0 \\ 0 & 0 & 0 & 0 & c_{55} & 0 \\ 0 & 0 & 0 & 0 & 0 & c_{66} \end{pmatrix}$$

We return now to the discussion of the generalized Hooke's law (21.2) for an arbitrary elastic medium that may or may not possess elastic symmetry of some kind. It is possible to show from energy considerations,[2] for example, that the 36 elastic coefficients c_{ij} are symmetric; that is,

(21.5) $\qquad\qquad c_{ij} = c_{ji}, \qquad (i, j = 1, 2, \cdots, 6).$

Thus, Hooke's law contains in its most general form 21 elastic coefficients. This symmetry has not been assumed, however, in establishing the forms of the arrays of coefficients (21.3) and (21.4), nor will it be used in the next section, where Hooke's law (21.2) is specialized to that for an isotropic medium.

It is worth noting that the statement of the law (21.2) is not devoid of inconsistency. In the process of formulating the notion of the components of strain e_{ij}, it was assumed that the components of displacement u_i are functions of the coordinates (x_1, x_2, x_3) of the body in its undeformed state; that is, Lagran-

---

[1] Note that elastic symmetry in the x_1x_2-plane and in the x_2x_3-plane implies elastic symmetry in the x_1x_3-plane.

[2] This is shown in Sec. 26, where the relation between Hooke's law and the strain-energy function is discussed. The strain energy is seen to be a homogeneous quadratic function of the strain components.

gian coordinates were used. On the other hand, Eulerian coordinates were employed in defining the components of the stress tensor τ_{ij}; that is, it was assumed that the τ_{ij} are functions of the coordinates (x_1', x_2', x_3') of the stressed (and hence deformed) medium. Of course, if the displacements u_i and their derivatives are small, then the values of $\tau_{ij}(x)$ and $\tau_{ij}(x')$ cannot differ by a great deal. As an indication of the order of approximation involved here, note that if $x_k' = x_k + u_k$, then

$$\frac{\partial \tau_{ij}}{\partial x_k} = \frac{\partial \tau_{ij}}{\partial x_l'} \frac{\partial x_l'}{\partial x_k} = \frac{\partial \tau_{ij}}{\partial x_l'} \left(\delta_{kl} + \frac{\partial u_l}{\partial x_k} \right) = \frac{\partial \tau_{ij}}{\partial x_k'} + \frac{\partial \tau_{ij}}{\partial x_l'} \frac{\partial u_l}{\partial x_k}.$$

Hence, in writing $\dfrac{\partial \tau_{ij}}{\partial x_k} = \dfrac{\partial \tau_{ij}}{\partial x_k'}$, we assume that the displacement derivatives are small compared with unity. In what follows, it will be assumed that both the components of strain e_{ij} and the components of stress τ_{ij} are functions of the initial coordinates (x_1, x_2, x_3).

References for Collateral Reading

A. E. H. Love: A Treatise on the Mathematical Theory of Elasticity, Cambridge University Press, London, Secs. 60–65, pp. 92–100.

Chap. VI of Love's treatise is given to a discussion of the equilibrium of nonisotropic elastic solids and contains further references on the subject. Voigt's Lehrbuch der Krystallphysik is a standard treatise on the subject.

L. Lecornu: Théorie mathématique de l'élasticité, Mémorial des sciences mathématiques, Gauthiers-Villars & Cie, Paris.

Contains a discussion of the theory of Poincaré regarding the number of elastic constants in the generalized Hooke's law. See pp. 12–18.

PROBLEMS

1. Are the principal axes of strain coincident with those of stress for an anisotropic medium with Hooke's law expressed by Eq. (21.2)? for a medium with one plane of elastic symmetry? for an orthotropic medium? *Hint:* Take the coordinate axes along the principal axes of strain so that $e_4 = e_5 = e_6 = 0$.

2. Show directly from the generalized Hooke's law, Eq. (21.2), that in an isotropic body the principal axes of strain coincide with those of stress. *Hint:* Take the coordinate axes along the principal axes of strain ($e_4 = e_5 = e_6 = 0$), and consider the effect on τ_{23} and τ_{31} of a rotation of axes by 180° about the x_3-axis.

3. In the case of an orthotropic medium, show that the array (21.4) of the coefficients c_{ij} is not simplified further if reflection in the x_1x_3-plane is considered.

22. Homogeneous, Isotropic Bodies.

We recall that a body is said to be isotropic if its elastic properties are independent of the orientation of the axes of coordinates. In this case, the array (21.4) of the elastic coefficients c_{ij} can be simplified, and we shall show that the number of essential elastic coefficients reduces to two, whenever the body is isotropic. If the medium is both homogeneous and isotropic, then the elastic coefficients are constants.

Since the medium is isotropic, the coefficients c_{ij} must be invariant relative to all possible choices of coordinate systems; in particular, they must remain unchanged when we introduce new coordinate axes x_1', x_2', x_3', obtained by rotating the x_1, x_2, x_3-system through a right angle about the x_1-axis. By considering the transformed stress components τ_i', in exactly the same way as was done in the preceding section, it is found that

$$c_{12} = c_{13}, \quad c_{31} = c_{21}, \quad c_{32} = c_{23}, \quad c_{33} = c_{22}, \quad c_{66} = c_{55}.$$

Similarly, a rotation of axes through a right angle about the x_3-axis leads to the relations

$$c_{21} = c_{12}, \quad c_{22} = c_{11}, \quad c_{23} = c_{13}, \quad c_{31} = c_{32}, \quad c_{55} = c_{44}.$$

We introduce, finally, the coordinate system x_1', x_2', x_3', got from the x_1, x_2, x_3-system by rotating the latter through an angle of $45°$ about the x_3-axis. In this case, we have

$$\tau_{12}' = -\tfrac{1}{2}\tau_{11} + \tfrac{1}{2}\tau_{22}, \quad e_{12}' = -\tfrac{1}{2}e_{11} + \tfrac{1}{2}e_{22},$$

or, noting the definitions (21.1),

$$\tau_6' = -\tfrac{1}{2}\tau_1 + \tfrac{1}{2}\tau_2, \quad e_6' = -e_1 + e_2.$$

From (21.4) and the relation $c_{66} = c_{44}$, we have

$$\tau_6 = c_{44}e_6.$$

When referred to the x_1', x_2', x_3'-axes, this becomes $\tau_6' = c_{44}e_6'$ or

(22.1) $$-\tfrac{1}{2}\tau_1 + \tfrac{1}{2}\tau_2 = c_{44}(-e_1 + e_2).$$

Now from (21.4)

$$\tau_1 = c_{11}e_1 + c_{12}e_2 + c_{13}e_3,$$
$$\tau_2 = c_{21}e_1 + c_{22}e_2 + c_{23}e_3,$$

and from the relations given above, namely,

$$c_{22} = c_{11}, \quad c_{23} = c_{13} = c_{21} = c_{12},$$

we get

$$-\tfrac{1}{2}\tau_1 + \tfrac{1}{2}\tau_2 = \tfrac{1}{2}(c_{11} - c_{12})(-e_1 + e_2).$$

Comparison of this equation with (22.1) yields the result

(22.2) $$c_{44} = \tfrac{1}{2}(c_{11} - c_{12}) \equiv \mu,$$

so that

$$\tau_6 = 2\mu e_6.$$

We shall find it convenient to write Hooke's law for an isotropic body in terms of the two constants λ and μ, where μ is defined by (22.2) and where we put

$$c_{12} = \lambda.$$

From (21.4) we can now write

$$\begin{aligned}\tau_{11} &= c_{11}e_{11} + c_{12}e_{22} + c_{12}e_{33} \\ &= c_{12}(e_{11} + e_{22} + e_{33}) + (c_{11} - c_{12})e_{11} \\ &= \lambda\vartheta + 2\mu e_{11}.\end{aligned}$$

Thus, the generalized Hooke's law for a homogeneous isotropic body can be written in the following form:

(22.3) $$\tau_{ij} = \lambda \delta_{ij}\vartheta + 2\mu e_{ij}, \quad (i, j = 1, 2, 3).$$

Equation (22.3) yields a simple relation connecting the invariants $\vartheta = e_{ii}$ and $\Theta = \tau_{ii}$.

Putting $j = i$ in (22.3) and noting that $\delta_{ii} = \delta_{11} + \delta_{22} + \delta_{33} = 3$, one finds that

$$\Theta \equiv \tau_{ii} = 3\lambda\vartheta + 2\mu e_{ii}$$

or

(22.4) $$\Theta = (3\lambda + 2\mu)\vartheta.$$

Equations (22.3) can now be solved easily for the strains e_{ij} in terms of the stresses τ_{ij}. We have

$$e_{ij} = \frac{-\lambda}{2\mu}\delta_{ij}\vartheta + \frac{1}{2\mu}\tau_{ij},$$

or

(22.5) $$e_{ij} = \frac{-\lambda \delta_{ij}}{2\mu(3\lambda + 2\mu)}\Theta + \frac{1}{2\mu}\tau_{ij}.$$

If the axes x_i are directed along the principal axes of strain, then $e_{23} = e_{31} = e_{12} = 0$. But from (22.3) we see that in this case τ_{23}, τ_{31}, and τ_{12} also vanish. Hence the axes x_i must lie along the principal axes of stress, and we have the result that

the principal axes of stress are coincident with the principal axes of strain if the medium is isotropic.

Henceforth no distinction will be made between the principal axes of strain and those of stress, and such axes will be referred to simply as the *principal axes*.

The constants λ and μ were introduced by G. Lamé and are called the *constants of Lamé*.

Since it was assumed that Eqs. (21.2) possess an inverse, it is clear that we must require that $\mu \neq 0$ and $3\lambda + 2\mu \neq 0$.

PROBLEM

Given Hooke's law for an isotropic body with coordinate axes along the principal axes:

$$\tau_\mathrm{I} = \lambda(e_\mathrm{I} + e_\mathrm{II} + e_\mathrm{III}) + 2\mu e_\mathrm{I}$$
$$\cdots\cdots\cdots\cdots\cdots\cdots$$

Use the invariance of $\tau_{ij} x_i x_j$, $e_{ij} x_i x_j$, and $x_i x_i$ to derive Eq. (22.3),

$$\tau_{ij} = \lambda \, \delta_{ij} \vartheta + 2\mu e_{ij}.$$

23. Elastic Moduli of Isotropic Bodies. In order to gain some insight into the significance of the elastic constants entering into the formulas (22.3) and (22.5), we shall consider some simple examples.

Assume that a cylinder, with its axis parallel to the x_1-axis, is subjected to the action of a force F, which gives rise to a tension T that is uniform throughout the cross section of the cylinder. Suppose that the lateral surface of the cylinder is free from traction. Then if the action of the stress is directed along the x_1-axis, one can write

(23.1) $\quad \tau_{11} = T = \text{const.}, \; \tau_{22} = \tau_{33} = \tau_{12} = \tau_{23} = \tau_{31} = 0.$

Substitution of these values in the formulas (22.5) gives the expressions for the strains associated with the state of stress described by (23.1), namely,

(23.2) $\quad \begin{cases} e_{11} = \dfrac{(\lambda + \mu)T}{\mu(3\lambda + 2\mu)}, & e_{22} = e_{33} = \dfrac{-\lambda T}{2\mu(3\lambda + 2\mu)}, \\ e_{12} = e_{23} = e_{31} = 0. \end{cases}$

Noting that
$$\frac{e_{22}}{e_{11}} = \frac{-\lambda}{2(\lambda + \mu)},$$
we introduce the abbreviations

(23.3) $\qquad \sigma \equiv \dfrac{\lambda}{2(\lambda + \mu)}, \qquad E \equiv \dfrac{\mu(3\lambda + 2\mu)}{\lambda + \mu}.$

Then Eqs. (23.2) can be written in the form

(23.4) $\qquad \begin{cases} e_{11} = \dfrac{1}{E} T, \quad e_{22} = e_{33} = \dfrac{-\sigma}{E} T = -\sigma e_{11}, \\ e_{12} = e_{23} = e_{31} = 0. \end{cases}$

If the stress T represents tension, so that $T > 0$, then a tensile stress will produce an extension in the direction of the axis of the cylinder and a contraction in its cross section. Accordingly, for $T > 0$, we have $e_{11} > 0$, $e_{22} < 0$, $e_{33} < 0$. It follows that E and σ are both positive.

Physical interpretations of the elastic moduli E and σ are easily obtained. It follows from the first of the formulas (23.4) that the quantity
$$E = \frac{T}{e_{11}}$$
represents the ratio of the tensile stress T to the extension e_{11} produced by the stress T. Again, from (23.4), it is seen that
$$\sigma = \left|\frac{e_{22}}{e_{11}}\right| = \left|\frac{e_{33}}{e_{11}}\right|;$$
thus σ denotes the ratio of the contraction of the linear elements perpendicular to the axis of the cylinder to the longitudinal extension of the rod. The quantity E is known as *Young's modulus*, and the number σ is called the *ratio of Poisson*.

It is easy to verify that one can express the constants λ and μ in terms of Young's modulus and Poisson's ratio as

(23.5) $\qquad \lambda = \dfrac{E\sigma}{(1+\sigma)(1-2\sigma)}, \qquad \mu = \dfrac{E}{2(1+\sigma)}.$

Consider next the state of pure shear characterized by the stress components
$$\tau_{23} = T = \text{const.}, \qquad \tau_{11} = \tau_{22} = \tau_{33} = \tau_{12} = \tau_{31} = 0.$$

Substituting these values in (22.5) yields

(23.6) $\quad e_{23} = \dfrac{1}{2\mu} T, \quad e_{11} = e_{22} = e_{33} = e_{12} = e_{31} = 0.$

These formulas show that a rectangular parallelepiped $OPQR$, whose faces are parallel to the coordinate planes, is sheared in the x_2x_3-plane (see Fig. 4) so that the right angle between the edges of the parallelepiped parallel to the x_2- and x_3-axes is diminished, for $T > 0$, by the angle $\alpha_{23} = 2e_{23}$. From (23.6) we have

$$\mu = \frac{T}{\alpha_{23}}.$$

Thus the number μ represents the ratio of the shearing stress T to the change in angle α_{23} produced by the shearing stress. For this reason the quantity μ is called the *modulus of rigidity* or the *shear modulus*. Since E and σ are both positive, it follows from the second of Eqs. (23.5) that μ is also positive.

Finally, consider the case of hydrostatic pressure characterized by the stress system

(23.7) $\quad \begin{cases} \tau_{11} = \tau_{22} = \tau_{33} = -p, \quad \tau_{12} = \tau_{23} = \tau_{31} = 0, \\ \Theta = \tau_{11} + \tau_{22} + \tau_{33} = -3p. \end{cases}$

The state of strain corresponding to this state of stress [see (22.5)] is

(23.8) $\quad e_{11} = e_{22} = e_{33} = \dfrac{-p}{3\lambda + 2\mu}, \quad e_{12} = e_{23} = e_{31} = 0.$

The cubical compression $\vartheta = e_{ii}$ can be obtained either from (23.8) or from the general relation (22.4) and (23.7). We get

$$\vartheta = e_{11} + e_{22} + e_{33} = \frac{-p}{\lambda + \dfrac{2}{3}\mu}.$$

which can be written as

$$\vartheta = \frac{-p}{k} \quad \text{or} \quad k = \frac{-p}{\vartheta}$$

by introducing the abbreviation

(23.9) $\quad k = \lambda + \tfrac{2}{3}\mu.$

Thus, the quantity k represents the ratio of the compressive stress to the cubical compression, and for this reason it is called the *modulus of compression*. Since for all physical substances a hydrostatic pressure tends to diminish the bulk, it is clear that k is positive. Substituting in (23.9) the expressions for λ and μ from (23.5) gives

$$k = \frac{E}{3(1 - 2\sigma)}.$$

Since k is positive for all physical substances, it follows that σ is less than one-half, and hence [see (23.5)] λ is positive. For most structural materials, the value of σ does not deviate much from one-third. If the material is highly incompressible (rubber, for example), σ is nearly one-half, and $\mu \doteq E/3$.

In the engineering literature, the modulus of shear is often denoted by G, and the reciprocal of Poisson's ratio σ is denoted by m; that is,

$$m = \frac{1}{\sigma}.$$

The stress-strain relations (22.5), when written by making the substitutions from (23.5), assume the simple form

$$(23.10) \qquad e_{ij} = \frac{1 + \sigma}{E} \tau_{ij} - \frac{\sigma}{E} \delta_{ij} \Theta,$$

where $\Theta = \tau_{ii}$. If we recall the notation of Sec. 14, these relations can also be given in the following form:

$$(23.11) \quad \begin{cases} e_{xx} = \frac{1}{E}[\tau_{xx} - \sigma(\tau_{yy} + \tau_{zz})], \\ e_{yy} = \frac{1}{E}[\tau_{yy} - \sigma(\tau_{zz} + \tau_{xx})], \\ e_{zz} = \frac{1}{E}[\tau_{zz} - \sigma(\tau_{xx} + \tau_{yy})], \\ e_{yz} = \frac{1 + \sigma}{E} \tau_{yz}, \quad e_{zx} = \frac{1 + \sigma}{E} \tau_{zx}, \quad e_{xy} = \frac{1 + \sigma}{E} \tau_{xy}. \end{cases}$$

The following table gives average values of E, μ, and σ for several elastic materials; the moduli E and μ are given in millions of pounds per square inch.

	E	μ	σ (Experimental)	$\sigma = \dfrac{E}{2\mu} - 1$
Carbon steels............	29.5	11.5	0.29	0.283
Wrought iron............	28.0	11.0	0.28	0.273
Cast iron................	16.5	6.5	0.25	0.269
Copper (hot-rolled).......	15.0	5.6	0.33	0.339
Brass, 2:1 (cold-drawn)....	13.0	4.9	0.33	0.327
Glass...................	8.0	3.2	0.25	0.250
Spruce (along the grain)...	1.5	0.08		

References for Collateral Reading

A. E. H. Love: A Treatise on the Mathematical Theory of Elasticity, Cambridge University Press, London, Secs. 69–71.

E. Trefftz: Handbuch der Physik, Verlag von Julius Springer, Berlin, vol. 6, Secs. 11–12.

S. Timoshenko: Theory of Elasticity, McGraw-Hill Book Company, Inc., New York, Sec. 6.

PROBLEMS

1. Show that Hooke's law in the form (23.11) can be obtained by the following argument. An elementary rectangular parallelepiped subjected to tensile stresses τ_{xx} on opposite faces will experience a longitudinal extension $e_{xx} = \tau_{xx}/E$ and lateral contractions $e_{yy} = e_{zz} = -\sigma e_{xx}$. Now consider the effect of stresses $\tau_{xx}, \tau_{yy}, \tau_{zz}$ and superpose the resulting strains to get Eq. (23.11).

2. Use Hooke's law to show that the stress invariant $\Theta = \tau_{jj}$ and the strain invariant $\vartheta = e_{jj}$ are connected by the relation $\Theta = 3k\vartheta$, where k is the modulus of compression.

3. Show that a stress vector cannot cross a free surface (one on which there is no external load). *Hint:* Let $\mathbf{v}$ be the normal to the free surface. Then $\overset{\nu}{T} = 0$ and from (16.1) $\overset{\nu'}{T} \cdot \mathbf{v} = \overset{\nu}{T} \cdot \mathbf{v}' = 0$.

4. Derive the following relations between the Lamé coefficients λ and μ, Poisson's ratio σ, Young's modulus E, and the bulk modulus k:

$$\lambda = \frac{2\mu\sigma}{1 - 2\sigma} = \frac{\mu(E - 2\mu)}{3\mu - E} = k - \frac{2}{3}\mu = \frac{E\sigma}{(1 + \sigma)(1 - 2\sigma)}$$

$$= \frac{3k\sigma}{1 + \sigma} = \frac{3k(3k - E)}{9k - E},$$

$$\mu = \frac{\lambda(1 - 2\sigma)}{2\sigma} = \frac{3}{2}(k - \lambda) = \frac{E}{2(1 + \sigma)} = \frac{3k(1 - 2\sigma)}{2(1 + \sigma)}$$

$$= \frac{3kE}{9k - E},$$

$$\sigma = \frac{\lambda}{2(\lambda + \mu)} = \frac{\lambda}{3k - \lambda} = \frac{E}{2\mu} - 1 = \frac{3k - 2\mu}{2(3k + \mu)}$$
$$= \frac{3k - E}{6k},$$
$$E = \frac{\mu(3\lambda + 2\mu)}{\lambda + \mu} = \frac{\lambda(1 + \sigma)(1 - 2\sigma)}{\sigma} = \frac{9k(k - \lambda)}{3k - \lambda}$$
$$= 2\mu(1 + \sigma) = \frac{9k\mu}{3k + \mu} = 3k(1 - 2\sigma),$$
$$k = \lambda + \frac{2}{3}\mu = \frac{\lambda(1 + \sigma)}{3\sigma} = \frac{2\mu(1 + \sigma)}{3(1 - 2\sigma)}$$
$$= \frac{\mu E}{3(3\mu - E)} = \frac{E}{3(1 - 2\sigma)}.$$

24. Equilibrium Equations for an Isotropic Elastic Solid. The complete system of equations of equilibrium of a homogeneous isotropic elastic solid is made up of the following equations:

a. Equations of Equilibrium. From (15.3)

(24.1) $\qquad \tau_{ij,j} + F_i = 0, \qquad (i, j = 1, 2, 3);$

b. Stress-Strain Relations. From (22.3)

(24.2) $\qquad \tau_{ij} = \lambda\, \delta_{ij}\vartheta + 2\mu e_{ij},$

where
$$\vartheta = e_{ii};$$
and [from (7.5)]

(24.3) $\qquad e_{ij} = \tfrac{1}{2}(u_{i,j} + u_{j,i}).$

The systems of Eqs. (24.1) and (24.2) must be satisfied at every interior point of the body τ, and on the surface Σ of the body τ the stresses must fulfill the equilibrium conditions (13.3)

(24.4) $\qquad \tau_{ij}\nu_j = \overset{\nu}{T}_i,$

where the ν_i are the direction cosines of the exterior normal ν to the surface Σ, and $\overset{\nu}{T}$ is the stress vector acting on the surface element with normal ν. To these equations one must adjoin the equations of compatibility [from (10.9)]

(24.5) $\qquad e_{ij,kl} + e_{kl,ij} - e_{ik,jl} - e_{jl,ik} = 0.$

It will be shown in Sec. 27 that the system of Eqs. (24.1) and (24.2), subject to the conditions of equilibrium on the surface

(24.4), is complete in the sense that if there exists a solution of the system, then that solution is unique. There are nine equations in the system on the set of nine unknown functions τ_{ij}, u_i, $(i, j = 1, 2, 3)$. Once the displacements u_i are determined, the strain components e_{ij} entering into (24.2) are readily calculated with the aid of the formulas (24.3). We have assumed that the displacements u_i are continuous functions of class $C^{(3)}$ throughout the region τ, and a reference to (24.2) shows that the components of stress τ_{ij} are continuous of class $C^{(2)}$ in the same region. The equations of equilibrium (24.1) contain the components F_i of the body force F, and they are assumed to be prescribed functions of the coordinates x_i of the undeformed body. Typical examples of the body forces F, occurring in practical applications, are centrifugal forces and forces of gravitation. Furthermore, the components $\overset{\nu}{T}_i$ of the external surface force $\overset{\nu}{T}$ are assumed to be prescribed functions of the coordinates x_i of the undeformed surface Σ of the body.

In order that the solution of the problem may exist, it is clear that one cannot prescribe the body force F and the surface force $\overset{\nu}{T}$ in a perfectly arbitrary manner, inasmuch as Eqs. (24.1) were established on the hypothesis that the body is in equilibrium. Hence one must demand that the distribution of the forces F and $\overset{\nu}{T}$, acting on the body τ, be such that the resultant force and the resultant moment vanish.

It is clear from physical considerations that instead of prescribing the distribution of the surface force $\overset{\nu}{T}$ acting on Σ, one could prescribe the displacements u_i on the surface Σ, and that the state of stress established in the interior of the body by deforming its surface Σ must also be characterized in a unique way. Thus, we are led to consider the following fundamental boundary-value problems of elasticity:

Problem I. *Determine the distribution of stress and the displacements in the interior of an elastic body in equilibrium when the body forces are prescribed and the distribution of the forces acting on the surface of the body is known.*

Problem II. *Determine the distribution of stress and the displacements in the interior of an elastic body in equilibrium when* *the body forces are prescribed and the displacements of the points on the surface of the body are prescribed functions.*

In many applications, it is important to consider a problem resulting from the combination of the problems stated above. Thus, one may have the displacements of the points on part of the surface prescribed and the distribution of forces specified over the remaining portion. Such a problem will be referred to as a mixed boundary-value problem.

The formulation of the fundamental boundary-value problems of elasticity given above suggests the desirability of expressing the differential equations for Prob. I entirely in terms of stresses and those for Prob. II entirely in terms of displacements. This is not difficult to do.

Let us first obtain the equations in terms of displacements u_i by substituting in (24.1) the expressions for stresses in terms of displacements. Making use of the formulas (24.3), we can write the system (24.2) in the form

$$(24.6) \qquad \tau_{ij} = \lambda \, \delta_{ij} u_{k,k} + \mu(u_{i,j} + u_{j,i}).$$

Substituting the values of the stress components (24.6) in the equilibrium equations (24.1) gives

$$\mu u_{i,jj} + (\lambda + \mu) u_{j,ji} + F_i = 0,$$

or

$$(24.7) \qquad \mu \nabla^2 u_i + (\lambda + \mu) \frac{\partial \vartheta}{\partial x_i} + F_i = 0$$

where

$$\vartheta = e_{ii} = u_{i,i} = \operatorname{div} \mathbf{u}.$$

Equations (24.7) are associated with the name of Navier.

Note that we need not adjoin the compatibility equations (10.9), for the only purpose of the latter is to impose restrictions on the strain components that shall ensure that the e_{ij} yield single-valued, continuous displacements u_i, when the region τ is simply connected.

It is clear that Prob. II is completely solved if one obtains the solution of the system (24.7) subject to the boundary conditions

$$u_i = f_i(x_1, x_2, x_3), \qquad (i = 1, 2, 3),$$

where the f_i are prescribed continuous functions on the boundary of the undeformed solid. From the knowledge of the functions

u_i, one can determine the strains, and hence the stresses by making use of the relations (24.2).

We now turn our attention to the first boundary-value problem. It was noted earlier that not every solution of the system of three equations of equilibrium (24.1) corresponds to a possible state of strain in an elastic body, because the components of strain, defined by the system of Eqs. (23.10), must satisfy the equations of compatibility (24.5). We proceed to derive the compatibility equations in terms of the stresses. If the expressions (23.10)

$$e_{ij} = \frac{1+\sigma}{E}\tau_{ij} - \frac{\sigma}{E}\delta_{ij}\Theta$$

are inserted in the compatibility equations (24.5)

$$e_{ij,kl} + e_{kl,ij} - e_{ik,jl} - e_{jl,ik} = 0,$$

we obtain

(24.8) $\tau_{ij,kl} + \tau_{kl,ij} - \tau_{ik,jl} - \tau_{jl,ik}$
$$= \frac{\sigma}{1+\sigma}(\delta_{ij}\Theta_{,kl} + \delta_{kl}\Theta_{,ij} - \delta_{ik}\Theta_{,jl} - \delta_{jl}\Theta_{,ik}).$$

Since the indices i, j, k, l assume values 1, 2, 3, there are $3^4 = 81$ equations in the system (24.5), but not all of these are independent, for an interchange of i and j or of k and l obviously does not yield new equations. Also for certain values of the indices (such as $i = j = k = l$), Eqs. (24.5) are identically satisfied and, as already noted in Sec. 10, the set of Eqs. (24.5) contains only six independent equations obtained by setting

$$k = l = 1, \quad i = j = 2;$$
$$k = l = 2, \quad i = j = 3;$$
$$k = l = 3, \quad i = j = 1;$$
$$k = l = 1, \quad i = 2, \quad j = 3;$$
$$k = l = 2, \quad i = 3, \quad j = 1;$$
$$k = l = 3, \quad i = 1, \quad j = 2.$$

Inasmuch as Eqs. (23.10) establish one-to-one correspondence between the e_{ij} and the τ_{ij}, the set of 81 equations (24.8) likewise contains only six independent equations. If we combine Eqs. (24.8) linearly by setting $k = l$ and summing with respect to the common index, we get

$$\tau_{ij,kk} + \tau_{kk,ij} - \tau_{ik,jk} - \tau_{jk,ik}$$
$$= \frac{\sigma}{1+\sigma}(\delta_{ij}\Theta_{,kk} + \delta_{kk}\Theta_{,ij} - \delta_{ik}\Theta_{,jk} - \delta_{jk}\Theta_{,ik}).$$

This is a set of nine equations of which only six are independent because of the symmetry in i and j. Consequently, in combining Eqs. (24.8) linearly, the number of independent equations is not reduced, and hence the resultant set of equations is equivalent to the original one.

Noting that
$$\tau_{ij,kk} = \nabla^2 \tau_{ij}$$
and
$$\tau_{kk} = \Theta,$$
the foregoing equations can be written as

(24.9) $\quad \nabla^2 \tau_{ij} + \Theta_{,ij} - \tau_{ik,jk} - \tau_{jk,ik}$
$$= \frac{\sigma}{1+\sigma}(\delta_{ij}\nabla^2\Theta + 3\Theta_{,ij} - 2\Theta_{,ij}),$$

if we make use of the continuity of the second derivatives of Θ.

Equations (24.9) can be written more neatly by utilizing the equations of equilibrium (24.1)
$$\tau_{ik,k} + F_i = 0.$$

Thus, differentiating (24.1) with respect to x_j, we get

(24.10) $\quad \tau_{ik,kj} = -F_{i,j}$

and, since $\tau_{ik,kj} = \tau_{ik,jk}$, we can rewrite (24.9) in the form

(24.11) $\quad \nabla^2 \tau_{ij} + \frac{1}{1+\sigma}\Theta_{,ij} - \frac{\sigma}{1+\sigma}\delta_{ij}\nabla^2\Theta = -(F_{i,j} + F_{j,i}).$

This set of six independent equations can be further simplified by expressing an invariant $\nabla^2\Theta$ in terms of the derivatives of the body force F. This may be done as follows.

If we set $k = i$ and $l = j$ in (24.8) and sum with respect to the common indices, we get

$$2\tau_{ij,ij} - \tau_{ii,jj} - \tau_{jj,ii} = \frac{\sigma}{1+\sigma}(2\,\delta_{ij}\Theta_{,ij} - \delta_{ii}\Theta_{,jj} - \delta_{jj}\Theta_{,ii}).$$

But
$$\tau_{ii} = \tau_{jj} = \Theta, \qquad \delta_{ij}\Theta_{,ij} = \Theta_{,ii} = \nabla^2\Theta,$$

§24 STRESS-STRAIN RELATIONS

and
$$\delta_{ii}\Theta_{,jj} = \delta_{jj}\Theta_{,ii} = 3\nabla^2\Theta.$$

The foregoing equation can be written as
$$\tau_{ij,ij} - \nabla^2\Theta = \frac{\sigma}{1+\sigma}(\nabla^2\Theta - 3\nabla^2\Theta)$$

or

(24.12) $$\tau_{ij,ij} = \frac{1-\sigma}{1+\sigma}\nabla^2\Theta.$$

The differentiation of the equilibrium equation
$$\tau_{ij,i} = -F_j$$
gives
$$\tau_{ij,ij} = -F_{j,j},$$

and inserting this in the left-hand member of (24.12) yields the formula

(24.13) $$\nabla^2\Theta = -\frac{1+\sigma}{1-\sigma}F_{j,j} \equiv -\frac{1+\sigma}{1-\sigma}\,\text{div}\,\mathbf{F}.$$

Substituting from (24.13) in (24.11) gives the final form of the compatibility equation in terms of stresses,

(24.14) $$\nabla^2\tau_{ij} + \frac{1}{1+\sigma}\Theta_{,ij} = -\frac{\sigma}{1-\sigma}\delta_{ij}\,\text{div}\,\mathbf{F} - (F_{i,j} + F_{j,i}).$$

Equations (24.14), when written out in unabridged notation, yield the following six equations of compatibility:

(24.15)
$$\begin{cases} \nabla^2\tau_{xx} + \dfrac{1}{1+\sigma}\dfrac{\partial^2\Theta}{\partial x^2} = -\dfrac{\sigma}{1-\sigma}\,\text{div}\,\mathbf{F} - 2\dfrac{\partial F_x}{\partial x}, \\[4pt] \nabla^2\tau_{yy} + \dfrac{1}{1+\sigma}\dfrac{\partial^2\Theta}{\partial y^2} = -\dfrac{\sigma}{1-\sigma}\,\text{div}\,\mathbf{F} - 2\dfrac{\partial F_y}{\partial y}, \\[4pt] \nabla^2\tau_{zz} + \dfrac{1}{1+\sigma}\dfrac{\partial^2\Theta}{\partial z^2} = -\dfrac{\sigma}{1-\sigma}\,\text{div}\,\mathbf{F} - 2\dfrac{\partial F_z}{\partial z}, \\[4pt] \nabla^2\tau_{yz} + \dfrac{1}{1+\sigma}\dfrac{\partial^2\Theta}{\partial y\,\partial z} = -\left(\dfrac{\partial F_y}{\partial z} + \dfrac{\partial F_z}{\partial y}\right), \\[4pt] \nabla^2\tau_{zx} + \dfrac{1}{1+\sigma}\dfrac{\partial^2\Theta}{\partial z\,\partial x} = -\left(\dfrac{\partial F_z}{\partial x} + \dfrac{\partial F_x}{\partial z}\right), \\[4pt] \nabla^2\tau_{xy} + \dfrac{1}{1+\sigma}\dfrac{\partial^2\Theta}{\partial x\,\partial y} = -\left(\dfrac{\partial F_x}{\partial y} + \dfrac{\partial F_y}{\partial x}\right), \end{cases}$$

Equations (24.15) were obtained by Michell in 1900 and, for the case when the body forces are absent, by Beltrami in 1892. They are known as the Beltrami-Michell *compatibility equations*. Thus, in order to determine the state of stress in the interior of an elastic body, one must solve the system of equations consisting of (24.1) and (24.15) subject to the boundary conditions (24.4).

The system of Eqs. (24.1) and (24.15) is equivalent to the system consisting of Eqs. (24.1), (24.2), and (24.5).

If the field of body force F is conservative, so that

$$F = \nabla \varphi$$

or

$$F_j = \varphi_{,j},$$

then

$$\operatorname{div} F \equiv F_{j,j} = \varphi_{,jj} \equiv \nabla^2 \varphi,$$

and

$$F_{i,j} = \varphi_{,ij}, \qquad F_{j,i} = \varphi_{,ji} = \varphi_{,ij},$$

so that (24.14) can be written as

$$(24.16) \qquad \nabla^2 \tau_{ij} + \frac{1}{1+\sigma} \Theta_{,ij} = -\frac{\sigma}{1-\sigma} \delta_{ij} \nabla^2 \varphi - 2\varphi_{,ij}.$$

We shall consider two particular cases of body forces, namely, the case in which F is a constant vector and that in which the potential function φ is harmonic (that is, $\operatorname{div} F = \nabla^2 \varphi = 0$).

If F is constant, then φ is a linear function. In this case the right-hand member of (24.16) vanishes, and we obtain the equations of Beltrami,

$$(24.17) \qquad \nabla^2 \tau_{ij} + \frac{1}{1+\sigma} \Theta_{,ij} = 0.$$

From (24.13) it follows that in this case

$$\nabla^2 \Theta = 0,$$

so that $\Theta = \tau_{ii}$ is a harmonic function. Equation (22.4) shows that the strain invariant $\vartheta = e_{ii}$ is also harmonic; that is,

$$\nabla^2 \vartheta = 0$$

whenever Θ is harmonic. From (24.17) it is seen that if the τ_{ij}

are of class $C^{(4)}$, the components of stress satisfy the *biharmonic* equation

$$\nabla^2 \nabla^2 \tau_{ij} \equiv \nabla^4 \tau_{ij} = 0,$$

and since the strain components e_{ij} are linear functions of the τ_{ij}, we have

$$\nabla^4 e_{ij} = 0.$$

A function V of class $C^{(4)}$, and satisfying the equation $\nabla^4 V = 0$, is called a *biharmonic function*.

If the body force F is derived from a harmonic potential function, so that

$$\operatorname{div} \boldsymbol{F} = \nabla^2 \varphi = 0,$$

then from (24.13) and (22.4) we see that

$$\nabla^2 \Theta = 0, \quad \text{and} \quad \nabla^2 \vartheta = 0.$$

We can thus enunciate a theorem.

THEOREM: *When the components of the body force F are constant, the invariants Θ and ϑ are harmonic functions, and the stress components τ_{ij} and strain components e_{ij} are biharmonic functions.*

When F is derived from a harmonic potential function, the invariants Θ and ϑ are also harmonic.

It will be shown, with the aid of some general theorems to be established in Sec. 26, that Probs. I and II have essentially unique solutions. Before proceeding to derive these theorems, however, we may note that on account of the linear character of Eqs. (24.1), (24.2), and (24.3), the principle of superposition is applicable to the fundamental problems of elasticity.

Thus, suppose that one finds a set of nine functions

$$\tau_{ij}^{(1)}, u_i^{(1)}, \qquad (i, j = 1, 2, 3),$$

which satisfy the systems (24.1) and (24.2) with prescribed body forces $F_i^{(1)}$. Also let a set of functions

(24.18) $$\tau_{ij}^{(2)}, u_i^{(2)}, \qquad (i, j = 1, 2, 3)$$

be the solutions of the systems corresponding to the choice of the body forces $F_i^{(2)}$. Then it is obvious that the solution

(24.19) $\quad \tau_{ij} = \tau_{ij}^{(1)} + \tau_{ij}^{(2)}, \quad u_i = u_i^{(1)} + u_i^{(2)}, \quad (i = j = 1, 2, 3)$

will correspond to the choice of the body force whose components are $F_i^{(1)} + F_i^{(2)}$. If the set of functions (24.18) represents the solution of the homogeneous system, that is, when $F_i^{(2)} = 0$, then the expressions (24.19) represent solution of the problem corresponding to the choice of the body force with components $F_i^{(1)}$.

PROBLEMS

1. Show that the following stress components are not the solution of a problem in elasticity, even though they satisfy the equations of equilibrium with zero body forces:

$$\tau_{xx} = c[y^2 + \sigma(x^2 - y^2)], \quad c \neq 0,$$
$$\tau_{yy} = c[x^2 + \sigma(y^2 - x^2)],$$
$$\tau_{zz} = c\sigma(x^2 + y^2),$$
$$\tau_{xy} = -2c\sigma xy,$$
$$\tau_{yz} = \tau_{zx} = 0.$$

2. The solutions of many problems in elasticity are either exactly or approximately independent of the value chosen for Poisson's ratio. This fact suggests that approximate solutions may be found by so choosing Poisson's ratio as to simplify the problem. Show that if one takes $\sigma = 0$, then

$$\lambda = 0, \quad \mu = \tfrac{1}{2}E, \quad k = \tfrac{1}{3}E,$$

and Hooke's law is expressed by

$$\tau_{ij} = E e_{ij} = \tfrac{1}{2}E(u_{i,j} + u_{j,i}).$$

Show by differentiation of these equations that

$$\tau_{ij,ij} = \tfrac{1}{2}(\tau_{ii,jj} + \tau_{jj,ii})$$

(no sum on repeated subscripts). That is, the six stress components are connected, in this case, by the three equilibrium equations

$$\tau_{ij,j} + F_i = 0$$

and by three compatibility equations, namely,

$$\frac{\partial^2 \tau_{xy}}{\partial x\, \partial y} = \frac{1}{2}\left(\frac{\partial^2 \tau_{xx}}{\partial y^2} + \frac{\partial^2 \tau_{yy}}{\partial x^2}\right),$$

and two similar equations obtained by cyclic interchange of x, y, z. Derive these compatibility conditions from Eq. (24.8) by setting $\sigma = 0$, $k = i, l = j$.

A and L. Föppl have discussed[1] the simplification of the equations of elasticity obtained by choosing for Poisson's ratio $\sigma = 0$ or $\sigma = \tfrac{1}{2}$.

[1] A. and L. Föppl, Drang und Zwang, vol. 1, Sec. 3.

Westergaard[1] has treated the problem of obtaining the general solution from a solution for a particular choice of Poisson's ratio.

3. Define the stress function S by

$$\tau_{ij} = S_{,ij} \equiv \frac{\partial^2 S}{\partial x_i \, \partial x_j}$$

and consider the case of zero body force. Show that if Poisson's ratio σ is assumed to vanish, then the equilibrium and compatibility equations given in the preceding problem reduce to

$$\nabla^2 S = \text{const.}$$

4. Show that if Poisson's ratio σ has the value $\tfrac{1}{2}$, then

$$\mu = \tfrac{1}{3}E, \qquad \lambda = \infty, \qquad k = \infty, \qquad \vartheta \equiv e_{ii} = u_{i,i} = 0.$$

Interpret physically the situation described by these elastic coefficients. From Hooke's law (23.10) deduce the relations

$$\begin{aligned}\tau_{ij} &= 2\mu e_{ij} + \tfrac{1}{3}\delta_{ij}\Theta \\ &= \mu(u_{i,j} + u_{j,i}) + \tfrac{1}{3}\delta_{ij}\Theta.\end{aligned}$$

Show that in this case

$$u_{j,ij} = \frac{\partial \vartheta}{\partial x_i} = 0$$

and that the equilibrium equations (24.1) can be written in the form

$$\nabla^2 u_i + \frac{1}{\mu}\left(\frac{1}{3}\Theta_{,i} + F_i\right) = 0.$$

That is, putting $u_1 = u$, $u_2 = v$, etc., the four functions u, v, w, Θ are to be determined from the four equations

$$\nabla^2 u + \frac{1}{\mu}\left(\frac{1}{3}\frac{\partial \Theta}{\partial x} + F_x\right) = 0,$$

$$\nabla^2 v + \frac{1}{\mu}\left(\frac{1}{3}\frac{\partial \Theta}{\partial y} + F_y\right) = 0,$$

$$\nabla^2 w + \frac{1}{\mu}\left(\frac{1}{3}\frac{\partial \Theta}{\partial z} + F_z\right) = 0,$$

$$\frac{\partial u}{\partial x} + \frac{\partial v}{\partial y} + \frac{\partial w}{\partial z} = 0.$$

This case ($\sigma = \tfrac{1}{2}$) has been discussed at length by A. and L. Föppl.[2]

[1] H. M. Westergaard, "Effects of a Change of Poisson's Ratio Analyzed by Twinned Gradients," *Journal of Applied Mechanics*, vol. 62 (1940), pp. A-113–A-116.

[2] A. and L. Föppl, *Drang und Zwang*, vol. 1, Sec. 3.

25. Dynamical Equations of an Isotropic Elastic Solid.

The differential equations of motion of an elastic solid can be obtained at once from the equations of equilibrium (24.1) by invoking the Principle of D'Alembert and adding the forces of inertia to the components F_i of the body force. If $\rho(x_1, x_2, x_3)$ is the density of the medium, then the components of the force of inertia acting on the mass contained within the volume element $d\tau$ are $-\rho \frac{\partial^2 u_i}{\partial t^2} d\tau$. Hence adding to the components F_i of the body force $\boldsymbol{F}$ in (24.1) the components of the force of inertia per unit volume gives the system of equations

$$(25.1) \qquad \tau_{ij,j} + F_i = \rho \ddot{u}_i,$$

where we write $\frac{\partial^2 u_i}{\partial t^2} \equiv \ddot{u}_i$.

Inasmuch as the stress-strain relations (24.2) do not involve body forces, they remain valid in this case also. The displacements u_i are now regarded as functions of the space variables x_i and of the time t.

It follows that the dynamical equations in terms of the displacements u_i can be written at once by referring to the set of Eqs. (24.7). Thus,

$$(25.2) \qquad \mu \nabla^2 u_i + (\lambda + \mu) \frac{\partial \vartheta}{\partial x_i} + F_i = \rho \ddot{u}_i.$$

To these equations it is necessary to adjoin the initial and the boundary conditions. Thus, at each point of the surface Σ of the undeformed medium, the surface forces $\overset{\nu}{T}_i$ or the displacements u_i must be prescribed. The functions u_i prescribed on the surface Σ, in general, are functions of the space coordinates x_i and of the time t. If the surface forces $\overset{\nu}{T}_i$ are prescribed as functions of x_i and t, then the components of stress must satisfy the usual equilibrium conditions (24.4) on the surface Σ, and in addition one must know the initial conditions on the displacements u_i and on their time derivatives. We set forth these conditions explicitly for the fundamental boundary-value problems of dynamical elasticity that correspond to the problems of equilibrium in Sec. 24.

Problem I. *Determine the displacements* $u_i(x_1, x_2, x_3, t)$ *that satisfy the system of Eqs.* (25.2) *and satisfy the conditions*

$$u_i = u_i^0(x_1, x_2, x_3), \frac{\partial u_i}{\partial t} = U_i^0(x_1, x_2, x_3), \quad \text{for } t = t_0 \text{ throughout } \tau,$$

and that satisfy on the surface Σ *of the region* τ *the boundary conditions*

$$\overset{\nu}{T}_i = f_i(x_1, x_2, x_3, t) \qquad \text{for } t \geq t_0.$$

Problem II. *Determine the displacements* $u_i(x_1, x_2, x_3, t)$ *that satisfy the system of Eqs.* (25.2) *and are such that on the surface* Σ *of* τ

$$u_i = U_i(x_1, x_2, x_3, t) \qquad \text{for } t \geq t_0.$$

As in Sec. 24, we may consider a mixed boundary-value problem in which the surface forces $\overset{\nu}{T}_i$ are prescribed functions of x_i and t over part of the surface and the displacements u_i are given functions of x_i and t over the rest of the surface. As an example of such a problem, consider an elastic plate clamped at the edges. Let the plate, initially at rest, be subjected to a normal load varying with time. In this case, the displacements are known on the edges of the plate (for $t \geq t_0$), while the surface forces are given functions of x_1, x_2, x_3, t.

References for Collateral Reading

A. E. H. Love: A Treatise on the Mathematical Theory of Elasticity, Cambridge University Press, London, Secs. 85, 86, 91, 92.

E. Trefftz: Handbuch der Physik, Verlag von Julius Springer, Berlin, vol. 6, Secs. 13–15.

S. Timoshenko: Theory of Elasticity, McGraw-Hill Book Company, Inc., New York, Secs. 61, 63.

26. The Strain-Energy Function and Its Connection with Hooke's Law.

We introduce the definition of the *unstrained* or *natural* state of a body as a standard state of uniform temperature and zero displacement, with reference to which all strains will be specified.

If the body is in the natural state at the instant of time $t = 0$, and if it is subjected to the action of external forces, then the latter may produce a deformation of the body and hence will do work. We shall be concerned with the rate at which work is done by the external body and surface forces. If (x_1, x_2, x_3)

denote the coordinates of an arbitrary material point P of the body in the unstrained state, then at any time t the coordinates of the same material point P will be $x_i + u_i(x_1, x_2, x_3, t)$. Since the displacement of the point P in the interval of time $(t, t + dt)$ is given by

$$\frac{\partial u_i}{\partial t} dt \equiv \dot{u}_i \, dt,$$

it follows that the work done in dt seconds by the body forces acting on the volume element $d\tau$ located at P is $F_i \dot{u}_i \, d\tau \, dt$. The work performed by the external surface forces in the same interval of time is $\overset{\nu}{T}_i \dot{u}_i \, d\sigma \, dt$, where $d\sigma$ is the element of surface. Denoting by $\mathcal{E}$ the total work done by the body and surface forces, we have the following expression for the rate of doing work on the matter originally occupying some region τ,

$$(26.1) \qquad \frac{d\mathcal{E}}{dt} = \int_\tau F_i \dot{u}_i \, d\tau + \int_\sigma \overset{\nu}{T}_i \dot{u}_i \, d\sigma;$$

here σ denotes the original surface of the unstrained region τ.

Now the surface integral appearing in (26.1) can be expressed as a volume integral by substituting for the components of the surface force $\overset{\nu}{T}$ their values from Eqs. (24.4) and by making use of the Divergence Theorem. We have

$$(26.2) \qquad \int_\sigma \overset{\nu}{T}_i \dot{u}_i \, d\sigma = \int_\sigma (\tau_{ij} \dot{u}_i) \nu_j \, d\sigma = \int_\tau (\tau_{ij} \dot{u}_i)_{,j} \, d\tau.$$

Carrying out the indicated differentiation in the integrand of the volume integral in (26.2) and recalling the formulas (7.5) give

$$\int_\sigma \overset{\nu}{T}_i \dot{u}_i \, d\sigma = \int_\tau \tau_{ij,j} \dot{u}_i \, d\tau + \int_\tau \tau_{ij} \dot{u}_{i,j} \, d\tau$$

$$= \int_\tau \tau_{ij,j} \dot{u}_i \, d\tau + \int_\tau \tau_{ij} \left(\frac{\dot{u}_{i,j} + \dot{u}_{j,i}}{2} + \frac{\dot{u}_{i,j} - \dot{u}_{j,i}}{2} \right) d\tau$$

$$= \int_\tau \tau_{ij,j} \dot{u}_i \, d\tau + \int_\tau (\tau_{ij} \dot{e}_{ij} + \tau_{ij} \dot{\omega}_{ij}) \, d\tau.$$

But $\omega_{ij} = -\omega_{ji}$, so that $\tau_{ij} \dot{\omega}_{ij} = 0$, and hence

$$(26.3) \qquad \int_\sigma \overset{\nu}{T}_i \dot{u}_i \, d\sigma = \int_\tau (\tau_{ij,j} \dot{u}_i + \tau_{ij} \dot{e}_{ij}) \, d\tau.$$

A reference to the dynamical equations (25.1) shows that we can write

$$\tau_{ij,j}\dot{u}_i = (\rho\ddot{u}_i - F_i)\dot{u}_i.$$

When this is inserted in Eq. (26.3) and the resulting expression used in (26.1), one obtains

(26.4) $$\frac{d\mathcal{E}}{dt} = \int_\tau \rho\ddot{u}_i\dot{u}_i\,d\tau + \int_\tau \tau_{ij}\dot{e}_{ij}\,d\tau.$$

The kinetic energy K of the body is defined as

$$K \equiv \tfrac{1}{2}\int_\tau \rho\dot{u}_i\dot{u}_i\,d\tau,$$

and for the rate of change of kinetic energy we have[1]

$$\frac{dK}{dt} = \int_\tau \rho\ddot{u}_i\dot{u}_i\,d\tau.$$

Hence Eq. (26.4) can be written in the form

(26.5) $$\frac{d\mathcal{E}}{dt} = \frac{dK}{dt} + \int_\tau \tau_{ij}\frac{\partial e_{ij}}{\partial t}\,d\tau.$$

We proceed to show that if the change of state (from the initial to the stressed state) is reversible and is either adiabatic (no transfer of heat) or isothermal (no change in temperature), then there exists a function W, of the independent variables e_1, e_2, $\cdots$, e_6, with the property that[2]

$$\tau_i = \frac{\partial W}{\partial e_i}, \quad (i = 1, 2, \cdots, 6),$$

where the definitions of τ_i and e_i were given in (21.1). From the first law of thermodynamics we have

(26.6) $$\frac{d\mathcal{E}}{dt} = \frac{dK}{dt} + \frac{dU}{dt} - \frac{dQ}{dt},$$

where Q is the amount of heat, and U is the internal energy associated with the region τ.

---

[1] It is assumed here that the variation of the density ρ with time is negligible.

[2] We recall that

$$\tau_1 = \tau_{11},\ \tau_2 = \tau_{22},\ \tau_3 = \tau_{33},\ \tau_4 = \tau_{23},\ \tau_5 = \tau_{31},\ \tau_6 = \tau_{12},$$
$$e_1 = e_{11},\ e_2 = e_{22},\ e_3 = e_{33},\ e_4 = 2e_{23},\ e_5 = 2e_{31},\ e_6 = 2e_{12}.$$

For an adiabatic change of state,

$$\frac{dQ}{dt} = 0;$$

this corresponds physically to a body performing small and rapid vibrations during which no heat transfer takes place. If we define the energy density W by the relation

$$U = \int_\tau W \, d\tau,$$

then[1]

$$\frac{dU}{dt} = \frac{d}{dt} \int_\tau W \, d\tau = \int_\tau \frac{\partial W}{\partial t} \, d\tau.$$

Comparison of (26.6) and (26.5) shows that for the adiabatic case we can write

$$\int_\tau \frac{\partial W}{\partial t} \, d\tau = \int_\tau \tau_{ij} \frac{\partial e_{ij}}{\partial t} \, d\tau.$$

Since the region τ may be chosen arbitrarily, it follows that

$$\frac{\partial W}{\partial t} = \tau_{ij}(e_{11}, \cdots, e_{12}) \frac{\partial e_{ij}}{\partial t}.$$

Hence the change in W in dt seconds (for fixed values of x_1, x_2, x_3) is

$$dW = \tau_{ij}(e_{11}, \cdots, e_{12}) \, de_{ij}, \quad (i, j = 1, 2, 3),$$

or, noting the definitions (21.1),

$$dW = \tau_i(e_1, \cdots, e_6) \, de_i, \quad (i = 1, 2, \cdots, 6),$$

where $de_i \equiv \dfrac{\partial e_i}{\partial t} dt$. Now if the e_i are considered to be the independent variables, then we can write

$$\tau_i = \frac{\partial W}{\partial e_i}, \quad (i = 1, 2, \cdots, 6).$$

If, on the other hand, the (reversible) change of state is isothermal, then we introduce the entropy S, the temperature T,

---

[1] It should be remembered that τ is the region occupied by the particles in the unstrained state, and hence the boundary σ of τ does not vary with time.

and write the second law of thermodynamics in the form

$$\frac{dQ}{dt} = T\frac{dS}{dt}, \quad \frac{dT}{dt} = 0.$$

The physical situation that is described here is one in which the passage from the initial to the final state is made slowly and in which the elastic body remains at the same temperature as the surrounding medium. Equation (26.6) now takes the form

$$\frac{d\mathcal{E}}{dt} = \frac{dK}{dt} + \frac{d}{dt}(U - TS).$$

In this case $U - TS$ plays the role of potential energy, and we define W, the energy of deformation per unit (initial) volume, by the formula

$$U - TS = \int_\tau W \, d\tau.$$

Then
$$\frac{d}{dt}(U - TS) = \int_\tau \frac{\partial W}{\partial t} \, d\tau,$$

and
$$\frac{d\mathcal{E}}{dt} = \frac{dK}{dt} + \int_\tau \frac{\partial W}{\partial t} \, d\tau.$$

Comparison of this equation with (26.5) shows that

$$\int_\tau \frac{\partial W}{\partial t} \, d\tau = \int_\tau \tau_{ij} \frac{\partial e_{ij}}{\partial t} \, d\tau,$$

and, as before, we have the result that

$$\tau_i = \frac{\partial W}{\partial e_i}, \quad (i = 1, 2, \cdots, 6).$$

We assume now that the *strain-energy function* W can be expanded as a power series in the strain quantities e_i, and we write

$$2W = c_0 + 2c_i e_i + c_{ij} e_i e_j + \cdots.$$

We shall neglect, in this theory of infinitesimal deformations, terms of order three and higher in the strains; the constant term c_0 can also be neglected, since we are interested only in the derivatives of W. Thus, we have

$$\tau_i = \frac{\partial W}{\partial e_i} = c_i + \frac{1}{2}(c_{ij} + c_{ji})e_j.$$

If the stresses are to vanish with the strains, then we must set $c_i = 0$, and we get

$$W = \tfrac{1}{2} c_{ij} e_i e_j$$

and hence

$$\tau_i = \frac{\partial W}{\partial e_i} = \frac{1}{2}(c_{ij} + c_{ji}) e_j.$$

It is thus seen that the coefficients in the generalized Hooke's law are symmetric.[1] That is, if we define $\bar{c}_{ij} \equiv \tfrac{1}{2}(c_{ij} + c_{ji})$, then $\bar{c}_{ij} = \bar{c}_{ji}$, and we can write (dropping the bars on the coefficients $\bar{c}_{ij}$)

(26.7)
$$\begin{cases} W = \tfrac{1}{2} c_{ij} e_i e_j = \tfrac{1}{2} \tau_i e_i, & (i, j = 1, 2, \cdots, 6), \\ = \tfrac{1}{2} \tau_{ij} e_{ij}, & (i, j = 1, 2, 3), \\ = \tfrac{1}{2}(\tau_{11} e_{11} + \tau_{22} e_{22} + \tau_{33} e_{33} + 2\tau_{23} e_{23} + 2\tau_{31} e_{31} \\ \phantom{W = \tfrac{1}{2}(} + 2\tau_{12} e_{12}), \\ \tau_i = \dfrac{\partial W}{\partial e_i} = c_{ij} e_j, & (i, j = 1, 2, \cdots, 6), \\ c_{ij} = c_{ji}. \end{cases}$$

It may prove instructive to consider the following simple physical argument, which is sometimes used to derive the expression for the strain-energy function $W = \tfrac{1}{2} \tau_{ij} e_{ij}$, in the case of an isotropic medium. Let the axes x_1, x_2, x_3 be the principal axes, and consider a parallelepiped, of edges dx_i, subjected to a normal stress τ_I acting on the faces normal to the x_1-axis. The work done in proceeding from a state of zero stress and zero strain to one of stress τ_I and extension e_I is the product of the average force $\tfrac{1}{2} \tau_\mathrm{I} \, dx_2 \, dx_3$ by the elongation $e_\mathrm{I} \, dx_1$. Thus, the strain energy accumulated per unit volume is $W = \tfrac{1}{2} \tau_\mathrm{I} e_\mathrm{I}$. If stresses τ_II and τ_III are also acting, then by superposition we get

(26.8) $$W = \tfrac{1}{2}(\tau_\mathrm{I} e_\mathrm{I} + \tau_\mathrm{II} e_\mathrm{II} + \tau_\mathrm{III} e_\mathrm{III}).$$

It is obvious that (26.8) is the form to which the expression $W = \tfrac{1}{2} \tau_{ij} e_{ij}$ specializes when the axes are taken along the principal directions.

The expression for the strain-energy function W, when the stresses and strains are referred to any coordinate system x_i, can be derived from (26.8) either by means of the laws of transforma-

---

[1] It will be recalled that the symmetry of elastic coefficients for a homogeneous isotropic medium was established in Sec. 21 without invoking energy considerations.

tion of the tensor components τ_{ij} and e_{ij} or by utilizing the strain invariants ϑ, ϑ_1 and ϑ_2. We prefer to insert in (26.7) the expressions for the τ_{ij} from (22.3) and get the following form for the strain-energy function in an isotropic medium:

$$W = \tfrac{1}{2}\tau_{ij}e_{ij} = \tfrac{1}{2}\lambda\vartheta e_{ii} + \mu e_{ij}e_{ij},$$

or

(26.9) $\quad W = \tfrac{1}{2}\lambda(e_{11} + e_{22} + e_{33})^2$
$\qquad + \mu(e_{11}^2 + e_{22}^2 + e_{33}^2 + 2e_{23}^2 + 2e_{31}^2 + 2e_{12}^2).$

A quadratic form that takes only positive values for every set of values of the independent variables, not all zero, is said to be a *positive definite* form. Equation (26.9) shows that the strain-energy function W is a positive definite form in the strain components e_{ij}, since both λ and μ are positive constants. This property of the function W will be used in the next section, where we establish the uniqueness of solution of the boundary-value problems of elasticity.

It follows from the relation $\tau_i = \dfrac{\partial W}{\partial e_i}$ that one can use the strain-energy function to define the stress-strain relations whenever the form of the function W is known.[1] If, on the other hand, the stress-strain relations can be found experimentally, then the strain-energy function W can be calculated from $W = \tfrac{1}{2}\tau_{ij}e_{ij}$.

As a consequence of the linear character of the stress-strain relations, the strain-energy function W must be expressible as a positive definite quadratic form in the stress components τ_{ij}. From $W = \tfrac{1}{2}\tau_{ij}e_{ij}$ and from Hooke's law in the form (23.10), we get, upon expanding,

(26.10) $\quad W = \dfrac{-\sigma}{2E}(\tau_{11} + \tau_{22} + \tau_{33})^2 + \dfrac{1+\sigma}{2E}(\tau_{11}^2 + \tau_{22}^2 + \tau_{33}^2)$
$\qquad\qquad + \dfrac{1+\sigma}{E}(\tau_{12}^2 + \tau_{23}^2 + \tau_{31}^2).$

---

[1] See in this connection Webster's Dynamics, Sec. 174, and some recent work by F. D. Murnaghan on finite strains ["Finite Deformations of an Elastic Solid," *American Journal of Mathematics*, vol. 59 (1937), pp. 235–260], in which the approach to Hooke's law is made by way of the energy function.

It is readily verified that

$$\frac{\partial W}{\partial \tau_i} = e_i, \quad (i = 1, 2, \cdots, 6).$$

It is clear that W, the energy of deformation per unit volume, has a physical meaning that is independent of the choice of the coordinate system. Since W is invariant relative to an orthogonal transformation of axes, and since it can be expressed as a quadratic function of the strains e_{ij}, it follows that the function W must involve the strains only through combinations of the strain invariants[1] ϑ, ϑ_2, ϑ_3. Since W is a homogeneous quadratic function, it cannot contain ϑ_3 and must involve only

$$\vartheta = e_\mathrm{I} + e_\mathrm{II} + e_\mathrm{III}$$
$$= e_{11} + e_{22} + e_{33},$$

and
$$\vartheta_2 = e_\mathrm{II} e_\mathrm{III} + e_\mathrm{III} e_\mathrm{I} + e_\mathrm{I} e_\mathrm{II}$$
$$= e_{22} e_{33} + e_{33} e_{11} + e_{11} e_{22} - e_{23}^2 - e_{31}^2 - e_{12}^2.$$

We have, in fact,

(26.11) $$W = (\tfrac{1}{2}\lambda + \mu)\vartheta^2 - 2\mu\vartheta_2.$$

Before proceeding to the proof of the uniqueness of solution of the fundamental problems of the theory of elasticity, we establish an important theorem concerning the potential energy of deformation.

THEOREM: *If a body is in equilibrium under a given system of body forces F_i and surface forces $\overset{\nu}{T}_i$, then the potential energy of deformation is equal to one-half the work that would be done by the external forces (of the equilibrium state) acting through the displacements u_i from the unstressed state to the state of equilibrium.*

The theorem asserts that

(26.12) $$\int_\tau F_i u_i \, d\tau + \int_\sigma \overset{\nu}{T}_i u_i \, d\sigma = 2 \int_\tau W \, d\tau.$$

Now the surface integral in (26.12) can be transformed in exactly the same way as was done in obtaining the formula (26.3). Making use of the equilibrium equations (15.3) and of the relation (26.7), we write

$$\int_\sigma \overset{\nu}{T}_i u_i \, d\sigma = \int_\tau (\tau_{ij,j} u_i + \tau_{ij} e_{ij}) \, d\tau = \int_\tau (-F_i u_i + 2W) \, d\tau.$$

---

[1] It can be shown that every invariant of a tensor e_{ij} can be expressed in terms of the principal invariants ϑ, ϑ_2, ϑ_3.

Then

$$\int_\tau F_i u_i \, d\tau + \int_\sigma \overset{\nu}{T}_i u_i \, d\sigma = 2 \int_\tau W \, d\tau,$$

and the theorem is proved. This formula will be utilized in establishing the uniqueness of solution of the problems of equilibrium of an elastic solid.

References for Collateral Reading

L. Brillouin: Les tenseurs en mécanique et en élasticité, Masson et Cie, Paris, Chap. X.

A. E. H. Love: A Treatise on the Mathematical Theory of Elasticity, Cambridge University Press, London, Secs. 60–65, 68, 120.

S. Timoshenko: Theory of Elasticity, McGraw-Hill Book Company, Inc., New York, Sec. 39.

E. Trefftz: Handbuch der Physik, Verlag von Julius Springer, Berlin, vol. 6, Secs. 16–18.

C. Schaefer: Einführung in die theoretische Physik, Walter de Gruyter & Company, Berlin, vol. 1, Secs. 118–121.

A. G. Webster: Dynamics of Particles and Rigid Bodies, Verlag von Julius Springer, Berlin, Secs. 172–174.

PROBLEMS

1. From $W = \dfrac{1}{2} c_{ij} e_i e_j$, $\tau = \dfrac{\partial W}{\partial e_i}$, and Euler's theorem on homogeneous functions (see I. S. Sokolnikoff, Advanced Calculus, Sec. 26), show that

$$W = \tfrac{1}{2}\tau_i e_i, \qquad (i = 1, 2, \cdots, 6).$$

2. Use Eq. (26.8) for the strain-energy function W when referred to the principal axes, Hooke's law as typified by

$$\tau_\mathrm{I} = \lambda(e_\mathrm{I} + e_\mathrm{II} + e_\mathrm{III}) + 2\mu e_\mathrm{I},$$

and the strain invariants ϑ, ϑ_2 to show that for an isotropic body one can write

$$W = \tfrac{1}{2}\lambda(e_\mathrm{I} + e_\mathrm{II} + e_\mathrm{III})^2 + \mu(e_\mathrm{I}^2 + e_\mathrm{II}^2 + e_\mathrm{III}^2)$$
$$= \tfrac{1}{2}(\lambda + 2\mu)\vartheta^2 - 2\mu\vartheta_2.$$

3. Consider the general homogeneous polynomial of degree two in the variables e_1, e_2, e_3,

$$f(e_1, e_2, e_3) = a_{ij} e_i e_j, \qquad a_{ij} = a_{ji}.$$

Show that if the polynomial is symmetric in the variables e_j, then it can be written as

$$f(e_1, e_2, e_3) = a_{11}(e_1^2 + e_2^2 + e_3^2) + 2a_{12}(e_1 e_2 + e_2 e_3 + e_3 e_1),$$

and in terms of the symmetric functions

$$\vartheta = e_1 + e_2 + e_3,$$
$$\vartheta_2 = e_1 e_2 + e_2 e_3 + e_3 e_1,$$

the polynomial takes the form

$$f(e_1, e_2, e_3) = a_{11}\vartheta^2 + 2(a_{12} - a_{11})\vartheta_2.$$

27. Uniqueness of Solution of the Boundary-value Problems of Elasticity. This section is concerned with the proof of uniqueness of the solutions of the boundary-value problems of the theory of elasticity for simply and multiply connected domains.[1]

It is clear that in order that the solutions of the equilibrium boundary-value problems (see Sec. 24) may exist, it is necessary to demand the vanishing of the resultant force and the resultant torque produced by the prescribed body and surface forces. This condition was implied in the derivation of the equilibrium equations (15.3).

In order to establish the uniqueness of solution of the boundary-value problems formulated in Sec. 24, assume that it is possible to obtain two solutions

(27.1) $$u_i^{(1)}, \tau_{ij}^{(1)}, \qquad (i, j = 1, 2, 3),$$

and

(27.2) $$u_i^{(2)}, \tau_{ij}^{(2)}, \qquad (i, j = 1, 2, 3).$$

Because of the linear character of the differential equations (see Sec. 24), it is clear that the set of functions defined by the formulas

$$u_i \equiv u_i^{(1)} - u_i^{(2)}, \qquad \tau_{ij} \equiv \tau_{ij}^{(1)} - \tau_{ij}^{(2)}$$

will satisfy Eqs. (24.1) with $F_i = 0$. Thus, for the "difference"

---

[1] It will not be possible to devote any space in this book to the question of existence of solutions of the fundamental boundary-value problems of elasticity. Suffice it to say that a proof of the existence of solutions of the first and second boundary-value problems has been given under quite general conditions. See A. Korn, "Über die Lösung des Grundproblems des Elastizitätstheorie," *Mathematische Annalen*, vol. 75 (1914), pp. 497–544; L. Lichtenstein, "Über die erste Randwertaufgabe der Elastizitätstheorie," *Mathematische Zeitschrift*, vol. 20 (1924), pp. 21–28; and D. I. Shermann, "A Special Static Problem of the Elasticity Theory," *Applied Mathematics and Mechanics*, Academy of Sciences, U.S.S.R., vol. 7 (1943), pp. 341–360.

u_i, τ_{ij} of the two solutions, we have from the formula (26.12)

$$\int_\sigma \overset{\nu}{T}_i u_i \, d\sigma = 2 \int_\tau W \, d\tau.$$

But since solutions (27.1) and (27.2) satisfy the boundary conditions, it follows that the components $\overset{\nu}{T}_i = \overset{\nu}{T}_i^{(1)} - \overset{\nu}{T}_i^{(2)}$ of the external surface forces vanish in the case of the first boundary-value problem, and the displacements $u_i = u_i^{(1)} - u_i^{(2)}$ vanish on the surface σ for the case of the second boundary-value problem. It is also obvious that the integrand of the surface integral will vanish in the case of the mixed problem. We thus have in all cases

$$\int_\tau W \, d\tau = 0.$$

But W is a positive definite quadratic form in the components of strain, and hence the integral can vanish only when $W = 0$; that is, when $e_{ij} = 0$, $(i, j = 1, 2, 3)$. But $e_{ij} = e_{ij}^{(1)} - e_{ij}^{(2)}$, and it follows that the components of the strain tensor for the two solutions must be identical, and hence the components of the stress tensor are also identical. As regards the uniqueness of displacements, we recall from Sec. 10 that they are determined to within the quantities representing rigid body motions. In the case of the second and mixed boundary-value problems, the displacements are determined uniquely, since they are prescribed at least over part of the surface of the body.

It is important to note that the foregoing proof assumes that the displacements u_i are single-valued functions, but imposes no restrictions on the connectivity of the region.

Consider now the dynamical case of Sec. 25, and assume that there are two solutions of the type (27.1) and (27.2) that satisfy the boundary conditions. Then, as above, the difference of two solutions

$$u_i, \tau_{ij}, \qquad (i, j = 1, 2, 3)$$

satisfies the differential equations when body forces are set equal to zero. We have in all cases the condition that

(27.3) $$\overset{\nu}{T}_i \frac{\partial u_i}{\partial t} = 0 \qquad \text{on } \sigma, \qquad t \geq t_0.$$

For in the case of the first dynamical problem, $\overset{\nu}{T_i} = 0$ on σ, $t \geq t_0$, and in the case of the second problem, the components of velocity $\dfrac{\partial u_i}{\partial t} = 0$ on σ, $t \geq t_0$, since $u_i = 0$ for all $t \geq t_0$.

Recalling that the displacements u_i correspond to the solution of Eqs. (25.2) when body forces are absent, and noting the expression (27.3), leads to the conclusion that both integrals in the expression (26.1) vanish, so that Eq. (26.5) becomes[1]

$$\frac{dK}{dt} + \frac{dU}{dt} = 0,$$

or

$$K + U = \text{const.}$$

But the constant of integration in the above formula must be zero, since the displacements u_i and the velocities $\dfrac{\partial u_i}{\partial t}$ vanish at the instant $t = t_0$. Hence

$$K + U = 0,$$

and since both the kinetic energy K and the function U are essentially positive, one has

$$K = U = 0 \qquad \text{for all } t \geq t_0.$$

It follows from these equations that

$$\frac{\partial u_j}{\partial t} = 0, \quad \text{and} \quad e_{ij} = 0, \quad (i, j = 1, 2, 3),$$

for all values of $t \geq t_0$. The first of the above-written relations states that we are dealing with a static case, and the second means that deformation of the body is not present, so that the solution (u_1, u_2, u_3) represents a rigid body motion. But the displacements (u_1, u_2, u_3) vanish at $t = t_0$, and hence rigid body motion cannot be present in our solution, or

$$u_1 = u_2 = u_3 = 0 \qquad \text{for all } t \geq t_0.$$

Thus, the two assumed solutions (27.1) and (27.2) are identical.

---

[1] In an isothermal case $U - TS = \displaystyle\int_\tau W \, d\tau$ plays the role of potential energy.

References for Collateral Reading

A. E. Love: A Treatise on the Mathematical Theory of Elasticity, Cambridge University Press, London, Secs. 118, 124.
C. Schaefer: Einführung in die theoretische Physik, Walter de Gruyter & Company, Berlin, vol. 1, Sec. 122.
S. Timoshenko: Theory of Elasticity, McGraw-Hill Book Company, Inc., New York, Sec. 64.
E. Trefftz: Handbuch der Physik, Verlag von Julius Springer, Berlin, vol. 6, Secs. 22–23.

28. Saint-Venant's Principle. It is obvious from the formulation of the fundamental boundary-value problems of the theory of elasticity that the exact solution of these problems is likely to present formidable mathematical difficulties because of the complicated form of the boundary conditions. Frequently it is possible to obtain a solution of the problem if the boundary conditions are slightly modified, and it is worth noting that in the technological applications of the theory of elasticity one can only approximate the mathematical formulation of the boundary conditions, so that the mathematical solution of the problem represents only an approximation to the actual situation. In 1855, B. de Saint-Venant expressed a principle that agrees well with the applications of the theory of elasticity to practical problems. The essence of the principle can be stated as follows:

If some distribution of forces acting on a portion of the surface of a body is replaced by a different distribution of forces acting on the same portion of the body, then the effects of the two different distributions on the parts of the body sufficiently far removed from the region of application of the forces are essentially the same, provided that the two distributions of forces are statically equivalent.

The phrase "statically equivalent" means that the two distributions of forces have the same resultant force and the same resultant moment.

To illustrate the meaning of the principle, consider a long beam, one end of which is fixed in a rigid wall, while the other is acted upon by a distribution of forces that gives rise to a resultant force F and a couple of moment M. Now there are infinitely many distributions of forces that may act on the end of the beam and that will have the same resultant F and the same resultant moment M. The principle of Saint-Venant asserts that while the distributions of stresses and strains near the region of application may differ greatly, the eccentricities of the local distribution will have no appreciable effect on the state of stress far enough from the points of application, so long as the systems of applied forces are statically equivalent. This principle is of great usefulness in practical applications, since it permits one to alter the boundary conditions and thus simplify the problem.

One would suspect from the generality of the statement of the principle that the latter is not easy to justify in all cases on purely mathematical grounds.[1] In specific instances, one can calculate the distribution of stresses produced by various statically equivalent systems of forces, and in problems on beams, for example, it is reasonable to assume that the local eccentricities are not felt at distances that are about ten times the greatest linear dimension of the area over which the forces are distributed.

References for Collateral Reading

A. E. H. LOVE: A Treatise on the Mathematical Theory of Elasticity, Cambridge University Press, London, Sec. 89.

S. TIMOSHENKO: Theory of Elasticity, McGraw-Hill Book Company, Inc., New York, Sec. 15.

R. V. SOUTHWELL: Theory of Elasticity for Engineers and Physicists, Oxford University Press, New York, Secs. 92–94.

---

[1] For various arguments as to the plausibility of the principle of Saint-Venant, based on considerations of the strain energy, see R. V. Southwell, "On Castigliano's Theorem of Least Work, and the Principle of St. Venant," *Philosophical Magazine*, ser. 6, vol. 45 (1923), p. 193; J. N. Goodier, "A General Proof of Saint-Venant's Principle," *Philosophical Magazine*, ser. 7, vol. 23 (1937), p. 607; and J. N. Goodier, "An Extension of Saint-Venant's Principle, with Applications," *Journal of Applied Physics*, vol. 13 (1942), p. 167. See also J. N. Goodier, "Supplementary Note on a General Proof of Saint-Venant's Principle," *Philosophical Magazine*, ser. 7, vol. 24 (1937), p. 325; O. Zanaboni, "Dimostrazione generale del principo del De Saint-Venant," *Atti della reale accademia nazionale dei lincei*, ser. 6, vol. 25 (1937), pp. 117–121, 595–601; and Richard v. Mises, "On Saint-Venant's Principle," *Bulletin of the American Mathematical Society*, vol. 51 (1945), pp. 555–562.

CHAPTER IV
EXTENSION, TORSION, AND FLEXURE OF HOMOGENEOUS BEAMS

29. Statement of Problem. This chapter is devoted to an analysis of the behavior of elastic beams bounded by a cylindrical surface (which is termed the *lateral surface* of the beam) and by a pair of planes normal to the lateral surface (which are called the *bases* of the cylinder). It contains a treatment of the technically important problem of torsion and flexure of cylinders and an account of the different methods of attack on the problems of the theory of elasticity concerned with a study of beams. An elegant method of solution of such problems, developed recently by N. I. Muschelišvili and others, will be considered in detail.[1] While it is not the purpose of this chapter to provide a compendium of the theory of beams, a number of problems will be worked out in detail, either because of their intrinsic importance in structural design, or for the sake of illustrating the methods of solution.

In dealing with special problems, no great saving of space is likely to result from the use of abridged notation; for this reason, we shall denote the variables x_1, x_2, and x_3 by x, y, and z, as was agreed in Sec. 7. We shall also write $\tau_{11} = \tau_{xx}$, $\tau_{23} = \tau_{yz}$, etc., for the components of the stress tensor and use the corresponding notation for the components of strain e_{ij}. The displacements u_i along the directions of the x, y, and z-axes will be labeled u, v, and w, and the components of body force F in the same directions will be written as F_x, F_y, and F_z.

Throughout this chapter, the z-axis of our coordinate system will be directed along the length of the beam parallel to the generators of the cylinder. The cylinder is assumed to be of length l, and one of its bases is taken to lie in the xy-plane, while the other is in the plane $z = l$. It is supposed that the lateral

---

[1] The development of several sections of this chapter follows along the lines of the prize-winning work by N. I. Muschelišvili, Nekotoriye Osnovniye Zadachi Matematicheskoi Teorii Uprugosti.

surface of the cylinder is free of external load and that the load on the beam is distributed over its bases, $z = 0$ and $z = l$, in a way that fulfills the equilibrium conditions of a rigid body.

The complete problem of equilibrium of an elastic beam can be formulated in the following way. Determine the components of stress τ_{ij} and the displacements u_i that, in the region τ occupied by the beam, satisfy the systems of equations

(29.1) $$\begin{cases} \dfrac{\partial \tau_{xx}}{\partial x} + \dfrac{\partial \tau_{xy}}{\partial y} + \dfrac{\partial \tau_{xz}}{\partial z} = -F_x, \\ \dfrac{\partial \tau_{yx}}{\partial x} + \dfrac{\partial \tau_{yy}}{\partial y} + \dfrac{\partial \tau_{yz}}{\partial z} = -F_y, \\ \dfrac{\partial \tau_{zx}}{\partial x} + \dfrac{\partial \tau_{zy}}{\partial y} + \dfrac{\partial \tau_{zz}}{\partial z} = -F_z, \end{cases}$$

(29.2) $$\begin{cases} \dfrac{\partial u}{\partial x} = \dfrac{1}{E}[\tau_{xx} - \sigma(\tau_{yy} + \tau_{zz})], \\ \dfrac{\partial v}{\partial y} = \dfrac{1}{E}[\tau_{yy} - \sigma(\tau_{zz} + \tau_{xx})], \\ \dfrac{\partial w}{\partial z} = \dfrac{1}{E}[\tau_{zz} - \sigma(\tau_{xx} + \tau_{yy})], \\ \dfrac{\partial v}{\partial x} + \dfrac{\partial u}{\partial y} = \dfrac{2(1+\sigma)}{E}\tau_{xy}, \\ \dfrac{\partial w}{\partial y} + \dfrac{\partial v}{\partial z} = \dfrac{2(1+\sigma)}{E}\tau_{yz}, \\ \dfrac{\partial u}{\partial z} + \dfrac{\partial w}{\partial x} = \dfrac{2(1+\sigma)}{E}\tau_{zx}, \end{cases}$$

and the boundary conditions

(29.3) $\tau_{zx}, \tau_{zy}, \tau_{zz}$, prescribed functions of x and y on the bases $z = 0$, $z = l$,

(29.4) $$\begin{cases} \tau_{xx}\nu_x + \tau_{xy}\nu_y = 0, \\ \tau_{yx}\nu_x + \tau_{yy}\nu_y = 0, \\ \tau_{zx}\nu_x + \tau_{zy}\nu_y = 0, \end{cases} \text{on the lateral surface of the cylinder.}$$

The functions τ_{ij}, naturally, must satisfy the Beltrami-Michell compatibility equations (24.15).

The problem, formulated with this degree of generality, presents formidable complications because of the difficulty of fulfilling the boundary conditions (29.3). In fact, the generality of formulation of the boundary conditions (29.3) is quite unnecessary from the practical point of view, since the actual distribution of applied stresses on the ends of the cylinder is rarely, if ever, known. A designer knows, more or less accurately, the resultant force T and the resultant moment M acting on the ends of the beam, and quite often the nature of the distribution of stresses over the ends of the beam, which give rise to the force T and the moment M, is a matter of indifference. On the other hand, if one accepts the principle of Saint-Venant and considers a beam whose length is large in comparison with the linear dimensions of its cross section, then the actual distribution of stresses over the ends has no appreciable influence on the character of the solution in portions of the beam sufficiently far removed from the ends. That is, one is free to prescribe any distribution of stresses at the end of the cylinder so long as the resultant forces and moments reduce to those given in the formulation of the problem. This principle will be applied throughout our discussion of beams. This means that the mathematical solution obtained will give, near the ends of the beam, either (1) the exact solution of the physical problem in which the applied stresses are distributed in the way specified by the solution, or (2) the approximate description of the physical situation in which the system of external forces and moments is statically equivalent to that assumed by the solution, but is distributed in some different manner.

One need be concerned with one of the bases only, for the specification of the resultant force T and of the resultant moment M on the base $z = l$ requires that the resultant force acting on the base $z = 0$ be $-T$ and that the resultant moment acting on the same base be so chosen as to satisfy the condition of static equilibrium. Let the point O' of intersection of the z-axis with the base $z = l$ be the center of gravity of the base, and suppose that a force T and a couple M are applied at O'. The force T can be resolved into two components, one in the direction of the z-axis, and the other in the plane of the base $z = l$. The component of force T_z in the direction of the z-axis will be responsible for tension or compression, while the other component T_B, lying in the plane of the base, will produce bending of the beam. The couple M, acting on the end of the beam, can likewise be decomposed into two couples, the moment of one of which is directed along the z-axis and hence will be responsible for twisting of the cylinder, while the moment of the other lies in the plane $z = l$ and will produce bending.

Thus, our problem can be solved, by utilizing the principle of superposition, if we succeed in solving the following four elementary problems:

1. Extension of a cylinder by longitudinal forces applied at the ends.
2. Bending of a cylinder by couples whose moments lie in the planes of the bases of the cylinder.
3. Torsion of a cylinder by couples whose moments are normal to the bases of the cylinder.
4. Flexure of a cylinder by a transverse force applied at one end of the cylinder, while on the other end there act a force equal in magnitude but oppositely directed to the transverse force and also a couple of such magnitude as to equilibrate the moment produced by the transverse forces.

Our general plan of attack upon the four elementary problems listed above is that of the Saint-Venant *semi-inverse method of solution*. This consists of making certain assumptions about the components of stress, strain, or displacement and yet leaving enough freedom in the quantities involved to satisfy the conditions of equilibrium and compatibility. In applying the semi-inverse method to problems on beams, we shall make one general assumption about the stress distribution in any beam; further assumptions regarding the stresses or the displacements will be introduced in the solution of each problem. These assumptions will be justified when it is shown that they lead in each case to a solution that satisfies the conditions of equilibrium and compatibility. Then the proof in Sec. 27 of the uniqueness of solution of the general boundary-value problems of elasticity assures us that the solution obtained is unique.

Now if we visualize the beam as made up of long filaments parallel to the axis of the cylinder, then it is sensible to assume that the action of forces and couples in the foregoing four problems may give rise to shearing stresses in the direction of the z-axis. These stresses act on the sides of the filaments and produce no stresses on the lateral surface of the filaments in the direction perpendicular to their lengths. Thus, let us assume tentatively that the system of stresses in all four problems is such that

$$\tau_{xx} = \tau_{yy} = \tau_{yx} = 0,$$

and let us investigate the consequences of this assumption.[1]

PROBLEMS

1. Show that if the stress components $\tau_{xx}, \tau_{xy}, \tau_{yy}$ and the body forces F_i vanish, then $\dfrac{\partial^2 \tau_{zz}}{\partial x^2} = \dfrac{\partial^2 \tau_{zz}}{\partial y^2} = \dfrac{\partial^2 \tau_{zz}}{\partial z^2} = 0$; that is, the stress component τ_{zz} is linear in x, in y, and in z. Write out the most general form of the function in this case.

2. Integrate the differential equations of equilibrium $\tau_{ij,j} + F_i = 0$ throughout the volume of an elastic solid, apply the Divergence Theorem, and show that the equations of static equilibrium

$$\int_\Sigma \overset{\nu}{T}_i \, d\sigma + \int_\tau F_i \, d\tau = 0$$

are satisfied and hence that the resultant force on the body vanishes.

3. Show with the help of the Divergence Theorem that if the following differential equations of equilibrium

$$\tau_{yj,j} + F_y = 0, \qquad \tau_{zj,j} + F_z = 0, \qquad (j = x, y, z)$$

are satisfied, then the following equation of static equilibrium also holds:

$$\int_\Sigma (y\overset{\nu}{T}_z - z\overset{\nu}{T}_y) \, d\sigma + \int_\tau (yF_z - zF_y) \, d\tau = 0,$$

thus expressing the vanishing of the x-component of the resultant moment on the body.

30. Extension of Beams by Longitudinal Forces. Let a force T, directed along the z-axis, be applied at the center of gravity

---

[1] One may equally well proceed by assuming that the distribution on the cross section of the stress constituting each component of the resultant force and couple is the same at all sections. This is equivalent to assuming that for Probs. 1, 2, and 3 we have

$$\frac{\partial \tau_{zx}}{\partial z} = \frac{\partial \tau_{zy}}{\partial z} = \frac{\partial \tau_{zz}}{\partial z} = 0,$$

while for Prob. 4 it is assumed that

$$\frac{\partial \tau_{zx}}{\partial z} = \frac{\partial \tau_{zy}}{\partial z} = \frac{\partial^2 \tau_{zz}}{\partial z^2} = 0.$$

See W. Voigt, *Abhandlungen der Gesellschaft der Wissenschaften zu Göttingen*, vol. 34 (1887), p. 53, and J. N. Goodier, *Philosophical Magazine*, ser. 7, vol. 23 (1937), p. 186.

of the area a of the cross section of the base $z = l$ of the cylinder. If the stresses giving rise to the force T are assumed to be uniformly distributed, then

$$\left. \begin{array}{c} \tau_{zz} = \dfrac{T}{a} = p \text{ (a const.)}, \\ \tau_{zx} = \tau_{zy} = 0, \end{array} \right\} \text{ on } z = l.$$

If we assume

$$\tau_{zz} = p, \qquad \tau_{xx} = \tau_{yy} = \tau_{xy} = \tau_{yz} = \tau_{zx} = 0,$$

throughout the cylinder, then the equilibrium equations (29.1) and (29.4) are obviously satisfied.[1] The Beltrami-Michell compatibility equations are also satisfied, since the components of the stress tensor are constants. The displacements u_i can be readily calculated. Thus, from (29.2),

$$\frac{\partial u}{\partial x} = -\frac{\sigma}{E}p, \qquad \frac{\partial v}{\partial y} = -\frac{\sigma}{E}p, \qquad \frac{\partial w}{\partial z} = \frac{p}{E},$$

$$\frac{\partial v}{\partial x} + \frac{\partial u}{\partial y} = 0, \qquad \frac{\partial w}{\partial y} + \frac{\partial v}{\partial z} = 0, \qquad \frac{\partial u}{\partial z} + \frac{\partial w}{\partial x} = 0,$$

and since the right-hand members of these equations are constants, one is justified in assuming that the solutions are linear functions of x, y, and z. A simple calculation gives

$$u = -\frac{\sigma p}{E}x, \qquad v = -\frac{\sigma p}{E}y, \qquad w = \frac{p}{E}z,$$

if one neglects the terms representing rigid motion of the beam as a whole. Of course one could obtain the displacement by making use of the general formula (10.6). This problem has already been discussed in Sec. 23.

PROBLEMS

1. Consider a bar of length l in., area of cross section a sq. in., Young's modulus E lb. per sq. in., and stretched by a force of T lb. applied at each end. Use both Eq. (26.7), $W = \frac{1}{2}\tau_{ij}e_{ij}$, and Eq. (26.12),

$$2\int_\tau W\, d\tau = \int_\tau F_i u_i\, d\tau + \int_\sigma \overset{\nu}{T}_i u_i\, d\sigma,$$

---

[1] The body forces F_i are assumed to vanish. The extension of a beam by gravitational forces is considered in the next section. The combined effect of extension of beams by longitudinal and gravitational forces can be obtained by applying the principle of superposition.

§30   EXTENSION, TORSION, AND FLEXURE OF BEAMS   103

to show that the strain-energy density W and the total strain energy $U = \int_\tau W \, d\tau$ stored in the bar are given by

$$W = \frac{T^2}{2a^2 E} \text{ in.-lb. per cu. in.,}$$

$$U = \frac{T^2 l}{2aE} \text{ in.-lb.}$$

2. Find the greatest amount of strain energy per unit volume that can be stored in a steel bar under tensile forces T without producing permanent set. Take the elastic limit to be 30×10^3 lb. per sq. in. and Young's modulus as 30×10^6 lb. per sq. in.

3. In Prob. 1 take $l = 10$ in., $a = 2$ sq. in., $T = 50 \times 10^3$ lb., $E = 30 \times 10^6$ lb. per sq. in. Find the strain-energy density W, and show numerically that the total strain energy U is one-half the product of the force T by the elongation of the rod.

4. Two gage marks 1 in. apart are made along the axis of a steel bar 10 in. long and of 2 sq. in. cross-sectional area. The bar is then subjected to a tensile force of 50,000 lb. Find the stress, strain, elongation between gage marks, and total elongation of bar. What is the total change of volume of the bar? What is the change in the cross-sectional area of the bar? Take $E = 30 \times 10^6$ lb. per sq. in., $\sigma = 0.3$.

5. Consider a beam stretched by a tensile force T applied at each end. The magnitude of the stress vector acting on a section with normal n is

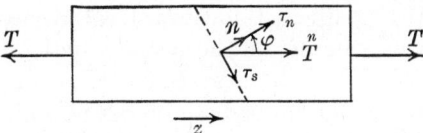

$\overset{n}{T} = T/(a \sec \varphi) = \tau_{zz} \cos \varphi$, where a is the area of the cross section. Resolve $\boldsymbol{T}$ into normal and shear stresses τ_n, τ_s, and show that

$$\tau_n = \tau_{zz} \cos^2 \varphi, \qquad \tau_s = \tau_{zz} \sin \varphi \cos \varphi.$$

Derive these results also from the formulas of Sec. 19b. Show that the maximum normal stress is τ_{zz} (at $\varphi = 0$) and the maximum shear stress is $\tfrac{1}{2}\tau_{zz}$ (at $\varphi = 45°$). Compare this with the theorem of Sec. 18. What are the inclinations of the cross sections on which the shear and normal stresses are equal in magnitude?

6. Find the maximum shear stress in the beam of Prob. 4. What is the normal stress on the planes on which the shear stress is a maximum?

7. Consider a rod under uniform longitudinal stress $\tau_{zz} = p$. Let the rod be so constrained that there is no lateral contraction in the x-direction ($e_{xx} = 0$), while the rod is free to contract laterally in the y-direction. Define the effective Young's modulus by $E' = \tau_{zz}/e_{zz}$ and the effective Poisson's ratio by $\sigma' = -e_{yy}/e_{zz}$, and show that due to the lateral constraint, one has

$$E' = \frac{E}{1 - \sigma^2}, \qquad \sigma' = \frac{\sigma}{1 - \sigma}.$$

What is the range of possible values for E'? for σ'?

8. Let the rod in the preceding problem be so constrained as to prevent any lateral contraction. Show that the effective Young's modulus has the value

$$E' = \frac{1 - \sigma}{(1 - 2\sigma)(1 + \sigma)} E.$$

What is the effective Poisson's ratio?

31. Beam Stretched by Its Own Weight. Before proceeding to the problem of bending of beams, we shall discuss one example of a problem requiring a consideration of the body force.

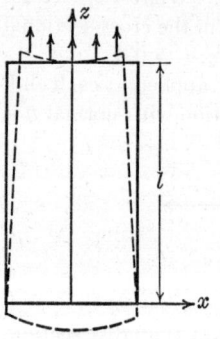

Fig. 15.

Let a beam of length l, shown in Fig. 15, be supported in a suitable manner at its upper base, and assume that the force of gravity, directed downward, is the only external force acting on the beam. If the xy-plane of the coordinate system is chosen to coincide with the lower base of the beam before deformation takes place and if the positive direction of the z-axis is vertically upward, then the stress components τ_{ij} satisfy the system of Eqs. (29.1) with $F_x = F_y = 0$ and $F_z = -\rho g$, where ρ is the density of the beam. The stresses acting on each cross section of the beam are produced by the weight of the lower part of the beam, and we shall suppose that the stresses are distributed uniformly. Thus, we assume the system of stresses

$$\tau_{zz} = \rho g z, \qquad \tau_{xx} = \tau_{yy} = \tau_{xy} = \tau_{yz} = \tau_{zx} = 0,$$

which obviously satisfies the equations of equilibrium and the compatibility equations (24.15). The conditions (29.4) that no forces are applied to the lateral surface of the beam are like-

§31 EXTENSION, TORSION, AND FLEXURE OF BEAMS

wise fulfilled. There are no tractions applied at the lower end; hence all components of stress vanish there, while at the upper end we have $\tau_{zz} = \rho g l$, which is directed vertically upward. Thus, the assumed distribution of stress requires that the upper end of the cylinder be supported in such a way as to yield a uniform distribution of stress.

In order to determine the displacements u_i, we note the relations (29.2), which yield

$$(31.1) \quad \frac{\partial u}{\partial x} = \frac{\partial v}{\partial y} = -\frac{\sigma \rho g z}{E}, \quad \frac{\partial w}{\partial z} = \frac{\rho g z}{E},$$

$$(31.2) \quad \frac{\partial v}{\partial x} + \frac{\partial u}{\partial y} = 0, \quad \frac{\partial w}{\partial y} + \frac{\partial v}{\partial z} = 0, \quad \frac{\partial u}{\partial z} + \frac{\partial w}{\partial x} = 0.$$

Integrating the last of Eqs. (31.1) gives

$$w = \frac{\rho g z^2}{2E} + w_0(x, y),$$

where w_0 is a function of x and y alone, and it follows from the last two of Eqs. (31.2) that

$$\frac{\partial u}{\partial z} = -\frac{\partial w_0}{\partial x}, \quad \text{and} \quad \frac{\partial v}{\partial z} = -\frac{\partial w_0}{\partial y}.$$

Hence

$$u = -z\frac{\partial w_0}{\partial x} + u_0(x, y), \quad \text{and} \quad v = -z\frac{\partial w_0}{\partial y} + v_0(x, y),$$

where u_0 and v_0 involve x and y only. Substituting the values of u and v just found in the first two of Eqs. (31.1) gives

$$(31.3) \quad \frac{\partial u_0}{\partial x} = 0, \quad \frac{\partial v_0}{\partial y} = 0, \quad \frac{\partial^2 w_0}{\partial x^2} = \frac{\sigma \rho g}{E}, \quad \frac{\partial^2 w_0}{\partial y^2} = \frac{\sigma \rho g}{E},$$

while the substitution of the same values in the first of Eqs. (31.2) yields

$$(31.4) \quad \frac{\partial^2 w_0}{\partial x \, \partial y} = 0, \quad \frac{\partial u_0}{\partial y} + \frac{\partial v_0}{\partial x} = 0.$$

It is clear from the first two of the differential equations (31.3) that

$$u_0 = F(y), \quad \text{and} \quad v_0 = G(x),$$

where F is a function of y alone, while G is a function of x. The functions F and G, as follows from the second of Eqs. (31.4), satisfy the equation

$$\frac{dF(y)}{dy} + \frac{dG(x)}{dx} = 0,$$

and this requires that $\frac{dF}{dy} = a$, $\frac{dG}{dx} = -a$, where a is a constant. Thus,

$$u_0 = F(y) = ay + b, \quad \text{and} \quad v_0 = G(x) = -ax + c.$$

The integration of the equations on w_0 is equally easy, and one finds

$$w_0 = \frac{\sigma \rho g}{2E}(x^2 + y^2) + a'x + b'y + c',$$

where a', b', and c' are constants.

Thus, the complete expression for the displacements is

$$u = -\frac{\sigma \rho g}{E} zx - a'z + ay + b,$$

$$v = -\frac{\sigma \rho g}{E} zy - b'z - ax + c,$$

$$w = \frac{\rho g}{2E}(z^2 + \sigma x^2 + \sigma y^2) + a'x + b'y + c'.$$

The linear part of the solution represents rigid body displacement.[1] If we prevent the point $(0, 0, l)$ from being displaced, then $u = v = w = 0$ for $x = 0$, $y = 0$, $z = l$. To prevent the possibility of rotation about the z-axis, we fix an element of area in the xz-plane and passing through the point $(0, 0, l)$; then $\frac{\partial v}{\partial x} = 0$ at $(0, 0, l)$. In order to eliminate rotation about the axes through $(0, 0, l)$ that are parallel to the x- and y-axes, we fix an element of the z-axis; then $\frac{\partial u}{\partial z} = 0$, $\frac{\partial v}{\partial z} = 0$, at $(0, 0, l)$. These six conditions enable us to eliminate the six constants a, b, c and a', b', c'. An elementary calculation shows that the displacement, in this case, is given by

(31.5) $\quad \begin{cases} u = -\dfrac{\sigma \rho g}{E} zx, \quad v = -\dfrac{\sigma \rho g}{E} zy, \\ w = \dfrac{\rho g}{2E}(z^2 + \sigma x^2 + \sigma y^2 - l^2). \end{cases}$

[1] See Prob. 4 at the end of this section.

It is seen from this solution that points on the z-axis are displaced vertically according to the law

$$w = -\frac{\rho g}{2E}(l^2 - z^2).$$

All other points of the beam have both vertical and horizontal displacements on account of the contraction in the transverse direction. The shape of the beam, after deformation, is indicated by the dotted lines in Fig. 15. Any cross section of the beam is shrunk laterally by an amount proportional to the distance from the lower end and is distorted into a paraboloid of revolution. This can be seen by noting that for a cross section $z = c$,

$$z' = c + w = c + \frac{\rho g(c^2 - l^2)}{2E} + \frac{\sigma \rho g}{2E}(x^2 + y^2).$$

The upper base of the cylinder is warped upward (see Fig. 15) because of the assumed uniform distribution of the stress component τ_{zz} over that face and the fixing of the point $(0, 0, l)$.

References for Collateral Reading

A. E. H. Love: A Treatise on the Mathematical Theory of Elasticity, Cambridge University Press, London, Sec. 86.

S. Timoshenko: Theory of Elasticity, McGraw-Hill Book Company, Inc., New York, Sec. 68.

PROBLEMS

1. Discuss the solution of the problem for the case where

$$\tau_{xx} = \tau_{yy} = \tau_{zz} = -p + \rho g z, \qquad \tau_{xy} = \tau_{yz} = \tau_{zx} = 0.$$

This state of stress corresponds to that found in a body immersed in a fluid whose density is the same as that of the body, where p is the pressure of the fluid at the level of the origin of coordinates.

2. Determine the displacements in a cylinder of length $2l$ and of density ρ when suspended in a fluid of density ρ'. Let the pressure of the fluid at the level of the center of gravity of the cylinder be p. Choose the origin of the coordinate system at the center of gravity of the cylinder, and let the z-axis be vertical. *Hint:* Assume a system of stresses

$$\tau_{xx} = \tau_{yy} = -p + \rho g z, \qquad \tau_{zz} = -p(\rho - \rho')gl + gz,$$
$$\tau_{xy} = \tau_{yz} = \tau_{zx} = 0.$$

3. Obtain the solution given in (31.5) from the general solution (10.6).

4. Show from Eq. (7.5) that the displacement components

$$u = \quad -ry + qz + a,$$
$$v = \quad rx \quad - pz + b,$$
$$w = -qx + py \quad + c$$

represent an (infinitesimal) rotation (p, q, r) and a translation (a, b, c).

5. Show that some of the results of Sec. 31 on a beam stretched by its own weight may be obtained readily by the procedure sketched below (used in strength of materials theory). As before, the stress on the faces of an element of cross-sectional area a and length dz is given by $\tau_{zz} = \rho g a z/a = \rho g z$. The elongation of this element is $\rho g z\, dz/E$. Integrate this expression over the length of the beam, and compare the result with that obtained from Sec. 31. Show that the total elongation in a beam stretched by its own weight W is the same as that produced by a load $\frac{1}{2}W$ applied at the end of the beam (with weight neglected).

32. Bending of Beams by Terminal Couples. In order to free the semi-inverse method of solution from elements of mystery that a beginner feels are involved in the usual statement: "Assume the system of stress defined by . . . ," we shall give first an intuitive picture of the probable state of affairs in a beam bent by a pair of couples applied at its ends. This picture will be of aid to us later on in considering the problem of deflection of an elastic plate and will bring into sharp focus the limitations of the approximate engineering theory of beams.

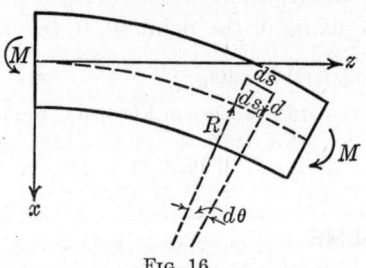

Fig. 16.

Let a pair of couples of magnitude M be applied to the ends of a beam as shown in Fig. 16. It is clear that the longitudinal filaments of which the beam may be thought to be composed will be contracted on the face of the beam toward the center of curvature, and those on the opposite face will be extended. We shall call the line passing through the centroids of the cross sections of the beam the *central line*. If we assume that the central line of the beam, indicated in the figure by a dotted curve, is unaltered in length, and if plane sections of the beam normal to the central line are assumed to remain plane and

§32   EXTENSION, TORSION, AND FLEXURE OF BEAMS

normal to the deformed central line, then it is easy to see that the magnitude of extension (or contraction) of the longitudinal filaments is given by the formula

$$e = \frac{d}{R}.$$

In this relation, d is the distance of the filament from the central plane drawn through the central line at right angles to the plane of the couple (the xz-plane in Fig. 16), and R is the radius of curvature of the central line. Now the length ds_0 of the portion of central filament subtended by an angle $d\theta$ is $ds_0 = R\,d\theta$, while the length of the element ds subtended by the same angle $d\theta$ and at a distance d from the central plane is

$$ds = (R + d)\,d\theta.$$

Hence the extension e is given by

$$e = \frac{ds - ds_0}{ds_0} = \frac{(R + d)\,d\theta - R\,d\theta}{R\,d\theta} = \frac{d}{R}.$$

This extension d/R of the longitudinal filaments can be thought to be produced by a longitudinal stress τ, which, from the third of Eqs. (29.2), is

$$\tau = \frac{E}{R}d.$$

Obviously τ denotes tension if the point in question is above the central line and compression if it is below. We choose the z-axis to coincide with the central line of the beam and take the x- and y-axes along the principal axes of inertia of the cross section A. From this choice of axes and from the definition of the central line as the line of centroids of the sections, we have

$$\int_A x\,d\sigma = \int_A y\,d\sigma = \int_A xy\,d\sigma = 0.$$

It follows that the distribution of stress in any section will be characterized by the formula

$$\tau_{zz} = -\frac{E}{R}x,$$

where the negative sign arises from our convention in regard to the signs of tensile and compressive stresses.

We shall now verify that the boundary conditions on the ends of the beam are satisfied, namely, that the resultant force and moment acting on the bases (or on any other cross section of the cylinder) reduce to a moment about the y-axis. The resultant force T acting on any section A has the components

$$T_x = \int_A \tau_{zx}\, d\sigma = 0, \qquad T_y = \int_A \tau_{zy}\, d\sigma = 0,$$

$$T_{z'} = \int_A \tau_{zz}\, d\sigma = -\frac{E}{R}\int_A x\, d\sigma = 0.$$

The resultant moment about the x-axis is

$$M_x = \int_A y\tau_{zz}\, d\sigma = -\frac{E}{R}\int_A xy\, d\sigma = 0,$$

while the moment about the y-axis is given by

$$M_y = -\int_A x\tau_{zz}\, d\sigma = \frac{E}{R}\int_A x^2\, d\sigma = \frac{EI_y}{R},$$

where I_y is the moment of inertia of the cross section about the y-axis.

Thus, the curvature of the central line of a beam bent by a couple of magnitude M is[1]

(32.1) $$R = \frac{EI}{M}.$$

Formula (32.1), connecting the curvature of the central line with the bending moment, is called the *Bernoulli-Euler law*. It will recur when we come to consider this problem in a rigorous way.

It appears from the foregoing discussion that the stress in a beam giving rise to a couple M is a longitudinal stress of magnitude

$$\tau_{zz} = -\frac{M}{I}x = -\frac{E}{R}x.$$

Under the action of the tensile stress τ_{zz}, the cross section of the beam will be deformed, and the amount of the transverse contrac-

---

[1] The subscript y on I and M has been dropped, since no confusion is likely to arise here.

§32 EXTENSION, TORSION, AND FLEXURE OF BEAMS

tion (or extension), from the definition of Poisson's ratio σ (see Sec. 23), is

$$\frac{\sigma x}{R} = \frac{\sigma M x}{EI}.$$

If the beam was initially of rectangular cross section $PQRS$ (Fig. 17), then, as will be shown in the rigorous discussion of this problem [see (32.10)], the parts RS and PQ of the boundary are each bent into a parabola whose radius of curvature is approximately R/σ. The *neutral plane* of the beam (that is, the plane in which there is no extension) and the faces of the beam that were originally parallel to the yz-plane are deformed into saddleshaped or anticlastic surfaces.

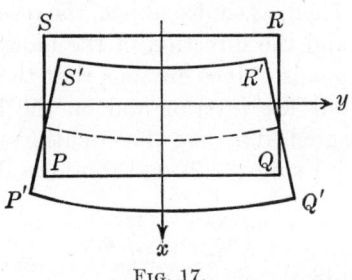

Fig. 17.

The experimental measurement of the principal curvatures of the anticlastic surfaces provides a method of determining Poisson's ratio[1] σ, while the measurement of the radius of curvature of the central line serves to determine Young's modulus E.

It is clear from formula (32.1) that a beam with a large value of EI will bend only slightly under the action of the couple M, and hence the magnitude of EI provides a measure of the rigidity

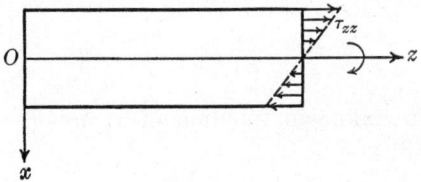

Fig. 18.

of the beam. For this reason the constant EI is called the *modulus of flexural rigidity*. In order to increase the flexural rigidity of a beam, one must design it so as to make the moment of inertia I as large as possible. This is one of the reasons for making beams with cross sections in the shape of the letters I, T, Z, etc.

[1] Those interested in the experimental determination of σ are referred to S. Timoshenko, Theory of Elasticity, p. 225, where further references on this subject will be found.

We are now ready to consider rigorously the problem of bending of a beam.

Assume the system of stresses

(32.2) $\quad \tau_{zz} = -\dfrac{M}{I} x, \quad \tau_{xx} = \tau_{yy} = \tau_{xy} = \tau_{yz} = \tau_{zx} = 0,$

and choose the axes of coordinates as before (see Fig. 18). Then, as shown above, the resultant force on any section vanishes, and the direction of the moment M of the couple is that of the y-axis. It is obvious that the equations of equilibrium throughout the interior and on the lateral surface of the cylinder are satisfied, as are the equations of compatibility.[1]

Using (32.2) and the formulas (29.2), we find

(32.3) $\quad \begin{cases} \dfrac{\partial u}{\partial x} = \dfrac{\sigma M}{EI} x, & \dfrac{\partial v}{\partial y} = \dfrac{\sigma M}{EI} x, & \dfrac{\partial w}{\partial z} = -\dfrac{M}{EI} x, \\ \dfrac{\partial v}{\partial x} + \dfrac{\partial u}{\partial y} = 0, & \dfrac{\partial w}{\partial y} + \dfrac{\partial v}{\partial z} = 0, & \dfrac{\partial u}{\partial z} + \dfrac{\partial w}{\partial x} = 0. \end{cases}$

The expressions for the displacements u_i can be obtained from the formulas (10.6) or by assuming u, v, and w to be functions of the second degree of x, y, and z with unknown coefficients and then determining the coefficients so as to satisfy Eqs. (32.3). We choose to integrate Eqs. (32.3) directly.

Thus, from the third of Eqs. (32.3), we obtain

$$w = -\frac{M}{EI} xz + w_0(x, y),$$

where w_0 is an unknown function of x and y. The fifth and sixth Eqs. (32.3) give

$$\frac{\partial u}{\partial z} = \frac{M}{EI} z - \frac{\partial w_0}{\partial x}, \quad \frac{\partial v}{\partial z} = -\frac{\partial w_0}{\partial y}.$$

Hence

(32.4) $\quad \begin{cases} u = \dfrac{M}{2EI} z^2 - z\dfrac{\partial w_0}{\partial x} + u_0(x, y), \\ v = -z\dfrac{\partial w_0}{\partial y} + v_0(x, y), \end{cases}$

---

[1] The body forces are assumed to vanish.

where u_0 and v_0 are unknown functions of x and y. Substituting these values in the first two of Eqs. (32.3) gives

(32.5)
$$\begin{cases} -z\dfrac{\partial^2 w_0}{\partial x^2} + \dfrac{\partial u_0}{\partial x} = \dfrac{\sigma M}{EI} x, \\ -z\dfrac{\partial^2 w_0}{\partial y^2} + \dfrac{\partial v_0}{\partial y} = \dfrac{\sigma M}{EI} x. \end{cases}$$

Since these equations are true for all values of z, it appears that

(32.6)
$$\frac{\partial^2 w_0}{\partial x^2} = 0, \qquad \frac{\partial^2 w_0}{\partial y^2} = 0,$$

and it follows from the integration of (32.5) that

$$u_0 = \frac{\sigma M}{2EI} x^2 + f_1(y), \qquad v_0 = \frac{\sigma M}{EI} xy + f_2(x).$$

Inserting these expressions in (32.4) and substituting the resulting values of u and v in the fourth of Eqs. (32.3) gives the condition

$$-2z\frac{\partial^2 w_0}{\partial x\, \partial y} + \frac{df_1(y)}{dy} + \frac{df_2(x)}{dx} + \frac{\sigma M}{EI} y = 0.$$

Since the last three terms in this equation are independent of z, it follows that

(32.7)
$$\frac{\partial^2 w_0}{\partial x\, \partial y} = 0,$$

and hence
$$\frac{df_1(y)}{dy} + \frac{df_2(x)}{dx} + \frac{\sigma M}{EI} y = 0.$$

Thus,

(32.8)
$$\frac{df_2}{dx} = -\alpha, \qquad \text{and} \qquad \frac{df_1}{dy} + \frac{\sigma M}{EI} y = \alpha,$$

where α is a constant. We note from (32.6) and (32.7) that w_0 is a linear function of x and y, say

$$w_0 = \beta x + \gamma y + c;$$

furthermore, integrating Eqs. (32.8) gives

$$f_2 = -\alpha x + b,$$
$$f_1 = -\frac{\sigma M}{2EI} y^2 + \alpha y + a,$$

where b and a are arbitrary constants. Thus, the expressions for the displacements become

(32.9) $$\begin{cases} u = \dfrac{M}{2EI}(z^2 + \sigma x^2 - \sigma y^2) & + \alpha y - \beta z + a, \\ v = \dfrac{M}{EI}\sigma xy & - \alpha x \qquad - \gamma z + b, \\ w = -\dfrac{M}{EI}xz & + \beta x + \gamma y \qquad + c. \end{cases}$$

The constants of integration appearing in the solution can be determined from the mode of fixing the beam. We can determine them in the same way as was done in Sec. 31, namely, by fixing the centroid of the left-hand face of the beam at the origin and by fixing an element of the z-axis and an element of the xz-plane at the origin. These conditions ensure that there is no rigid body motion of translation or rotation about the origin. They can be formulated explicitly as follows:

$$u = v = w = \frac{\partial u}{\partial z} = \frac{\partial v}{\partial z} = \frac{\partial v}{\partial x} = 0 \quad \text{at } (0, 0, 0).$$

It follows from these relations that

$$\alpha = \beta = \gamma = a = b = c = 0.$$

The vanishing of the constants of integration also follows readily from Eqs. (3.5), from which it is seen that a, b, c represent a rigid body translation, while α, β, γ characterize a rigid body rotation about the origin. The solution can now be written in the form

(32.10) $$\begin{cases} u = \dfrac{M}{2EI}(z^2 + \sigma x^2 - \sigma y^2), \\ v = \dfrac{M}{EI}\sigma xy, \\ w = -\dfrac{M}{EI}xz. \end{cases}$$

It is clear from (32.10) that the filaments lying in the central plane $x = 0$ suffer no extensions; that is, the plane $x = 0$ is the neutral plane of the beams. The longitudinal filaments on one side of the central plane ($x > 0$) are contracted, whereas those on the other side ($x < 0$) are extended. Points which, prior to deformation, had the coordinates x_i go into points with coordi-

§32 EXTENSION, TORSION, AND FLEXURE OF BEAMS

nates x_i', where $x_i' = x_i + u_i$. Hence the points on the z-axis (that is, the points on the central line of the beam) go into points

$$(32.11) \quad x' = \frac{M}{2EI} z^2 = \frac{M}{2EI} z'^2, \quad y' = 0, \quad z' = z.$$

The *plane of bending* is defined to be the plane containing the deformed central line of the beam. Equation (32.11) shows that in this example, the plane of bending coincides with the plane ($y = 0$) of the couple M. The curve defined by (32.11) is a parabola whose radius of curvature R is given by the formula

$$\frac{1}{R} = \frac{\frac{d^2 x'}{dz'^2}}{\left[1 + \left(\frac{dx'}{dz'}\right)^2\right]^{3/2}},$$

which is nearly equal to $\frac{d^2 x'}{dz'^2}$ if $\frac{dx'}{dz'}$ is small. It follows from (32.11) that for small deflections

$$\frac{1}{R} = \frac{M}{EI},$$

which is the Bernoulli-Euler law, discovered earlier from rough geometrical considerations. This formula states that the magnitude of the bending moment M is proportional to the curvature of the central line of the beam. The Bernoulli-Euler law forms the point of departure for all considerations in the technical theory of beams.[1]

Consider a cross section of the beam made by the plane $z = c$. After deformation, points in this cross section will lie in the plane

$$z' = c + w = c - \frac{M}{EI} xc = c\left(1 - \frac{x}{R}\right).$$

If the curvature is small, we can replace x by x' and obtain

$$z' = c\left(1 - \frac{x'}{R}\right),$$

which is the equation of a plane normal to the deformed central line. Hence the assumption that the normal sections remain plane after deformation (made at the beginning of this section) is valid.

[1] See Sec. 61.

In order to see how the cross sections of the beam are deformed, consider a beam of rectangular cross section. The sides $y = \pm b$ of the beam will go into

$$y' = \pm b + v = \pm b \left(1 + \frac{\sigma}{R} x\right),$$

and for small values of $1/R$ this is nearly the same as

$$y' = \pm b \left(1 + \frac{\sigma}{R} x'\right).$$

Thus, the vertical sides become inclined, as shown in Fig. 17. The points in the section $z = c$, which lie on the upper and lower faces $x = \pm a$ of the beam, will go into points

$$x' = \pm a + u = \pm a + \frac{1}{2R}[c^2 + \sigma(a^2 - y^2)].$$

Hence, for small values of $1/R$, we have

$$x' = \pm a + \frac{1}{2R}[c^2 + \sigma(a^2 - y'^2)],$$

which is the equation of a parabola whose curvature at any point of the section is nearly σ/R.

It may be remarked in conclusion that if the moment M of the couple is not directed along one of the principal axes of inertia of the section, then the couple can be resolved into two couples, each of which has moments directed along the principal axes. Then the foregoing considerations become applicable to each of the couples, and the solution of the problem can be obtained by superposition. It turns out that in this general case the plane of the couple is also perpendicular to the neutral (undeformed) surface, although the plane of the couple does not necessarily coincide with the plane of bending.[1]

References for Collateral Reading

A. E. H. LOVE: A Treatise on the Mathematical Theory of Elasticity, Cambridge University Press, London, Secs. 86–88.

S. TIMOSHENKO: Theory of Elasticity, McGraw-Hill Book Company, Inc., New York, Secs. 67, 70.

---

[1] See in this connection Secs. 52–61, dealing with the flexure problem.

E. Trefftz: Handbuch der Physik, Verlag von Julius Springer, Berlin, vol. 6, Sec. 29.

R. V. Southwell: Theory of Elasticity for Engineers and Physicists, Oxford University Press, New York, Secs. 46, 125, 164–170.

PROBLEMS

1. Define the strain deviations e'_{ij} by

$$e_{ij} = \tfrac{1}{3}\vartheta \delta_{ij} + e'_{ij};$$

that is,

$$e_{xx} = \tfrac{1}{3}\vartheta + e'_{xx}, \quad e_{yy} = \tfrac{1}{3}\vartheta + e'_{yy}, \quad e_{zz} = \tfrac{1}{3}\vartheta + e'_{zz},$$
$$e_{xy} = e'_{xy}, \quad e_{yz} = e'_{yz}, \quad e_{zx} = e'_{zx},$$

where $\tfrac{1}{3}\vartheta \equiv \tfrac{1}{3}(e_{xx} + e_{yy} + e_{zz})$ is the mean extension.

Show that the cubical dilatation

$$\vartheta' \equiv e'_{xx} + e'_{yy} + e'_{zz} = 0,$$

and hence that the strain deviation tensor represents a change in shape without a change in volume.

Show that the principal strains e_i and the principal strain deviations e'_i are connected by the relations

$$e'_I = e_I - \tfrac{1}{3}\vartheta, \quad e'_{II} = e_{II} - \tfrac{1}{3}\vartheta, \quad e'_{III} = e_{III} - \tfrac{1}{3}\vartheta.$$

Hint: The principal strain deviations are the roots of the determinantal equation $|e'_{ij} - e' \delta_{ij}| = 0$.

2. Define the stress deviation τ'_{ij} by

$$\tau_{ij} = \tfrac{1}{3}\Theta \delta_{ij} + \tau'_{ij},$$

or

$$\tau_{xx} = \tfrac{1}{3}\Theta + \tau'_{xx}, \quad \tau_{yy} = \tfrac{1}{3}\Theta + \tau'_{yy}, \quad \tau_{zz} = \tfrac{1}{3}\Theta + \tau'_{zz},$$
$$\tau_{xy} = \tau'_{xy}, \quad \tau_{yz} = \tau'_{yz}, \quad \tau_{zx} = \tau'_{zx},$$

where $\tfrac{1}{3}\Theta \equiv \tfrac{1}{3}(\tau_{xx} + \tau_{yy} + \tau_{zz})$ is the mean normal stress. Show that the stress deviation and strain deviation tensors are related by

$$\tau'_{ij} = 2\mu e'_{ij},$$

and that

$$\Theta' \equiv \tau'_{xx} + \tau'_{yy} + \tau'_{zz} = 0.$$

Show that the principal stress deviations τ'_i and the principal stresses τ_i are connected by the relations

$$\tau'_I = \tau_I - \tfrac{1}{3}\Theta, \quad \tau'_{II} = \tau_{II} - \tfrac{1}{3}\Theta, \quad \tau'_{III} = \tau_{III} - \tfrac{1}{3}\Theta.$$

3. Verify the identity

$$\vartheta^2 \equiv (e_{xx} + e_{yy} + e_{zz})^2$$
$$= 3(e_{xx}^2 + e_{yy}^2 + e_{zz}^2) - (e_{xx} - e_{yy})^2 - (e_{yy} - e_{zz})^2 - (e_{zz} - e_{xx})^2,$$

and show that the strain-energy density can be written in the form

$$W = W_1 + W_2$$

where

$W_1 = 3k\vartheta^2,$
$W_2 = \tfrac{1}{3}\mu[(e_{xx} - e_{yy})^2 + (e_{yy} - e_{zz})^2 + (e_{zz} - e_{xx})^2 + 6(e_{xy}^2 + e_{yz}^2 + e_{zx}^2)]$
$\quad = \tfrac{1}{3}\mu[(e'_{xx} - e'_{yy})^2 + (e'_{yy} - e'_{zz})^2 + (e'_{zz} - e'_{xx})^2 + 6(e'^2_{xy} + e'^2_{yz} + e'^2_{zx})],$

and where e'_{ij} is the strain-deviation tensor. Show that W_1 depends only on the change of volume, while W_2 is that part of the strain-energy density arising from a change of shape. We call W_1 the strain-energy density of dilatation and W_2 the strain-energy density of distortion.

4. From $\Theta = 3k\vartheta$ and $\tau'_{ij} = 2\mu e'_{ij}$ (see Prob. 2), show that the strain-energy density W can be written as the sum of the strain-energy density of dilatation W_1 and the strain-energy density of distortion W_2, where

$$W_1 = \frac{1}{3k}\Theta^2,$$

$$W_2 = \frac{1}{12\mu}[(\tau_{xx} - \tau_{yy})^2 + (\tau_{yy} - \tau_{zz})^2 + (\tau_{zz} - \tau_{xx})^2$$
$$+ 6(\tau_{xy}^2 + \tau_{yz}^2 + \tau_{zx}^2)]$$
$$= \frac{1}{12\mu}[(\tau'_{xx} - \tau'_{yy})^2 + (\tau'_{yy} - \tau'_{zz})^2 + (\tau'_{zz} - \tau'_{xx})^2$$
$$+ 6(\tau'^2_{xy} + \tau'^2_{yz} + \tau'^2_{zx})].$$

The strain-energy density of distortion has been used as a criterion for failure of the material. See S. Timoshenko, Strength of Materials, vol. 2 (or Theory of Elasticity, p. 137), and A. and L. Föppl, Drang und Zwang, vol. 1, Sec. 6.

5. Consider a beam stretched by a longitudinal stress p uniformly distributed over the end sections. Show that the strain-energy density of dilatation W_1 and that of distortion W_2 are given by

$$W_1 = \frac{p^2}{3k}, \qquad W_2 = \frac{p^2}{6\mu} = \frac{1+\sigma}{3(1-2\sigma)}W_1.$$

Show that in the torsion of a cylindrical shaft we have

$$W_1 = 0, \qquad W_2 = \frac{r^2}{2\mu} = \frac{1}{2\mu}(\tau_{zx}^2 + \tau_{zy}^2).$$

6. Show that the strain energy stored in a beam of length l (parallel to the x-axis) bent by end couples M_z can be written as

$$U = \int_\tau W\, d\tau = \int_0^l \frac{M_z^2}{2EI_z}\, dx = \frac{M_z^2 l}{2EI_z}.$$

33. Torsion of a Circular Shaft.

In the preceding section, we formed a physical picture of the distribution of stress in a beam bent by couples from a consideration of the extension of a longitudinal filament. In this section we shall be guided by the displacements and shall deduce the stresses from the functions u_i.

Consider a circular cylinder, of length l, with one of its bases fixed in the xy-plane, while the other base (in the plane $z = l$) is acted upon by a couple whose moment lies along the z-axis. Under the action of the couple, the beam will be twisted, and the generators of the cylinder will be deformed into helical curves.

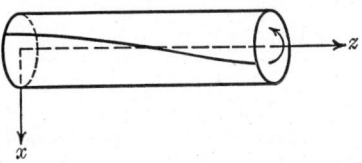

Fig. 19.

On account of the symmetry of the cross section, it is reasonable to suppose that sections of the cylinder by planes normal to the z-axis will remain plane after deformation and that the action of the couple will merely rotate each section through some angle θ. The amount of rotation will clearly depend on the distance of the section from the base $z = 0$, and since the deformations are small, it is sensible to assume that the amount of rotation θ is proportional to the distance of the section from the fixed base. Thus,

$$\theta = \alpha z,$$

where α is the twist per unit length; that is, the relative angular displacement of a pair of cross sections that are unit distance apart.

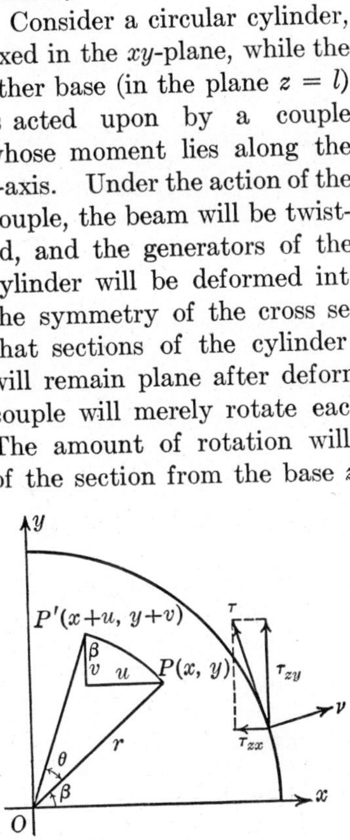

Fig. 20.

If the cross sections of the cylinder remain plane after deformation, then the displacement w, along the z-axis, is zero. The displacements u and v are readily calculated. Thus, consider any point $P(x, y)$ in the circular cross section, which, before deformation, occupied the position shown in Fig. 20. After deformation, the point P will occupy a new position $P'(x + u, y + v)$. In terms of the angular displacement θ of the point P, we have

$$u = r \cos(\beta + \theta) - r \cos \beta = x(\cos \theta - 1) - y \sin \theta,$$
$$v = r \sin(\beta + \theta) - r \sin \beta = x \sin \theta + y(\cos \theta - 1),$$

where β is the angle between the radius vector r and the x-axis so that $x = r \cos \beta$, $y = r \sin \beta$. If the angle θ is small, we can write

$$u = -\theta y, \qquad v = \theta x.$$

Since $\theta = \alpha z$, we have for the displacements of any point with coordinates x, y, z

(33.1) $\qquad u = -\alpha z y, \qquad v = \alpha z x, \qquad w = 0.$

The system of stresses associated with the displacements (33.1) is given at once by the formulas (24.6). We thus have

(33.2) $\qquad \tau_{zy} = \mu \alpha x, \qquad \tau_{zx} = -\mu \alpha y, \qquad \tau_{xx} = \tau_{yy} = \tau_{zz} = \tau_{xy} = 0,$

which obviously satisfy the equations of equilibrium (with no body forces acting) and the equations of compatibility. The boundary conditions on the lateral surface are likewise satisfied. The first two of Eqs. (29.4) are identically satisfied, and the last one gives

$$\tau_{zx} \cos(x, \nu) + \tau_{zy} \cos(y, \nu)$$
$$= -\mu \alpha y \cos(x, \nu) + \mu \alpha x \cos(y, \nu) \equiv 0,$$

since for a circle of radius a, $\cos(x, \nu) = x/a$ and $\cos(y, \nu) = y/a$.

The only nonvanishing component of the couple M produced by the distribution of stresses (33.2) over the end of the cylinder is M_z, which is easily calculated. Thus,

$$M_z = \iint (x \tau_{zy} - y \tau_{zx}) \, dx \, dy$$
$$= \mu \alpha \iint (x^2 + y^2) \, dx \, dy = \mu \alpha I_0,$$

where $I_0 = \pi a^4 / 2$ is the polar moment of inertia of the circular cross-section of radius a.

The resultant force acting on the end of the cylinder vanishes, and it follows from Saint-Venant's principle that whatever be the distribution of forces over the end of the cylinder that gives rise to the couple of magnitude M_z, the distribution of stress sufficiently far from the ends of the cylinder is essentially that specified by (33.2).

The stress vector[1]

$$T = i\tau_{zx} + j\tau_{zy} + k\tau_{zz} = \mu\alpha(-iy + jx),$$

acting at a point (x, y) on any cross section z-constant, lies in the plane of the section and is normal to the radius vector r joining the point (x, y) with the origin $(0, 0)$. The magnitude of T is

(33.3) $\qquad T = \sqrt{\tau_{zx}^2 + \tau_{zy}^2} = \mu\alpha\sqrt{x^2 + y^2} = \mu\alpha r.$

From this we see that the maximum stress is a tangential stress that acts on the boundary of the cylinder and has the magnitude $\mu\alpha a$, where a is the radius of the cylinder.

34. Torsion of Cylindrical Bars. Consider a cylindrical bar subjected to no body forces and free from external forces on its lateral surface. One end of the bar is fixed in the plane $z = 0$, while the other end, in the plane $z = l$, is twisted by a couple of magnitude M whose moment is directed along the axis of the bar.

Navier, being guided by Coulomb's solution of the torsion problem for a circular shaft, assumed that in the general case of torsion of noncircular bars, the sections of the bar perpendicular to the z-axis will remain plane. This assumption led him to erroneous conclusions. The fact that the displacements characterized by formulas (33.1) cannot be valid for bars whose sections are not circular can be seen from the boundary conditions (29.4). A substitution of the stresses (33.2) in the third of the boundary conditions yields

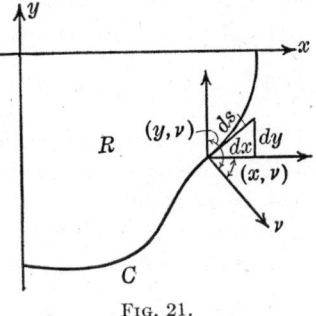

Fig. 21.

(34.1) $\qquad -\mu\alpha y \cos(x, \nu) + \mu\alpha x \cos(y, \nu) = 0,$

where ν, as always, denotes the exterior normal to the boundary C of the cross section R of the beam. But from Fig. 21 it is seen that

---

[1] We denote the unit base vectors along the x-, y-, and z-axes by i, j, and k, respectively.

$$(34.2) \quad \begin{cases} \dfrac{dx}{ds} = \cos(x, s) = \sin(x, \nu) = -\cos(y, \nu), \\ \dfrac{dy}{ds} = \sin(x, s) = \cos(x, \nu), \end{cases}$$

so that, upon dividing out the nonvanishing factor $\mu\alpha$, Eq. (34.1) becomes

$$x\, dx + y\, dy = 0.$$

This is the differential equation of a family of circles. Thus, circular cylinders are the only bodies whose lateral surfaces can be expected to be free from applied external forces if the state of stress characterized by the formulas (33.2) obtains in the interior.

A natural modification of Navier's assumption is to suppose that for cylinders other than circular ones, cross sections do not remain plane but are warped, and that each section is warped in the same way. This leads us to assume displacements of the form

$$(34.3) \quad u = -\alpha z y, \quad v = \alpha z x, \quad w = \alpha \varphi(x, y),$$

where $\varphi(x, y)$ is some function of x and y, and α, as before, is the angle of twist per unit length of the bar. The function $\varphi(x, y)$ must be so determined as to satisfy the differential equations (29.1) and the boundary conditions (29.4).

A simple calculation of the stresses corresponding to the displacements (34.3) gives

$$(34.4) \quad \begin{cases} \tau_{yz} = \mu\alpha\left(\dfrac{\partial \varphi}{\partial y} + x\right), \quad \tau_{zx} = \mu\alpha\left(\dfrac{\partial \varphi}{\partial x} - y\right), \\ \tau_{xy} = \tau_{xx} = \tau_{yy} = \tau_{zz} = 0. \end{cases}$$

A substitution of these values in the equilibrium equations (29.1) shows that the equilibrium equations will be satisfied if $\varphi(x, y)$ satisfies the equation

$$(34.5) \quad \nabla^2 \varphi \equiv \frac{\partial^2 \varphi}{\partial x^2} + \frac{\partial^2 \varphi}{\partial y^2} = 0$$

throughout the cross section of the cylinder. Furthermore, if the system of stresses is to satisfy the boundary conditions

(29.4) on the lateral surface of the cylinder, we see that

$$\left(\frac{\partial \varphi}{\partial x} - y\right) \cos (x, \nu) + \left(\frac{\partial \varphi}{\partial y} + x\right) \cos (y, \nu) = 0 \quad \text{on } C,$$

where C is the boundary of the cross section R of the cylinder (Fig. 19). But

$$\frac{\partial \varphi}{\partial x} \cos (x, \nu) + \frac{\partial \varphi}{\partial y} \cos (y, \nu) \equiv \frac{d\varphi}{d\nu},$$

so that the boundary condition can be written in the form

(34.6) $$\frac{d\varphi}{d\nu} = y \cos (x, \nu) - x \cos (y, \nu) \quad \text{on } C.$$

It follows from (34.5) that $\varphi(x, y)$ must be a harmonic function throughout the region R bounded by the curve C, and that on the boundary C the normal derivative of $\varphi(x, y)$ must assume the value given by (34.6). Since the displacements are single-valued functions, it follows from (34.3) that $\varphi(x, y)$ must also be a single-valued function. Thus, the problem of determining the *torsion function* $\varphi(x, y)$ is a special case of the second boundary-value problem of Potential Theory. This latter problem is associated with the name of Neumann, and consists of determining a function Φ that is harmonic in a given region, and whose normal derivative is prescribed on the boundary of the region. We shall meet this problem again in our study of several problems of elasticity. At this time we shall simply remark that the harmonic function φ is determined by the boundary condition (34.6) to within an arbitrary constant.[1] The substitution of $\varphi +$ constant in formulas (34.4) obviously does not alter the stresses, and it is clear from (34.3) that the addition of a constant to φ means a shift of the cylinder as a whole in the direction of the z-axis. Thus, the additive constant in the solution of the Problem of Neumann is immaterial in our case.

The condition for the existence of a solution Φ of the Problem of Neumann is that the integral of the normal derivative of the function Φ, calculated over the entire boundary C, vanish. This

---

[1] See Sec. 42.

follows from the identity

$$\int_C \frac{d\Phi}{d\nu} ds = \iint_R \text{div } (\nabla \Phi) \, d\sigma = \iint_R \nabla^2 \Phi \, d\sigma$$

and from the fact that $\nabla^2 \Phi = 0$. This condition is satisfied in our case, for [see (34.6)]

$$\int_C \frac{d\varphi}{d\nu} ds = \int_C [y \cos (x, \nu) - x \cos (y, \nu)] \, ds$$
$$= \int_C (y \, dy + x \, dx) = 0,$$

since the integrand is the exact differential of the function $\frac{1}{2}(x^2 + y^2) +$ constant.

It is easy to show that the distribution of stresses given by Eqs. (34.4) is equivalent to a torsional couple applied at the end $z = l$ of the cylinder and that the resultant force acting on the end of the cylinder vanishes. Now the resultant force in the x-direction is given by

$$\iint_R \tau_{zx} \, dx \, dy = \mu\alpha \iint_R \left(\frac{\partial \varphi}{\partial x} - y \right) dx \, dy,$$

and this can be written as

$$(34.7) \quad \mu\alpha \iint_R \left\{ \frac{\partial}{\partial x} \left[x \left(\frac{\partial \varphi}{\partial x} - y \right) \right] + \frac{\partial}{\partial y} \left[x \left(\frac{\partial \varphi}{\partial y} + x \right) \right] \right\} dx \, dy,$$

since φ satisfies the differential equation (34.5). Green's Theorem is directly applicable to the integral (34.7), and we get

$$(34.8) \quad \mu\alpha \int_C x \left[\frac{d\varphi}{d\nu} - y \cos (x, \nu) + x \cos (y, \nu) \right] ds,$$

where the line integral is evaluated over the boundary C of the region R. The integral (34.8) vanishes on account of the boundary condition (34.6). It is shown in a similar way that

$$\iint_R \tau_{zy} \, dx \, dy = 0,$$

so that the resultant force acting on the end of the cylinder vanishes.

It remains to show that the system of stresses defined by Eqs. (34.4) is statically equivalent to a torsional couple. The resultant moment of the external forces applied to the end of the beam is

$$M_z = \int\int_R (x\tau_{zy} - y\tau_{zx})\, dx\, dy$$

(34.9)
$$= \mu\alpha \int\int_R \left(x^2 + y^2 + x\frac{\partial\varphi}{\partial y} - y\frac{\partial\varphi}{\partial x}\right) dx\, dy.$$

The integral appearing in (34.9) depends on the torsion function φ and hence on the cross section R of the beam. Setting

(34.10) $$D = \mu \int\int_R \left(x^2 + y^2 + x\frac{\partial\varphi}{\partial y} - y\frac{\partial\varphi}{\partial x}\right) dx\, dy,$$

we have

(34.11) $$M = D\alpha.$$

The formula (34.11) shows that the twisting moment M is proportional to the angle of twist per unit length, α, so that the constant D provides a measure of the rigidity of a beam subjected to torsion. For this reason the constant D (depending on the modulus of rigidity μ and on the shape of the cross section only) is called the *torsional rigidity* of the beam.

It follows from the foregoing that the torsion problem for a beam of any cross section is completely solved once the function $\varphi(x, y)$ is determined. For the torsional rigidity D is determined by φ from (34.10), and the moment M required to produce the angle of twist per unit length, α, can be calculated from (34.11).

In carrying out the foregoing calculations, no assumptions were made regarding the location of the origin O or concerning the orientation of the axes x, y. Inasmuch as the first two of the formulas (34.3) represent the infinitesimal rotation of any cross section of the beam as a whole about the z-axis, it may seem at first glance that a different choice of the axis of rotation parallel of the axis of z may yield a different solution of the problem. For instance, if the axis z' is chosen parallel to the z-axis, and if it intersects the xy-plane at some point (x_1, y_1), then the displacements u_1, v_1, and w_1 will be

$$u_1 = -\alpha z(y - y_1), \quad v_1 = \alpha z(x - x_1), \quad w_1 = \alpha\varphi_1(x, y),$$

and there is no *a priori* reason why the functions $\varphi_1(x, y)$ and $\varphi(x, y)$ should be identical.

Calculating stresses that correspond to displacements (u_1, v_1, w_1) yields

(34.12)
$$\begin{cases} \tau_{zy} = \mu\alpha\left(\dfrac{\partial\varphi_1}{\partial y} + x - x_1\right), \\ \tau_{zx} = \mu\alpha\left(\dfrac{\partial\varphi_1}{\partial x} - y + y_1\right), \\ \tau_{xy} = \tau_{xx} = \tau_{yy} = \tau_{zz} = 0, \end{cases}$$

and the substitution of these values in the equations of equilibrium (29.1) shows that the function φ_1 likewise satisfies the equation

$$\nabla^2\varphi_1 \equiv \frac{\partial^2\varphi_1}{\partial x^2} + \frac{\partial^2\varphi_1}{\partial y^2} = 0.$$

Moreover, the third of the boundary conditions (29.4) demands that

$$\left(\frac{\partial\varphi_1}{\partial x} + y_1\right)\cos(x,\nu) + \left(\frac{\partial\varphi_1}{\partial y} - x_1\right)\cos(y,\nu)$$
$$= y\cos(x,\nu) - x\cos(y,\nu)$$

or

$$\frac{d}{d\nu}(\varphi_1 + y_1 x - x_1 y) = y\cos(x,\nu) - x\cos(y,\nu).$$

But the function $\varphi_1 + y_1 x - x_1 y$ is harmonic, and since it satisfies the same boundary condition as the function φ [see (34.6)], it follows from the uniqueness of solution of the problem of Neumann[1] that the two can differ only by a constant. Thus,

$$\varphi_1 = \varphi(x, y) - y_1 x + x_1 y + \text{const.}$$

A simple calculation making use of the formulas (34.12) shows that the system of stresses obtained by using the function $\varphi_1(x, y)$ is identical with that obtained by using the function $\varphi(x, y)$. It follows that the displacement in the two cases can differ only by a rigid body displacement. Thus, we see that the position of the origin of coordinates is immaterial in this problem.

References for Collateral Reading

O. D. KELLOGG: Foundations of Potential Theory, Verlag von Julius Springer, Berlin, Chap. IX, Sec. 5; Chap. XI, Sec. 12.

[1] See O. D. Kellogg, Foundations of Potential Theory, Chap. XI, Sec. 12.

A. E. H. Love: A Treatise on the Mathematical Theory of Elasticity, Cambridge University Press, London, Sec. 215.
S. Timoshenko: Theory of Elasticity, McGraw-Hill Book Company, Inc., New York, Secs. 69, 72, 73.
R. V. Southwell: Theory of Elasticity for Engineers and Physicists, Oxford University Press, New York, Secs. 155, 336–341.
E. Trefftz: Handbuch der Physik, Verlag von Julius Springer, Berlin, vol. 6, Sec. 42.
A. G. Webster: Dynamics of Particles and Rigid Bodies, Verlag von Julius Springer, Berlin, Sec. 184.

35. Stress Function. Since the torsion function $\varphi(x, y)$ is harmonic in the region R representing the cross section of the beam, one can construct the analytic function[1] $\varphi + i\psi$, where $\psi(x, y)$ is the conjugate harmonic function, which is related to $\varphi(x, y)$ by the Cauchy-Riemann equations, namely,

$$(35.1) \qquad \frac{\partial \varphi}{\partial x} = \frac{\partial \psi}{\partial y}, \qquad \frac{\partial \varphi}{\partial y} = -\frac{\partial \psi}{\partial x}.$$

Since the function $\varphi + i\psi$ is an analytic function of the complex variable $x + iy$, it is clear that the function $\psi(x, y)$ is determined by the formula

$$(35.2) \quad \psi(x, y) = \int_C \left(\frac{\partial \psi}{\partial x} dx + \frac{\partial \psi}{\partial y} dy \right)$$
$$= \int_{P_0(x_0, y_0)}^{P(x, y)} \left(-\frac{\partial \varphi}{\partial y} dx + \frac{\partial \varphi}{\partial x} dy \right),$$

where the integral is evaluated over an arbitrary path joining some point $P_0(x_0, y_0)$ with an arbitrary point $P(x, y)$ belonging to the region R. If the region R is simply connected, then the function $\psi(x, y)$ will be single-valued; otherwise, $\psi(x, y)$ may turn out to be multiple-valued. For the time being, we shall be primarily concerned with simply connected regions, and the discussion in this section will be confined to such regions.

It is not difficult to phrase the torsion problem in terms of the conjugate function $\psi(x, y)$. Thus, noting the relations (34.2), one can write the expression for the normal derivative $\frac{d\varphi}{d\nu}$ with the aid of the tangential derivatives $\frac{dx}{ds}$ and $\frac{dy}{ds}$, so that

---

[1] Some basic results of the theory of analytic functions of a complex variable may be found in I. S. and E. S. Sokolnikoff, Higher Mathematics for Engineers and Physicists, Chap. X, pp. 440–491.

$$\frac{d\varphi}{d\nu} = \frac{\partial \varphi}{\partial x} \cos (x, \nu) + \frac{\partial \varphi}{\partial y} \cos (y, \nu)$$
$$= \frac{\partial \varphi}{\partial x} \frac{dy}{ds} - \frac{\partial \varphi}{\partial y} \frac{dx}{ds}.$$

Making use of the Cauchy-Riemann equations (35.1), we have

$$\frac{d\varphi}{d\nu} = \frac{\partial \psi}{\partial x} \frac{dx}{ds} + \frac{\partial \psi}{\partial y} \frac{dy}{ds} \equiv \frac{d\psi}{ds}.$$

Moreover, the boundary condition (34.6) can be written as

$$\frac{d\varphi}{d\nu} = y \cos (x, \nu) - x \cos (y, \nu)$$
$$= x \frac{dx}{ds} + y \frac{dy}{ds} = \frac{1}{2} \frac{d}{ds} (x^2 + y^2).$$

Hence

$$\frac{d\psi}{ds} = \frac{1}{2} \frac{d}{ds} (x^2 + y^2),$$

so that

(35.3) $\qquad \psi = \frac{1}{2}(x^2 + y^2) + \text{const.} \qquad$ on C.

It will be recalled that the torsion function φ is determined to within a nonessential arbitrary constant; the derivatives of φ and hence those of ψ [see (35.1)] are determined uniquely, and the function ψ is determined by means of (35.2) to within a constant depending on the choice of $P_0(x_0, y_0)$. Accordingly, we are free to assign any value to the constant of integration in (35.3), since this choice will not affect stresses, and the two sets of displacements that correspond to two different choices of the arbitrary constant will differ from one another by a rigid body motion.

Thus, instead of solving a problem of Neumann, we can equally well solve a problem of Dirichlet by determining a function that is harmonic in a given region, and by assuming prescribed values on the boundary of the region.

On account of the remarks just made, our problem consists of determining a function ψ that satisfies the equation

$$\frac{\partial^2 \psi}{\partial x^2} + \frac{\partial^2 \psi}{\partial y^2} = 0 \qquad \text{in } R,$$

and that satisfies the boundary condition

(35.4) $$\psi = \tfrac{1}{2}(x^2 + y^2).$$

It is known that the solution of this problem is unique,[1] and there are general methods that permit one to construct solutions of the problem of Dirichlet. We shall consider some of them in detail in succeeding sections.

We shall now formulate the torsion problem in terms of the function Ψ, which is defined as follows:

(35.5) $$\Psi = \psi(x, y) - \tfrac{1}{2}(x^2 + y^2).$$

We have

$$\frac{\partial \Psi}{\partial x} = \frac{\partial \psi}{\partial x} - x, \qquad \frac{\partial \Psi}{\partial y} = \frac{\partial \psi}{\partial y} - y,$$

and, upon recalling the formulas (34.4) and (35.1), it follows that

(35.6) $$\tau_{zx} = \mu\alpha \frac{\partial \Psi}{\partial y}, \qquad \tau_{zy} = -\mu\alpha \frac{\partial \Psi}{\partial x}.$$

Since the stress components τ_{zx} and τ_{zy} are obtained from the function $\Psi(x, y)$ by differentiation, the latter is called the *stress function*. It is readily checked that the stress function Ψ satisfies Poisson's equation

(35.7) $$\nabla^2 \Psi \equiv \frac{\partial^2 \Psi}{\partial x^2} + \frac{\partial^2 \Psi}{\partial y^2} = -2 \qquad \text{in } R,$$

and on the boundary C of the region R [cf. (35.3) and (35.5)] assumes the value

$$\Psi = \text{const.}$$

Consider a family of curves, in the plane of the cross section of the beam, obtained by setting

(35.8) $$\Psi(x, y) = \text{const.}$$

The slope $\dfrac{dy}{dx}$ of the tangent line to any curve of the family defined by (35.8) is determined from the formula

$$\frac{\partial \Psi}{\partial x} + \frac{\partial \Psi}{\partial y} \frac{dy}{dx} = 0,$$

---

[1] See O. D. Kellogg, Foundations of Potential Theory, Chap. IX, Sec. 5.

and, upon noting the relations (35.6), we obtain

$$\frac{dy}{dx} = \frac{\tau_{zy}}{\tau_{zx}}.$$

Thus, at each point of the curve $\Psi(x, y) = $ const., the stress vector

$$\boldsymbol{\tau} = \boldsymbol{i}\tau_{zx} + \boldsymbol{j}\tau_{zy}$$

is directed along the tangent to the curve. The curves

$$\Psi(x, y) = \text{const.}$$

are called the *lines of shearing stress*. The magnitude τ of the tangential stress is

$$\tau = \sqrt{\tau_{zy}^2 + \tau_{zx}^2} = \mu\alpha\sqrt{\left(\frac{\partial \Psi}{\partial x}\right)^2 + \left(\frac{\partial \Psi}{\partial y}\right)^2}.$$

Recalling that for a circular cylinder the magnitude of the tangential stress is given by

$$\tau = \sqrt{\tau_{zx}^2 + \tau_{zy}^2} = \mu\alpha\sqrt{x^2 + y^2},$$

we see that in this case the maximum shearing stress occurs on the boundary of the section. It is not difficult to prove that in the general case the points at which maximum shearing stress occurs lie on the boundary C of the section, so that elastic failure of material in shear is to be expected on the lateral surface of the beam. In order to prove the assertion, we refer to a theorem.

THEOREM: *Let a function Φ, continuous of class $C^{(2)}$, and not identically equal to a constant, satisfy the inequality $\nabla^2 \Phi \geqq 0$ in the region R; then this function attains its maximum on the boundary C of the region R.*

The proof of this theorem follows at once from the well-known property of subharmonic functions. It will be recalled that a function $\Phi(x, y)$ is called subharmonic in the region R if at every point (x_1, y_1) of the region

(1) $$\Phi(x_1, y_1) \leq \frac{1}{2\pi r}\int_\gamma \Phi(x, y)\, ds,$$

where the integral is evaluated over the circle γ of sufficiently small radius r and with center at (x_1, y_1). But it is known[1] that whenever $\Phi(x, y)$ is continuous of class $C^{(2)}$ in R, then the necessary and sufficient condition for the function Φ to be subharmonic in R is that $\nabla^2 \Phi \geq 0$ at every point of R. Since $\Phi(x, y)$ is continuous in the closed region R, the Theorem of Weierstrass gives an assurance that $\Phi(x, y)$ attains its maximum somewhere in the region. Assume that the maximum value is attained not on the boundary C but at some interior point (x_1, y_1); then, since $\Phi(x, y)$ is not identically equal to a constant,

$$(2) \qquad \Phi(x_1, y_1) > \frac{1}{2\pi r} \int_\gamma \Phi(x, y)\, ds,$$

where r is a sufficiently small radius of the circle γ, whose center is at (x_1, y_1). An inspection of the inequalities (1) and (2) leads one to conclude that the assumption that the maximum is attained in the interior of the region R is untenable.

Since

$$\tau^2 = \mu^2 \alpha^2 \left[\left(\frac{\partial \Psi}{\partial x}\right)^2 + \left(\frac{\partial \Psi}{\partial y}\right)^2 \right],$$

a simple calculation shows that

$$\nabla^2(\tau^2) = 2\mu^2 \alpha^2 \left[\left(\frac{\partial^2 \Psi}{\partial x^2}\right)^2 + 2\left(\frac{\partial^2 \Psi}{\partial x\, \partial y}\right)^2 + \left(\frac{\partial^2 \Psi}{\partial y^2}\right)^2 \right],$$

where we make use of the equation satisfied by Ψ. The expression for $\nabla^2(\tau^2)$ is positive, and it follows from the foregoing theorem that τ attains its maximum on the boundary of the region. Since the strength of the beam to resist torsion depends on the maximum shearing stress, practical rules for the design of beams carrying torsional loads are expressed in terms of the safe maximum shearing stress τ.

The expression (34.10) for the torsional rigidity D can be written in an interesting way in terms of the stress function Ψ. This expression will prove valuable to us in considering a method of solution of torsion problems by the membrane analogy introduced by L. Prandtl.

[1] See, for example, O. D. Kellogg, Foundations of Potential Theory, pp. 315–317; I. I. Privaloff, Subharmonic Functions, p. 39 and p. 54; Tibor Radó, Subharmonic Functions, pp. 12–13.

We first recall the formula (34.11),
$$M = D\alpha,$$
where
$$M = \iint_R (x\tau_{zy} - y\tau_{zx})\, dx\, dy.$$

Since[1]
$$\tau_{zx} = \mu\alpha\frac{\partial \Psi}{\partial y}, \qquad \tau_{zy} = -\mu\alpha\frac{\partial \Psi}{\partial x},$$

we have
$$M = -\mu\alpha \iint_R \left(x\frac{\partial \Psi}{\partial x} + y\frac{\partial \Psi}{\partial y}\right) dx\, dy,$$

so that

$$(35.9) \quad D = -\mu \iint_R \left(x\frac{\partial \Psi}{\partial x} + y\frac{\partial \Psi}{\partial y}\right) dx\, dy$$
$$= -\mu \iint_R \left[\frac{\partial (x\Psi)}{\partial x} + \frac{\partial (y\Psi)}{\partial y}\right] dx\, dy + 2\mu \iint_R \Psi\, dx\, dy.$$

The first of the double integrals in the foregoing can be transformed by Green's Theorem so that (35.9) reads

$$D = -\mu \int_C \Psi[x \cos(x, \nu) + y \cos(y, \nu)]\, ds + 2\mu \iint_R \Psi\, dx\, dy.$$

But we can choose [see (35.4) and (35.5)]
$$\Psi = 0 \qquad \text{on } C,$$
and the foregoing expression becomes

$$(35.10) \quad D = 2\mu \iint_R \Psi\, dx\, dy.$$

It is obvious from (35.10) that the torsional rigidity of a beam whose cross section R is bounded by the contour C is twice the product of the shear modulus μ and the volume enclosed by the surface $z = \Psi(x, y)$ and the plane $z = 0$. We shall see in a later section that a homogeneous, uniformly stretched membrane subjected to a uniform pressure is distorted into a surface whose differential equation is of the same form as that for the stress function Ψ. The connection between the surface of the loaded membrane and the stress function Ψ is utilized in the experimental

[1] See formulas (35.6).

§35 EXTENSION, TORSION, AND FLEXURE OF BEAMS

determination of the magnitude of stresses in cylinders whose cross sections are such as to make a mathematical determination of the torsion function very difficult.

Before proceeding to a consideration of specific examples, we note that our solution requires that the tangential stresses τ_{zx} and τ_{zy} be distributed over the ends of the beam in a manner specified by (34.4). In practical applications, this particular distribution of stress may not correspond to the actual physical situation, but on the basis of Saint-Venant's principle, we can assert that sufficiently far from the ends of the beam, the stress will depend on the magnitude of the couple M and will be quite independent of the mode of distribution of tractions over the ends of the beam.

We have seen that the torsion problem can be reduced to the problem of finding a function $\psi(x, y)$ that is harmonic in the region R and takes the values $\frac{1}{2}(x^2 + y^2)$ on the boundary C of R. Some special methods of solving the torsion problem will be considered in the following sections. In the next two sections, our plan of attack will be to consider a particular harmonic function ψ that contains some undetermined coefficients. These undetermined coefficients will be chosen in such a way that on the boundary of a certain region, ψ takes on the values $\frac{1}{2}(x^2 + y^2)$. In Sec. 26, a solution in the form of an infinite series will be obtained for rectangular and triangular prisms. The general solution of the torsion problem for a beam of arbitrary cross section R is then given by mapping the region R upon the interior of a circle and then considering the solutions of the problems of Dirichlet and Neumann for the circular region.

References for Collateral Reading

A. E. H. LOVE: A Treatise on the Mathematical Theory of Elasticity, Cambridge University Press, London, Secs. 216–220.

W. D. MACMILLAN: The Theory of the Potential, McGraw-Hill Book Company, Inc., New York, Sec. 127.

O. D. KELLOGG: Foundations of Potential Theory, Verlag von Julius Springer, Berlin, Chap. IX, Sec. 3; Chap. XI, Secs. 1, 12.

PROBLEMS

1. Consider a circular shaft of length l, radius a, and shear modulus μ, twisted by a couple M. Show that the greatest angle of twist θ and the maximum shear stress $T = \sqrt{\tau_{zx}^2 + \tau_{zy}^2}$ are given by

$$\theta_{\max} = \frac{2Ml}{\pi\mu a^4},$$

$$T_{\max} = \frac{2M}{\pi a^3}.$$

2. A steel shaft of circular cross section 2 in. in diameter and 5 ft. long is twisted by end couples. Find the maximum twisting moment and angle of twist if the greatest shear stress is not to exceed 10,000 lb. per sq. in. Take $E = 30 \cdot 10^6$ lb. per sq. in., $\sigma = 0.3$.

3. The shaft of the preceding problem is not to be twisted more than $1°$. What is the corresponding maximum shear stress?

4. Derive the expression

$$M_z = \frac{63{,}000}{n} H$$

for the torque M_z on a solid circular shaft transmitting H hp. at a speed n r.p.m. *Hint:* Let the radius of the shaft (or pulley) be r in., and let $T = M_z/r$ be the tension in the belt. Calculate the work done in each minute against M_z. (1 hp. = 33,000 ft.-lb. per min.)

5. Derive the expressions

$$\alpha = \frac{32 M_z}{\mu \pi d^4} = \frac{2(\tau_s)_{\max}}{\mu d} = \frac{640{,}000}{\mu n d^4} H$$

for the twist per inch length α (radians) in a solid circular shaft of diameter d in., transmitting H hp. at n r.p.m. against a torque of M_z in.-lb.

6. How much torque can be transmitted by a solid circular shaft 3 in. in diameter if the allowable shear stress is 10,000 lb. per sq. in.? What is the angle of twist per foot of length? Use $\mu = 12 \times 10^6$ lb. per sq. in.

36. Torsion of Elliptical Cylinder. It was shown above that the solution of the torsion problem for a cylinder of any cross section is completely determined if one obtains the harmonic function ψ that on the boundary C of the cross section assumes the value

(36.1) $$\psi = \tfrac{1}{2}(x^2 + y^2).$$

Consider the harmonic function

(36.2) $$\psi = c^2(x^2 - y^2) + k^2,$$

where c and k are constants. The function defined by (36.2) will enable us to solve the torsion problem for some region R

§36 EXTENSION, TORSION, AND FLEXURE OF BEAMS

on the boundary of which (36.2) reduces to (36.1). Hence points of the boundary C of the region R are determined by equating (36.1) and (36.2). Thus,

$$c^2(x^2 - y^2) + k^2 = \tfrac{1}{2}(x^2 + y^2),$$

or

(36.3) $$(\tfrac{1}{2} - c^2)x^2 + (\tfrac{1}{2} + c^2)y^2 = k^2.$$

The curve defined by Eq. (36.3) is an ellipse

$$\frac{x^2}{a^2} + \frac{y^2}{b^2} = 1,$$

if we choose $c^2 < \tfrac{1}{2}$ and

$$a = \frac{k}{\sqrt{\tfrac{1}{2} - c^2}}, \qquad b = \frac{k}{\sqrt{\tfrac{1}{2} + c^2}}.$$

or

$$c^2 = \frac{1}{2}\frac{a^2 - b^2}{a^2 + b^2}, \qquad k^2 = \frac{a^2 b^2}{a^2 + b^2}.$$

Substituting the values of c and k in terms of a and b in (36.2), we obtain the solution of our boundary-value problem for an ellipse with semiaxes a and b, namely,

(36.4) $$\psi = \frac{1}{2}\frac{a^2 - b^2}{a^2 + b^2}(x^2 - y^2) + \frac{a^2 b^2}{a^2 + b^2}.$$

The components of stress (34.4) can be expressed directly in terms of the function ψ by noting the Cauchy-Riemann equations (35.1). Thus,

$$\tau_{zx} = \mu\alpha\left(\frac{\partial \psi}{\partial y} - y\right), \qquad \tau_{zy} = \mu\alpha\left(-\frac{\partial \psi}{\partial x} + x\right).$$

Hence

(36.5) $$\tau_{zx} = \frac{-2\mu\alpha a^2 y}{a^2 + b^2}, \qquad \tau_{zy} = \frac{2\mu\alpha b^2 x}{a^2 + b^2}.$$

The torsion moment M is

$$M = \int\!\!\int_R (x\tau_{zy} - y\tau_{zx})\,dx\,dy$$
$$= \frac{2\mu\alpha}{a^2 + b^2}\left(b^2 \int\!\!\int_R x^2\,dx\,dy + a^2 \int\!\!\int_R y^2\,dx\,dy\right)$$
$$= \frac{2\mu\alpha}{a^2 + b^2}(a^2 I_x + b^2 I_y),$$

where I_x and I_y are the moments of inertia of the elliptical section about the x- and y-axes. Recalling that

$$I_x = \frac{\pi a b^3}{4}, \qquad I_y = \frac{\pi a^3 b}{4},$$

we have

$$M = \frac{\pi \mu \alpha a^3 b^3}{a^2 + b^2},$$

so that the torsional rigidity

$$D = \frac{\pi \mu a^3 b^3}{a^2 + b^2}.$$

It was shown in the preceding section that the maximum shearing stress on any cross section occurs on the boundary of the

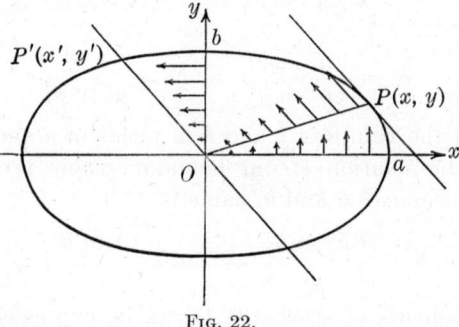

Fig. 22.

section. The location of the points on the boundary at which the greatest stress $\tau_{\max}$ occurs can be determined[1] by maximizing the expression for τ that has been obtained as a function of a single variable by utilizing the equation of the boundary C. In the case of an elliptical cylinder, the points of greatest shearing stress can be found easily from some simple geometrical considerations.

Consider an elliptical section, shown in Fig. 22, and draw from the center of the ellipse a semidiameter OP to an arbitrary point $P(x, y)$ of the boundary. Since the diameter of the ellipse conjugate to the diameter through P is parallel to the tangent line[2] at $P(x, y)$, it follows that the conjugate semidiameter OP'

---

[1] See Prob. 1 at the end of this section.
[2] See, for example, W. F. Osgood and W. C. Graustein, Plane and Solid Analytic Geometry, Chap. XIV.

intersects the curve at the point $P'(x', y')$, where

$$x' = -\frac{ay}{b}, \qquad y' = \frac{bx}{a}.$$

When the stresses at $P(x, y)$ are written in terms of the coordinates x', y' of the point P', we have

$$\tau_{zx} = \frac{2\mu\alpha ab}{a^2 + b^2} x', \qquad \tau_{zy} = \frac{2\mu\alpha ab}{a^2 + b^2} y',$$

so that the direction of stress at the point P is parallel to the conjugate semidiameter OP'. Furthermore, the magnitude τ of the tangential stress at $P(x, y)$ is

$$\tau = \sqrt{\tau_{zx}^2 + \tau_{yz}^2} = \frac{2\mu\alpha ab}{a^2 + b^2} \sqrt{x'^2 + y'^2} = \frac{2\mu\alpha ab}{a^2 + b^2} r',$$

where r' is the distance OP'. Since the conjugate semidiameter is of maximum length when the point P is at an extremity of the minor axis, it follows that

$$\tau_{\max} = \frac{2\mu\alpha a^2 b}{a^2 + b^2}.$$

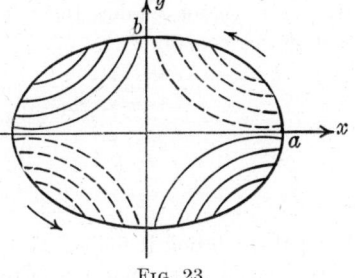

Fig. 23.

Thus, the maximum stress occurs at the extremities of the minor axis of the ellipse, contrary to an intuitive expectation that the maximum stress would be at the points of maximum curvature.

It is easy to verify that the conjugate harmonic function φ, apart from a nonessential constant, is[1]

(36.6) $$\varphi = -\frac{a^2 - b^2}{a^2 + b^2} xy.$$

This function determines the warping of the cross sections of the cylinder, for the displacement along the z-axis is given by $w = \alpha\varphi(x, y)$. The contour lines, obtained by setting $\varphi = $ constant, are the hyperbolas shown in Fig. 23. The dotted lines indi-

[1] See Eq. (36.1).

cate the portions of the section that become concave, and the solid those that become convex, when the cylinder is twisted by a couple in the directions shown in the figure by arrows.

The lines of shearing stress are determined by drawing the contour lines for the surface $z = \Psi(x, y)$. Setting $\Psi(x, y) = $ constant gives,[1] in this case, a family of concentric ellipses,

$$\frac{x^2}{a^2} + \frac{y^2}{b^2} - 1 = \frac{-c'(a^2 + b^2)}{a^2 b^2},$$

similar to the ellipse $x^2/a^2 + y^2/b^2 = 1$.

The displacement of the points of the cylinder is given at once by the formulas (34.3). The results obtained in Sec. 33 for a beam of circular cross section follow at once from the formulas of this section upon setting $b = a$.

PROBLEMS

1. Show that in the torsion of an elliptical cylinder, the magnitude of the stress vector τ takes the following value on the boundary of the section $z = $ constant:

$$\tau = 2\mu\alpha \frac{ab}{a^2 + b^2} \sqrt{a^2 - e^2 x^2},$$

$$e = \frac{1}{a} \sqrt{a^2 - b^2}.$$

From this relation it follows that the maximum shearing stress occurs at the ends of the minor axis of the ellipse.

2. Derive the expression (36.6) from

$$\varphi(x, y) = \int_{P_0}^{P} \left(\frac{\partial \psi}{\partial y} dx - \frac{\partial \psi}{\partial x} dy \right) + \text{const.},$$

and evaluate the line integral over the path consisting of the straight-line segments from $P_0(x_0, y_0)$ to $Q(x, y_0)$ and from $Q(x, y_0)$ to $P(x, y)$.

3. Show that the stress function for an elliptical section can be written as

$$\Psi = \frac{-a^2 b^2}{a^2 + b^2} \left(\frac{x^2}{a^2} + \frac{y^2}{b^2} - 1 \right)$$

---

[1] See Prob. 3 at the end of this section.

and is thus proportional to the function appearing in the equation of the boundary of the section. The problem of determining the sections for which this proportionality holds has been treated by Leibenson.[2]

37. Simple Solutions of the Torsion Problem. Effect of Grooves. The method of solution of the torsion problem illustrated in the preceding section was used by Saint-Venant, who selected a number of simple polynomial solutions of the equation

$$(37.1) \qquad \nabla^2 \psi = 0,$$

and determined the equation of the boundary of the cross section of the cylinder on which the function ψ reduces to $\frac{1}{2}(x^2 + y^2)$. Inasmuch as the real and imaginary parts of every analytic function of a complex variable $x + iy$ satisfy Eq. (37.1), we can build up a list of functions ψ, and by working, so to speak, backwards, can determine the equations of the contours for which these functions ψ represent the solution of the torsion problem. For example, if we consider the function $(x + iy)^n$, then by choosing $n = 2$, we get two solutions, $x^2 - y^2$ and $2xy$, of Eq. (37.1). The first of these solutions was utilized in the preceding section to solve the torsion problem for an elliptical cylinder. If n is set equal to 3, we obtain the harmonic functions $x^3 - 3xy^2$ and $3x^2y - y^3$. Now consider the harmonic function

$$(37.2) \qquad \psi = c(x^3 - 3xy^2) + k,$$

where c and k are constants. The function ψ determines the solution of the torsion problem for a cylinder whose cross section has the equation

$$(37.3) \qquad c(x^3 - 3xy^2) + k = \frac{1}{2}(x^2 + y^2).$$

By altering the values of the parameters in (37.3), we obtain various cross sections, some of which may be of technical interest. If we set $c = -1/6a$ and $k = 2a^2/3$, then (37.3) can be written in the factored form as

$$(x - a)(x - y\sqrt{3} + 2a)(x + y\sqrt{3} + 2a) = 0,$$

so that the boundary of the region is the equilateral triangle of altitude $3a$ (see Fig. 24).

---

[2] L. LEIBENSON, "Über den Zusammenhang zwischen der Spannungsfunktion bei Torsion und der Konturgleichung eines Prismenquerschnittes," *Wissenschaftliche Berichte der Moskauer Universität*, vol. 2 (1934), pp. 99–102 (in Russian with a German summary).

Making use of the formulas (35.6), we find

$$\tau_{zx} = \frac{\mu\alpha}{a} y(x - a), \qquad \tau_{zy} = \frac{\mu\alpha}{2a} (x^2 + 2ax - y^2).$$

We see from these formulas that the x-component of the shearing stress vanishes along the x-axis, while the y-component becomes

$$(\tau_{zy})_{y=0} = \frac{\mu\alpha}{2a} x(x + 2a).$$

The distribution of stress along the x-axis is indicated in Fig. 25. The shearing stress is a maximum at the midpoints of the sides

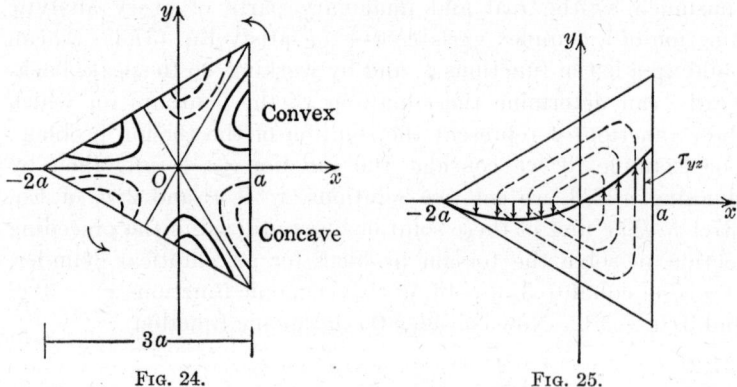

FIG. 24.   FIG. 25.

of the triangle, and its value is

$$\tau_{\max} = \tfrac{3}{2} a\mu\alpha.$$

The stress vanishes at the corners and at the origin O. The direction of the lines of shearing stress is along the curves

$$\psi - \tfrac{1}{2}(x^2 + y^2) = \text{const.}$$

a few of which are indicated in Fig. 25 by dotted lines. It is easily checked, with the aid of (34.9), that the torsional couple has the magnitude

$$M = \tfrac{3}{5} \mu\alpha I_0,$$

where $I_0 = 3\sqrt{3}\, a^4$ is the polar moment of inertia of the triangle. The nature of the distortion of the initially plane sections is indicated in Fig. 24, where the contour lines of the surface $\varphi(x, y) \equiv (3x^2 y - y^3)/6a = \text{constant}$ are shown.

§37 EXTENSION, TORSION, AND FLEXURE OF BEAMS

It appears from this example, and from that of the preceding section, that a circular shaft of the same cross-sectional area as an elliptical beam or a triangular prism has the greatest torsional rigidity.[1] It was shown by Saint-Venant that, for simply connected regions, the torsional rigidity for a given cross-sectional area of a beam increases as the polar moment of inertia of the cross section decreases. One can also prove that if the region is simply connected, then, for a given moment M and for a given cross-sectional area, the smallest maximum stress will be found in a circular beam.

The effect of grooves or slots in the beam on the maximum shearing stress can be discussed in an elementary way by studying an example due to C. Weber.[2]

Consider a pair of harmonic functions,

$$x \quad \text{and} \quad \frac{x}{x^2 + y^2},$$

and introduce the polar coordinates defined by the equations $x = r \cos \theta$, $y = r \sin \theta$. We can construct a harmonic function ψ,

$$\psi = a\left(x - b^2 \frac{x}{x^2 + y^2}\right) + \frac{1}{2} b^2 = a\left(r \cos \theta - \frac{b^2 \cos \theta}{r}\right) + \frac{1}{2} b^2,$$

where a and b are constants.

On the boundary C of the cross section, ψ must reduce to

$$\psi = \tfrac{1}{2}(x^2 + y^2) = \tfrac{1}{2} r^2,$$

so that the equation of the boundary for which the function ψ solves the torsion problem is

$$a\left(r \cos \theta - \frac{b^2 \cos \theta}{r}\right) + \frac{1}{2} b^2 = \frac{1}{2} r^2,$$

or

$$r^2 - b^2 - 2a(r^2 - b^2) \frac{\cos \theta}{r} = 0.$$

Factoring this expression gives

$$(r^2 - b^2)\left(1 - \frac{2a \cos \theta}{r}\right) = 0.$$

---

[1] See Prob. 1 at the end of this section.

[2] This problem is discussed in S. Timoshenko, Theory of Elasticity, pp. 238–239, where further references will be found.

Thus, the boundary is made up of two circles

$$r = b \quad \text{and} \quad r = 2a \cos \theta,$$

which are shown in Fig. 26.

Since the function ψ is known, one can easily calculate the stresses τ_{zx} and τ_{zy}. It turns out that the maximum shearing stress is at the point A and has the value[1]

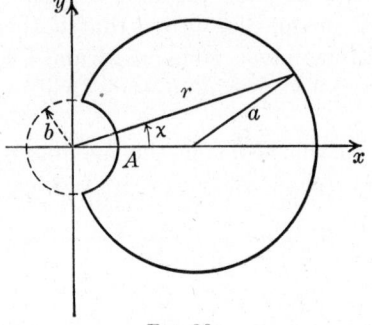

Fig. 26.

$$\tau_{\max} \doteq 2\mu\alpha a,$$

which is twice as great as the peripheral stress in a circular shaft of radius a. This example indicates the importance of considering stresses in slots and keyways of shafts.

References for Collateral Reading

A. E. H. Love: A Treatise on the Mathematical Theory of Elasticity, Cambridge University Press, London, Secs. 221–223.

S. Timoshenko: Theory of Elasticity, McGraw-Hill Book Company, Inc., New York, Secs. 74–75.

J. W. Geckeler: Handbuch der Physik, Verlag von Julius Springer, Berlin, vol. 6, p. 143.

PROBLEMS

1. Let D_0 be the torsional rigidity of a circular cylinder, D_e that of an elliptical cylinder, and D_t that of a beam whose cross section is an equilateral triangle. Show that

$$D_e = kD_0, \qquad D_t = \frac{2\pi \sqrt{3}}{15} D_0,$$

where

$$k = \frac{2ab}{a^2 + b^2} \leq 1,$$

and a, b are the semiaxes of the elliptical section.

2. Consider a circular shaft of radius a with a circular groove of radius b along a generator of the shaft. (See Sec. 37.) Show that on the groove the shearing stresses are

---

[1] See Prob. 2 at the end of this section.

$$\tau_{zx} = \mu\alpha(2a\cos\theta - b)\sin\theta,$$
$$\tau_{zy} = -\mu\alpha(2a\cos\theta - b)\cos\theta,$$
$$\tau = \sqrt{\tau_{zx}^2 + \tau_{zy}^2} = \mu\alpha(2a\cos\theta - b),$$

while on the shaft we have

$$\tau_{zx} = \frac{\mu\alpha}{4a}(b^2 - 4a^2\cos^2\theta)\frac{\sin 2\theta}{\cos^2\theta},$$
$$\tau_{zy} = -\frac{\mu\alpha}{4a}(b^2 - 4a^2\cos^2\theta)\frac{\cos 2\theta}{\cos^2\theta},$$
$$\tau = \mu\alpha\left(a - \frac{b^2}{4a}\sec^2\theta\right).$$

Find the magnitude of the shearing stress at the point (see Fig. 26) where the groove enters the shaft.

38. Torsion of a Rectangular Beam and of a Triangular Prism. Consider a beam of rectangular cross section, and let one side of the cross section, of length a, be parallel to the x-axis and that of length b be parallel to the y-axis. It will be supposed that $b \geq a$ and that the z-axis passes through the center of the cross section.

The torsion problem will be solved if we succeed in determining the function $\psi(x, y)$ that is harmonic in the region bounded by $x = \pm a/2$, $y = \pm b/2$, and assumes on the boundary of the region the values $\frac{1}{2}(x^2 + y^2)$. In this case, the boundary conditions can be written as

(38.1) $\quad \psi\left(\pm\frac{a}{2}, y\right) = \frac{a^2}{8} + \frac{y^2}{2}, \quad \psi\left(x, \pm\frac{b}{2}\right) = \frac{b^2}{8} + \frac{x^2}{2}.$

The boundary conditions (38.1) are somewhat complicated, and it will simplify our search for the function ψ if we introduce a function $f(x, y)$, defined by the formula

(38.2) $\quad f(x, y) = \frac{\partial^2 \psi}{\partial x^2} + 1.$

The function $f(x, y)$ is obviously harmonic. Since the function $\psi(x, y)$ satisfies the equation

$$\frac{\partial^2 \psi}{\partial x^2} + \frac{\partial^2 \psi}{\partial y^2} = 0,$$

we can also write

(38.3) $$f(x, y) = -\frac{\partial^2 \psi}{\partial y^2} + 1.$$

By differentiating Eqs. (38.1), we see that

$$\frac{\partial^2 \psi}{\partial y^2} = 1 \quad \text{on } x = \pm \frac{a}{2},$$

$$\frac{\partial^2 \psi}{\partial x^2} = 1 \quad \text{on } y = \pm \frac{b}{2},$$

and from (38.2) and (38.3) it follows that the boundary values of the harmonic function $f(x, y)$ are

(38.4) $$\begin{cases} f(x, y) = 0 & \text{on } x = \pm \frac{a}{2}, \\ f(x, y) = 2 & \text{on } y = \pm \frac{b}{2}. \end{cases}$$

The function $f(x, y)$ satisfies the equation

(38.5) $$\frac{\partial^2 f}{\partial x^2} + \frac{\partial^2 f}{\partial y^2} = 0,$$

and we seek a solution of this equation in the form of an infinite series

$$f(x, y) = \sum_{n=0}^{\infty} c_n X_n(x) Y_n(y),$$

where each term of the series satisfies the differential equation (38.5), and where $X_n(x)$ and $Y_n(y)$ are, respectively, functions of x alone and of y alone. Substituting $X_n(x) Y_n(y)$ in (38.5), and denoting the derivatives by primes, we get

$$X_n''(x) Y_n(y) + X_n(x) Y_n''(y) = 0,$$

or

$$\frac{X_n''(x)}{X_n(x)} = -\frac{Y_n''(y)}{Y_n(y)}.$$

Since the left-hand member of this expression is a function of x alone and the right-hand member depends only on y, the equality can be fulfilled only if each member is equal to a constant, say $-k_n^2$. We are thus led to a pair of ordinary differential equations

$$\frac{d^2 X_n}{dx^2} + k_n^2 X_n = 0 \quad \text{and} \quad \frac{d^2 Y_n}{dy^2} - k_n^2 Y_n = 0,$$

§38 EXTENSION, TORSION, AND FLEXURE OF BEAMS

whose linearly independent solutions are

$$X_n = \begin{cases} \cos k_n x, \\ \sin k_n x, \end{cases} \quad Y_n = \begin{cases} \cosh k_n y, \\ \sinh k_n y. \end{cases}$$

Since our solutions must satisfy the boundary conditions (38.4), we reject the terms involving the odd functions $\sin k_n x$ and $\sinh k_n y$, and choose the product $X_n Y_n$ of the form

$$\cos k_n x \cosh k_n y,$$

where

$$k_n = \frac{(2n+1)\pi}{a}.$$

Thus, each term of the series

(38.6) $$f(x, y) = \sum_{n=0}^{\infty} c_n \cos k_n x \cosh k_n y$$

satisfies the first of the boundary conditions (38.4), and it remains to satisfy the conditions on the edges $y = \pm b/2$. Substituting $y = \pm b/2$ in (38.6) yields the equation

(38.7) $$2 = \sum_{n=0}^{\infty} c_n \cosh \frac{k_n b}{2} \cos k_n x,$$

from which it follows that the coefficients c_n can be formally determined by utilizing the scheme used in expanding functions in Fourier series. If we multiply both members of (38.7) by $\cos (2m+1)\pi x/a$, and integrate term by term with respect to x between the limits $-a/2$ and $a/2$, then because of the orthogonal property of trigonometric functions, namely,

$$\int_{-a/2}^{a/2} \cos k_n x \cos k_m x \, dx = \begin{cases} 0 \text{ if } m \neq n, \\ \dfrac{a}{2} \text{ if } m = n, \end{cases}$$

we get

$$\int_{-a/2}^{a/2} 2 \cos k_m x \, dx = \frac{a}{2} c_m \cosh \frac{k_m b}{2}.$$

Upon evaluating the integral, we see that

$$c_m = \frac{8(-1)^m}{\pi(2m+1)} \cdot \frac{1}{\cosh k_m b/2},$$

so that the formal solution is

$$(38.8) \qquad f(x, y) = \frac{8}{\pi} \sum_{n=0}^{\infty} \frac{(-1)^n}{2n + 1} \frac{\cosh k_n y}{\cosh k_n b/2} \cos k_n x.$$

The stresses τ_{zx} and τ_{zy} are given by the formulas

$$(38.9) \qquad \tau_{zx} = \mu\alpha \left(\frac{\partial \psi}{\partial y} - y \right), \qquad \tau_{zy} = \mu\alpha \left(-\frac{\partial \psi}{\partial x} + x \right),$$

and since

$$\frac{\partial^2 \psi}{\partial x^2} = f(x, y) - 1,$$

and

$$\frac{\partial^2 \psi}{\partial y^2} = -f(x, y) + 1,$$

we see that in order to evaluate stresses, we must integrate the series (38.8) with respect to x and y. Integrating, and making use of the fact that $\tau_{zx} = 0$ on $x = \pm a/2$ and $\tau_{zy} = 0$ on $y = \pm b/2$, we obtain

$$(38.10) \quad \begin{cases} \dfrac{\partial \psi}{\partial x} = -x + \dfrac{8a}{\pi^2} \sum_{n=0}^{\infty} \dfrac{(-1)^n}{(2n+1)^2} \dfrac{\cosh k_n y}{\cosh k_n b/2} \sin k_n x, \\[2mm] \dfrac{\partial \psi}{\partial y} = y - \dfrac{8a}{\pi^2} \sum_{n=0}^{\infty} \dfrac{(-1)^n}{(2n+1)^2} \dfrac{\sinh k_n y}{\cosh k_n b/2} \cos k_n x. \end{cases}$$

Hence the stresses τ_{zx} and τ_{zy} can be calculated from the series

$$(38.11) \quad \begin{cases} \tau_{zx} = \dfrac{-8a\mu\alpha}{\pi^2} \sum_{n=0}^{\infty} \dfrac{(-1)^n}{(2n+1)^2} \dfrac{\sinh k_n y}{\cosh k_n b/2} \cos k_n x, \\[2mm] \tau_{zy} = \mu\alpha \left(2x - \dfrac{8a}{\pi^2} \sum_{n=0}^{\infty} \dfrac{(-1)^n}{(2n+1)^2} \dfrac{\cosh k_n y}{\cosh k_n b/2} \sin k_n x \right). \end{cases}$$

The solutions (38.11) are formal, but the series converge so rapidly that there is no serious difficulty in justifying the term-by-term differentiation to show that the equilibrium equations are satisfied. The x-component of shear obviously vanishes

§38   EXTENSION, TORSION, AND FLEXURE OF BEAMS   147

when $y = 0$, while the y-component at the midpoint of the longer side is equal to

$$(38.12) \quad \tau_{zy}\bigg|_{\substack{x=a/2 \\ y=0}} = \mu\alpha a \left(1 - \frac{8}{\pi^2} \sum_{n=0}^{\infty} \frac{1}{(2n+1)^2} \operatorname{sech} \frac{k_n b}{2}\right).$$

It is not difficult to prove that (38.12) gives the maximum value of the shearing stress, by taking note of the fact that the term $2x$ in the parenthesis of (38.11) dominates the series. Now in the most unfavorable case (for convergence) of a square beam ($b = a$),

$$(38.13) \quad \tau_{\max} = \mu\alpha a \left\{1 - \frac{8}{\pi^2}\left[\operatorname{sech}\frac{\pi}{2} \right.\right.$$
$$\left.\left. + \sum_{n=1}^{\infty} \frac{1}{(2n+1)^2} \operatorname{sech}(2n+1)\frac{\pi}{2}\right]\right\}.$$

But

$$\sum_{n=1}^{\infty} \frac{1}{(2n+1)^2} \operatorname{sech}(2n+1)\frac{\pi}{2} < \frac{1}{9}\sum_{n=1}^{\infty} \frac{2e^{-(2n+1)\frac{\pi}{2}}}{1 + e^{-(2n+1)\pi}}$$

$$< \frac{2}{9}\sum_{n=1}^{\infty} e^{-(2n+1)\frac{\pi}{2}} = \frac{2}{9}\frac{e^{-\frac{3\pi}{2}}}{1 - e^{-\pi}} = 0.002.$$

Since $\operatorname{sech} \pi/2 = 0.4$, it follows that the first term in the bracket gives the value of the entire bracket with the accuracy of ½ per cent. Hence, for practical calculations, the value of $\tau_{\max}$ can be assumed to be given by the formula

$$\tau_{\max} \doteq \mu\alpha a \left(1 - \frac{8}{\pi^2} \operatorname{sech} \frac{\pi b}{2a}\right).$$

The twisting moment

$$M = \int_{-b/2}^{b/2} \int_{-a/2}^{a/2} (x\tau_{zy} - y\tau_{zx})\, dx\, dy$$

is calculated by making use of the series (38.11). The result of the calculations is

$$M = \frac{\mu\alpha b a^3}{6} + \frac{16\mu\alpha a^3 b}{\pi^4} \sum_{n=0}^{\infty} \frac{1}{(2n+1)^4} - \frac{64\mu\alpha a^4}{\pi^5} \sum_{n=0}^{\infty} \frac{\tanh k_n b/2}{(2n+1)^5},$$

and since
$$\sum_{n=0}^{\infty} \frac{1}{(2n+1)^4} = \frac{\pi^4}{96},$$
we have the formula

(38.14) $$M = \frac{\mu\alpha b a^3}{3} - \frac{64\mu\alpha a^4}{\pi^5} \sum_{n=0}^{\infty} \frac{\tanh k_n b/2}{(2n+1)^5}.$$

Now the series in (38.14) can be written as
$$\tanh \frac{\pi b}{2a} + \sum_{n=1}^{\infty} \frac{\tanh k_n b/2}{(2n+1)^5},$$
and we note that $\sum_{n=1}^{\infty} \frac{\tanh k_n b/2}{(2n+1)^5}$ is less than
$$\sum_{n=1}^{\infty} \frac{1}{(2n+1)^5} = 0.0046,$$
while $\tanh \pi b/2a \geqq 0.917$. Thus, the first term of the series gives the value of the sum to within ½ per cent, and one can use, for practical purposes, the approximate formula
$$M \doteq \frac{\mu\alpha b a^3}{3} - \frac{64\mu\alpha a^4}{\pi^5} \tanh \frac{\pi b}{2a}.$$

Inasmuch as the partial derivatives $\frac{\partial \psi}{\partial x}$ and $\frac{\partial \psi}{\partial y}$ are known from (38.10), it is a straightforward matter to compute the torsion function φ. Noting formula (35.1), it is found that
$$\varphi(x, y) = xy - \frac{8a^2}{\pi^3} \sum_{n=0}^{\infty} \frac{(-1)^n}{(2n+1)^3} \frac{\sinh k_n y}{\cosh k_n b/2} \sin k_n x.$$

Accordingly, the displacement w is given by $w = \alpha\varphi(x, y)$. The contour lines of the surface $\varphi(x, y) = $ constant for the case $b = a$ are shown in Fig. 27. The section is divided into eight triangular regions, which are warped as shown by the contour lines in Fig. 27. The function $\psi(x, y)$ can be determined by

§38 EXTENSION, TORSION, AND FLEXURE OF BEAMS

integrating Eqs. (38.10) and recalling the boundary condition $\psi(a/2, y) = \frac{1}{2}(x^2 + y^2) = a^2/8 + y^2/2$; the result is

$$(38.15) \quad \psi(x, y) = \frac{a^2}{4} + \frac{1}{2}(y^2 - x^2)$$

$$- \frac{8a^2}{\pi^3} \sum_{n=0}^{\infty} \frac{(-1)^n}{(2n+1)^3} \frac{\cosh k_n y}{\cosh k_n b/2} \cos k_n x.$$

The solution of the torsion problem for a prism whose cross section is an isosceles right triangle (Fig. 28) can be obtained

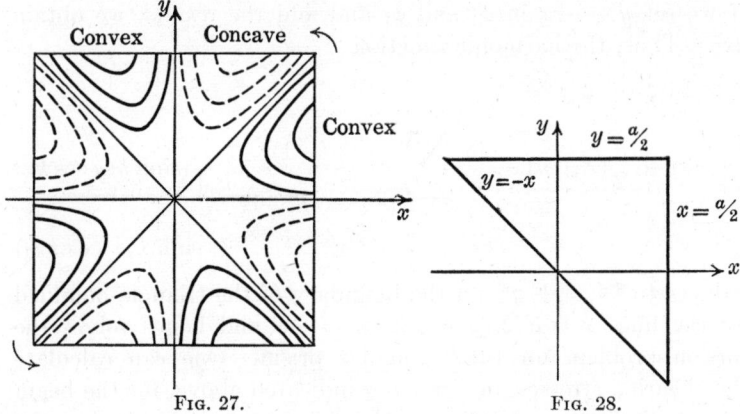

Fig. 27.   Fig. 28.

from the foregoing solution for a rectangular prism.[1] We construct the harmonic functions

$$\psi_1 = \frac{a^2}{4} + \frac{1}{2}(y^2 - x^2) - \left(x - \frac{a}{2}\right)\left(y - \frac{a}{2}\right)$$

$$- \frac{8a^2}{\pi^3} \sum_{n=0}^{\infty} \frac{(-1)^n}{(2n+1)^3} \frac{\sinh k_n y}{\sinh k_n a/2} \cos k_n x,$$

and

$$\psi_2 = \frac{a^2}{4} + \frac{1}{2}(x^2 - y^2) - \left(x - \frac{a}{2}\right)\left(y - \frac{a}{2}\right)$$

$$- \frac{8a^2}{\pi^3} \sum_{n=0}^{\infty} \frac{(-1)^n}{(2n+1)^3} \frac{\sinh k_n x}{\sinh k_n a/2} \cos k_n y.$$

[1] See B. G. Galerkin, *Bulletin de l'académie des sciences de Russie*, 1919, p. 111, and G. Kolossoff, *Comptes rendus de l'académie des sciences* (Paris), vol. 178 (1924), p. 2057.

A comparison with the expression (38.15) for ψ shows that the function ψ_1 reduces to $\frac{1}{2}(x^2 + y^2)$ on the sides $x = a/2$ and $y = a/2$. That ψ_2 also satisfies the boundary conditions on these sides can be shown either from considerations of symmetry or by direct calculation of the boundary values and by noting the expansion

$$\frac{a^2}{4} - x^2 = \frac{8a^2}{\pi^3} \sum_{n=0}^{\infty} \frac{(-1)^n}{(2n+1)^3} \cos \frac{(2n+1)\pi x}{a}.$$

If we set $y = -x$ in ψ_1 and ψ_2 and add the results, we obtain $2x^2$. Thus, the harmonic function

$$\psi = \frac{1}{2}(\psi_1 + \psi_2)$$

$$= -xy + \frac{a}{2}(x+y) - \frac{4a^2}{\pi^3} \sum_{n=0}^{\infty} \frac{(-1)^n}{(2n+1)^3 \sinh \frac{k_n a}{2}} (\sinh k_n y \cos k_n x$$

$$+ \sinh k_n x \cos k_n y)$$

reduces to $\frac{1}{2}(x^2 + y^2)$ on the boundary of the triangle bounded by the lines $x = a/2$, $y = a/2$, $y = -x$ and hence solves the torsion problem for the triangular prism. One can calculate the shearing stresses, in a manner indicated above, for the beam of rectangular cross section, and it is possible to show that the maximum shearing stress is at the middle of the hypotenuse.

References for Collateral Reading

A. E. H. LOVE: A Treatise on the Mathematical Theory of Elasticity, Cambridge University Press, London, Secs. 221–225.

S. TIMOSHENKO: Theory of Elasticity, McGraw-Hill Book Company, Inc., New York, Secs. 78, 79.

RIEMANN-WEBER: Die Differential- und Integralgleichungen der Physik, Friedrich Vieweg & Sohn, Brunswick, vol. 2, Chap. XVIII, Sec. 2, or 1930 ed., by Frank and v. Mises, Mary S. Rosenberg, New York, vol. 2, Chap. VIII, Sec. 2.

B. G. GALERKIN: "Torsion of Parabolic Prisms," *Messenger of Mathematics*, vol. 54 (1924), pp. 97–110.

39. Complex Form of Fourier Series. The discussion of the torsion of a rectangular beam in the preceding section utilized the expansion of a certain function in a trigonometric series. We shall have occasion to make frequent use of Fourier series

§39 EXTENSION, TORSION, AND FLEXURE OF BEAMS

expansions, and it is the purpose of this section to recall some facts about Fourier series and to give a representation of Fourier series in complex form. Sufficient conditions for the expansion of an arbitrary function in a Fourier series are given by the following theorem.

Theorem: *Let $f(\theta)$ be a real, single-valued function defined arbitrarily in the interval $0 \leq \theta \leq 2\pi$, and outside this interval defined by the equation $f(\theta + 2\pi) = f(\theta)$. If $f(\theta)$ has at most a finite number of points of ordinary discontinuity and a finite number of maxima and minima in the interval $0 \leq \theta \leq 2\pi$, then it can be represented by the series*

$$(39.1) \qquad \frac{a_0}{2} + \sum_{k=1}^{\infty} (a_k \cos k\theta + b_k \sin k\theta)$$

with

$$(39.2) \quad a_n = \frac{1}{\pi} \int_0^{2\pi} f(t) \cos nt \, dt, \qquad b_n = \frac{1}{\pi} \int_0^{2\pi} f(t) \sin nt \, dt,$$

and the series converges at every point $\theta = \theta_0$ to the value

$$\tfrac{1}{2}[f(\theta_0+) + f(\theta_0-)].$$

The symbols $f(\theta_0+)$ and $f(\theta_0-)$ stand for the right- and left-hand limits of $f(\theta)$ as $\theta \to \theta_0$.

The restrictions imposed upon the function $f(\theta)$ in this theorem are known as the *Dirichlet conditions*.[1] We assume that the reader is familiar with this theorem.

If $f(\theta)$ not only satisfies the conditions of Dirichlet but is continuous in the closed interval $(0, 2\pi)$,[2] then one can show that the Fourier series converges *uniformly* in the closed interval $(0, 2\pi)$.

We also have the following theorem concerning the bounds on the coefficients in Fourier series.

Theorem: *If the function $f(\theta)$ is periodic and is such that its pth derivative satisfies the condition of Dirichlet in the interval*

---

[1] The restrictions imposed on the function $f(\theta)$ can be relaxed, and it is sufficient to demand that $f(\theta)$ be a function of bounded variation.

[2] In this case, the requirement of periodicity imposes the condition $f(0) = f(2\pi)$.

$(0, 2\pi)$, then the Fourier coefficients for $f(\theta)$ satisfy the inequalities

$$|a_n| < \frac{M}{n^{p+1}}, \quad \text{and} \quad |b_n| < \frac{M}{n^{p+1}},$$

where M is a positive number independent of n.

An important conclusion follows directly from this theorem. Let the function $f(\theta)$ have the first derivative $f'(\theta)$, which satisfies the conditions of Dirichlet. Then the Fourier series for such a function has coefficients of order $1/n^2$, so that

$$|a_n \cos n\theta + b_n \sin n\theta| \leq |a_n \cos n\theta| + |b_n \sin n\theta|$$
$$\leq |a_n| + |b_n| < \frac{M}{n^2},$$

where M is a positive number independent of n. Since the series of positive constants

$$\sum_{n=1}^{\infty} \frac{M}{n^2} = M \sum_{n=1}^{\infty} \frac{1}{n^2}$$

converges, it follows from the Weierstrass M test that the Fourier series for a function whose first derivative satisfies the conditions of Dirichlet is absolutely and uniformly convergent, and hence can be integrated term by term.[1]

Since the coefficients of the series obtained by differentiating the series term by term are of the order na_n and nb_n, it is clear that in order to ensure the convergence of the derived series, it is sufficient to demand that the second derivative $f''(\theta)$ fulfill the conditions of Dirichlet in the interval $(0, 2\pi)$.

The Fourier series (39.1) can be written in an equivalent form

(39.3) $$f(\theta) = c_0 + \sum_{k=1}^{\infty} c_k e^{ik\theta} + \sum_{k=1}^{\infty} c_{-k} e^{-ik\theta}$$
$$\equiv \sum_{k=-\infty}^{\infty} c_k e^{ik\theta},$$

where

(39.4) $$c_n = \frac{1}{2\pi} \int_0^{2\pi} f(t) e^{-int} \, dt, \quad (n = 0, \pm 1, \pm 2, \cdots).$$

---

[1] As a matter of fact, every Fourier series can be integrated term by term.

In order to establish the identity of the representation (39.3) with (39.1), it is merely necessary to recall the Euler formula

$$e^{iu} = \cos u + i \sin u,$$

and verify that the formula (39.4) gives for $n > 0$

(39.5) $\quad c_n = \dfrac{a_n}{2} - i\dfrac{b_n}{2}, \quad c_{-n} = \dfrac{a_n}{2} + i\dfrac{b_n}{2}, \quad c_0 = \dfrac{a_0}{2}.$

Then

$$\sum_{k=-\infty}^{\infty} c_k e^{ik\theta} = \frac{a_0}{2} + \sum_{k=1}^{\infty} \left(\frac{a_k}{2} - i\frac{b_k}{2}\right)(\cos k\theta + i \sin k\theta)$$

$$+ \sum_{k=1}^{\infty} \left(\frac{a_k}{2} + i\frac{b_k}{2}\right)(\cos k\theta - i \sin k\theta)$$

$$= \frac{a_0}{2} + \sum_{k=1}^{\infty} (a_k \cos k\theta + b_k \sin k\theta).$$

Let $f_1(\theta)$ and $f_2(\theta)$ be a pair of real functions, each of which can be expanded in Fourier series in the interval $(0, 2\pi)$, and form the complex function $f_1(\theta) + if_2(\theta)$. Then

$$f_1(\theta) + if_2(\theta) = \sum_{k=-\infty}^{\infty} c_k e^{ik\theta},$$

where

$$c_n = \frac{1}{2\pi} \int_0^{2\pi} [f_1(t) + if_2(t)]e^{-int}\, dt, \quad (n = 0, \pm 1, \pm 2, \cdots).$$

If we set

$$c_n = \gamma_n + i\,\delta_n,$$

where γ_n and δ_n are real numbers, then

$$f_1(\theta) + if_2(\theta) = \sum_{k=-\infty}^{\infty} (\gamma_k + i\,\delta_k)(\cos k\theta + i \sin k\theta)$$

$$= \sum_{k=-\infty}^{\infty} (\gamma_k \cos k\theta - \delta_k \sin k\theta)$$

$$+ i \sum_{k=-\infty}^{\infty} (\delta_k \cos k\theta + \gamma_k \sin k\theta)$$

$$= \gamma_0 + \sum_{k=1}^{\infty} [(\gamma_k + \gamma_{-k}) \cos k\theta - (\delta_k - \delta_{-k}) \sin k\theta]$$
$$+ i \delta_0$$
$$+ i \sum_{k=1}^{\infty} [(\delta_k + \delta_{-k}) \cos k\theta + (\gamma_k - \gamma_{-k}) \sin k\theta].$$

Hence

$$f_1(\theta) = \tfrac{1}{2}a_0 + \sum_{k=1}^{\infty} (a_k \cos k\theta + b_k \sin k\theta),$$

$$f_2(\theta) = \tfrac{1}{2}a_0' + \sum_{k=1}^{\infty} (a_k' \cos k\theta + b_k' \sin k\theta),$$

where

$$\tfrac{1}{2}a_0 = \gamma_0, \quad a_k = \gamma_k + \gamma_{-k}, \quad b_k = -\delta_k + \delta_{-k},$$
$$\tfrac{1}{2}a_0' = \delta_0, \quad a_k' = \delta_k + \delta_{-k}, \quad b_k' = \gamma_k - \gamma_{-k},$$
$$(k = 1, 2, 3, \cdots).$$

It follows from these formulas that the representation of a complex function $f_1(\theta) + if_2(\theta)$ in a series of the type (39.3) is unique, since the representation of the functions $f_1(\theta)$ and $f_2(\theta)$ in series of the type (39.1) is unique.

References for Collateral Reading

D. JACKSON: Fourier Series and Orthogonal Polynomials, Mathematical Association of America, Chicago, Chap. I.

I. S. SOKOLNIKOFF: Advanced Calculus, McGraw-Hill Book Company, Inc., New York, Chap. XI.

E. T. WHITTAKER and G. N. WATSON: Modern Analysis, Cambridge University Press, London, p. 170.

K. KNOPP: Theory and Application of Infinite Series, Blackie & Son, Ltd., Glasgow, p. 356.

H. S. CARSLAW: Fourier Series and Integrals, The Macmillan Company, New York.

R. V. CHURCHILL: Fourier Series and Boundary Value Problems, McGraw-Hill Book Company, Inc., New York.

40. Summary of Some Results of the Complex Variable Theory. We shall need in our subsequent work some theorems from the theory of functions of a complex variable. In this section, some of the more familiar results will be stated without proof, and the proofs of the less familiar ones will be outlined. A detailed discussion can be found in the reference books listed at the end of the next section.

It will be recalled that a single-valued function
$$f(\mathfrak{z}) = u(x, y) + iv(x, y)$$
of a complex variable $\mathfrak{z} = x + iy$ is called *analytic* or *holomorphic* in a given region R if it possesses a unique derivative at every point of the region R. Points at which the function $f(\mathfrak{z})$ ceases to have a derivative are termed the singular points of the analytic function. The necessary and sufficient conditions for the analyticity of the function $f(\mathfrak{z})$ are given by the well-known Cauchy-Riemann equations

$$(40.1) \qquad \frac{\partial u}{\partial x} = \frac{\partial v}{\partial y}, \qquad \frac{\partial v}{\partial x} = -\frac{\partial u}{\partial y},$$

where it is assumed that the partial derivatives involved are continuous functions of x and y. It is known that if $f(\mathfrak{z})$ is analytic in the region R, then not only do the first partial derivatives of u and v exist in R, but also those of all higher orders. It follows from this observation and from (40.1) that the real and imaginary parts of an analytic function satisfy the equation of Laplace; that is,

$$\nabla^2 u = 0, \qquad \nabla^2 v = 0.$$

The following theorem is basic to all considerations of the theory of analytic functions.

CAUCHY'S INTEGRAL THEOREM: *If $f(\mathfrak{z})$ is continuous in the closed region*[1] *R bounded by a simple closed contour C, and if $f(\mathfrak{z})$ is analytic at every interior point of R, then*

$$\int_C f(\mathfrak{z}) \, d\mathfrak{z} = 0.$$

This theorem can easily be extended to the case of multiply connected regions to yield another.

THEOREM: *If $f(\mathfrak{z})$ is continuous in the closed, multiply connected region R bounded by the exterior simple contour C and by the interior simple contours $C_1, C_2, \cdots, C_n$, then the integral of $f(\mathfrak{z})$ over the exterior contour C is equal to the sum of the integrals over the interior contours, whenever $f(\mathfrak{z})$ is analytic in the interior of R. The integration over all the contours is performed in the same direction.*

---

[1] The term *continuous in a closed region* is used to mean that the function is continuous up to and on the boundary.

The following numerical results are worth noting.

If n is an integer and $\mathfrak{z} = a$ is a fixed point that lies either within or without the simple closed contour C, then

$$\int_C (\mathfrak{z} - a)^n \, d\mathfrak{z} = 0 \qquad \text{if } n \neq -1.$$

If the point a is outside the contour C, then the truth of the formula follows from Cauchy's Theorem, whatever be the value of n; if it is within, then the result follows from elementary calculations. If the point a is within the contour, then an elementary calculation gives

$$\int_C \frac{d\mathfrak{z}}{\mathfrak{z} - a} = 2\pi i.$$

This latter formula, in conjunction with Cauchy's Integral Theorem, can be used to establish Cauchy's Integral Formula.

CAUCHY'S INTEGRAL FORMULA: *If $\mathfrak{z} = a$ is an interior point of the region R, then*

$$(40.2) \qquad \frac{1}{2\pi i} \int_C \frac{f(\mathfrak{z}) \, d\mathfrak{z}}{\mathfrak{z} - a} = f(a),$$

whenever $f(\mathfrak{z})$ is continuous in the closed region R bounded by a contour C and analytic at every interior point of R.

If the variable of integration in (40.2) is denoted by ζ, and if $\mathfrak{z}$ is any point interior to R, then (40.2) becomes

$$(40.3) \qquad f(\mathfrak{z}) = \frac{1}{2\pi i} \int_C \frac{f(\zeta) \, d\zeta}{\zeta - \mathfrak{z}}.$$

Calculation of the derivative from the formula (40.3) yields

$$f'(\mathfrak{z}) = \frac{1}{2\pi i} \int_C \frac{f(\zeta) \, d\zeta}{(\zeta - \mathfrak{z})^2},$$

and in general,

$$f^{(n)}(\mathfrak{z}) = \frac{n!}{2\pi i} \int_C \frac{f(\zeta) \, d\zeta}{(\zeta - \mathfrak{z})^{n+1}}.$$

The Integral Formula of Cauchy can be used to establish the fact that an analytic function $f(\mathfrak{z})$ can be expanded in Taylor's series, so that

$$f(\mathfrak{z}) = f(a) + f'(a)(\mathfrak{z} - a) + \cdots + \frac{f^{(n)}(a)}{n!}(\mathfrak{z} - a)^n + \cdots.$$

§40 EXTENSION, TORSION, AND FLEXURE OF BEAMS

This series converges to $f(\mathfrak{z})$ at every point $\mathfrak{z}$ interior to any circle γ that lies within the region R, and whose center is at a. Moreover, the representation of $f(\mathfrak{z})$ in Taylor's series is unique.

Consider now the region R bounded by two concentric circles C_1 and C_2, and let $\mathfrak{z} = a$ be the center of the circles. If $f(\mathfrak{z})$ is continuous in the closed annular region formed by C_1 and C_2, and if it is analytic at every interior point of the ring, then one can represent $f(\mathfrak{z})$ by the Laurent series

$$f(\mathfrak{z}) = \sum_{k=-\infty}^{\infty} b_k (\mathfrak{z} - a)^k,$$

where

$$b_k = \frac{1}{2\pi i} \int_C \frac{f(\mathfrak{z}) \, d\mathfrak{z}}{(\mathfrak{z} - a)^{k+1}}, \quad (k = 0, \pm 1, \pm 2, \cdots),$$

and where C is an arbitrary path drawn in R that encloses C_1. It is obvious that the series of Laurent reduces to Taylor's series whenever the function $f(\mathfrak{z})$ is analytic throughout the region bounded by the circle C_1.

If $f(\mathfrak{z})$ has a pole of order m at $\mathfrak{z} = a$, then the Laurent series about the point $\mathfrak{z} = a$ takes the form

$$f(\mathfrak{z}) = \frac{b_{-m}}{(\mathfrak{z}-a)^m} + \cdots + \frac{b_{-2}}{(\mathfrak{z}-a)^2} + \frac{b_{-1}}{\mathfrak{z}-a} + b_0 + b_1(\mathfrak{z}-a) + b_2(\mathfrak{z}-a)^2 + \cdots.$$

If we set $\mathfrak{z} - a = \zeta$ and integrate around a curve C enclosing $\mathfrak{z} = a$ and no other singularity of $f(\mathfrak{z})$, then

$$\int_C f(\mathfrak{z}) \, d\mathfrak{z} = \int_C \left[\frac{b_{-m}}{\zeta^m} + \cdots + \frac{b_{-2}}{\zeta^2} + \frac{b_{-1}}{\zeta} + b_0 + b_1 \zeta + b_2 \zeta^2 + \cdots \right] d\zeta$$

$$= b_{-1} \int_C \frac{d\zeta}{\zeta} = 2\pi i b_{-1}.$$

The quantity b_{-1} is called the *residue* of $f(\mathfrak{z})$ at the pole $\mathfrak{z} = a$. If

$$f(\mathfrak{z}) = \frac{g(\mathfrak{z})}{(\mathfrak{z}-a) h(\mathfrak{z})}$$

has a *simple* pole at $\mathfrak{z} = a$, $(m = 1)$, then the residue at $\mathfrak{z} = a$ is $g(a)/h(a)$. In general, when C encloses n poles at $\mathfrak{z} = a_1$, $\mathfrak{z} = a_2, \cdots, \mathfrak{z} = a_n$, the last equation is replaced by

$$\int_C f(\mathfrak{z})\, d\mathfrak{z} = 2\pi i \times (\text{sum of residues at poles}).$$

If the Laurent expansion of $f(\mathfrak{z})$ at each pole is known, then to evaluate

$$\frac{1}{2\pi i} \int_C f(\mathfrak{z})\, d\mathfrak{z}$$

we have merely to add the coefficients of

$$\frac{1}{\mathfrak{z} - a_1},\ \frac{1}{\mathfrak{z} - a_2},\ \cdots$$

in the several expansions.

The evaluation of residues may often be simplified by observing that if $f(\mathfrak{z})$ and $g(\mathfrak{z})$ are analytic at $\mathfrak{z} = a$, and if $\mathfrak{z} - a$ is a nonrepeated factor of $g(\mathfrak{z})$, then the residue at a of $f(\mathfrak{z})/g(\mathfrak{z})$ is $f(a)/g'(a)$. The residue at a multiple pole can be found from the theorem that the residue at $\mathfrak{z} = a$ of $f(\mathfrak{z})/(\mathfrak{z} - a)^n$ is $f^{(n-1)}(a)/(n-1)!$; it is assumed that $f(\mathfrak{z})$ is analytic at a and that n is a positive integer.

It should be noted that the function $f(\zeta)$ in the formula (40.3) of Cauchy represents the values of an analytic function $f(\mathfrak{z})$ on the boundary C of the region R. Now if we consider the integral

$$(40.4) \qquad \Phi_1(\mathfrak{z}) = \frac{1}{2\pi i} \int_C \frac{F(\zeta)}{\zeta - \mathfrak{z}}\, d\zeta,$$

where $F(\zeta)$ is *any continuous function* defined on the boundary C, then this integral defines some function of $\mathfrak{z}$, and it is easy to verify that $\Phi_1(\mathfrak{z})$ has the derivative $\Phi_1'(\mathfrak{z})$, which is given by the formula

$$\Phi_1'(\mathfrak{z}) = \frac{1}{2\pi i} \int_C \frac{F(\zeta)\, d\zeta}{(\zeta - \mathfrak{z})^2},$$

and in general,

$$\Phi_1^{(n)}(\mathfrak{z}) = \frac{n!}{2\pi i} \int_C \frac{F(\zeta)\, d\zeta}{(\zeta - \mathfrak{z})^{n+1}}.$$

§40   EXTENSION, TORSION, AND FLEXURE OF BEAMS

Thus, the function $\Phi_1(\mathfrak{z})$ is analytic for every value of $\mathfrak{z}$ that is interior to the region R bounded by C. Again, if $\mathfrak{z}$ is some point exterior to the region R, then the integral

$$(40.5) \qquad \Phi_2(\mathfrak{z}) = \frac{1}{2\pi i} \int_C \frac{F(\zeta)\, d\zeta}{\zeta - \mathfrak{z}}$$

defines some function $\Phi_2(\mathfrak{z})$, and it is easy to see that $\Phi_2(\mathfrak{z})$ likewise has derivatives of all orders and hence is analytic. Thus, the integrals (40.4) and (40.5) of Cauchy's type define two analytic functions that in general will be distinct. The situation here is the same even when $F(\zeta)$ represents the boundary values of some analytic function $f(\mathfrak{z})$. For, by (40.3), if $\mathfrak{z}$ is interior to the contour, the value of the integral is precisely equal to $f(\mathfrak{z})$, and if $\mathfrak{z}$ is outside the contour, then the integral defines the function 0, since the integrand $f(\zeta)/(\zeta - \mathfrak{z})$ is an analytic function of ζ throughout R. It should be observed that as $\mathfrak{z}$ tends to some definite point ζ on the contour from inside and from outside C, the difference between the two limiting values in this example is $f(\zeta) - 0 = f(\zeta)$.

One can raise a similar question regarding the connection of the limiting values of the functions $\Phi_1(\mathfrak{z})$ and $\Phi_2(\mathfrak{z})$ with the *density function* $F(\zeta)$. If we place no restrictions on the function $F(\zeta)$ beyond continuity on the contour, then the problem becomes an exceedingly difficult one. If, however, some further restrictions on $F(\zeta)$ are imposed, then it is possible to establish a definite connection[1] of the density function $F(\zeta)$ with the limits $\lim_{\mathfrak{z}\to\zeta}\Phi_1(\mathfrak{z})$ and $\lim_{\mathfrak{z}\to\zeta}\Phi_2(\mathfrak{z})$.

We shall make use of integrals of Cauchy's type to represent analytically some functions that are useful in the theory of elasticity. However, it must be noted that such representation is not unique, so that the same function can be represented by different integrals of Cauchy's type. As an illustration, consider a contour C that contains in the interior the point $\mathfrak{z} = 0$, and let us determine the analytic function that vanishes at every point of the region R enclosed by C. If we choose in (40.4) the density function $F(\zeta) = 0$, then $\Phi_1(\mathfrak{z}) \equiv 0$. Also, if we choose

$$F(\zeta) = \frac{1}{\zeta},$$

---

[1] See E. Picard, Leçons sur quelques types simples d'équations aux dérivées partielles.

then

$$\Phi_1(\mathfrak{z}) = \frac{1}{2\pi i}\int_C \frac{d\zeta}{\zeta(\zeta-\mathfrak{z})} = 0$$

for every position of the point $\mathfrak{z}$ in the region R.[1] Hence if we add this integral to an integral of Cauchy's type that defines an analytic function $\Psi(\mathfrak{z})$, we shall obtain another integral of Cauchy's type that defines the same analytic function $\Psi(\mathfrak{z})$. It follows from these remarks that no conclusion can be drawn concerning the equality of the density functions $F_1(\zeta)$ and $F_2(\zeta)$ from the equality of the two integrals

$$\frac{1}{2\pi i}\int_C \frac{F_1(\zeta)}{\zeta-\mathfrak{z}}\,d\zeta = \frac{1}{2\pi i}\int_C \frac{F_2(\zeta)}{\zeta-\mathfrak{z}}\,d\zeta$$

for all values of $\mathfrak{z}$ in the interior of C. We shall see, however, that if some additional restrictions are imposed on the density functions and on the contour C, then the equality will obtain. This is the subject of the next section, which contains a discussion of the Theorem of Harnack.

41. Theorem of Harnack.[2] In considering the applications of the theory of functions of a complex variable to problems in elasticity, we shall most frequently deal with the region bounded by the unit circle; that is, the region $|\mathfrak{z}| \leq 1$. In order to avoid a possible misinterpretation of the formulas, we shall draw the unit circle in the complex ζ-plane, where $\zeta = \xi + i\eta$ (ξ and η being real). The boundary of the unit circle $|\zeta| \leq 1$ will be denoted by the letter γ, and the points on the boundary γ by $\sigma = e^{i\theta}$.* All functions of the argument θ will be assumed to be periodic, so that $f(\theta + 2\pi) = f(\theta)$.

---

[1] For $\dfrac{1}{\zeta(\zeta-\mathfrak{z})} = \dfrac{1}{\mathfrak{z}(\zeta-\mathfrak{z})} - \dfrac{1}{\mathfrak{z}\zeta}$; hence

$$\frac{1}{2\pi i}\int_C \frac{d\zeta}{\zeta(\zeta-\mathfrak{z})} = \frac{1}{2\pi i\mathfrak{z}}\int_C \frac{d\zeta}{\zeta-\mathfrak{z}} - \frac{1}{2\pi i\mathfrak{z}}\int_C \frac{d\zeta}{\zeta} = \frac{1}{\mathfrak{z}} - \frac{1}{\mathfrak{z}} = 0.$$

[2] The proofs contained in this and the following section are due to N. I. Muscheliŝvili, Nekotoriye Osnovniye Zadachi Matematicheskoi Teorii Uprugosti.

* The letter σ was used earlier for Poisson's ratio and in the expression $d\sigma$ for the element of area, but the distinction is so obvious that no complications should arise.

§41 EXTENSION, TORSION, AND FLEXURE OF BEAMS

THEOREM: *Let $f(\theta)$ and $\varphi(\theta)$ be continuous, real functions of the argument θ (defined on the unit circle γ); if*

$$(41.1) \quad \frac{1}{2\pi i} \int_\gamma \frac{f(\theta)\, d\sigma}{\sigma - \zeta} = \frac{1}{2\pi i} \int_\gamma \frac{\varphi(\theta)\, d\sigma}{\sigma - \zeta}$$

for all values of ζ interior to γ, then

$$f(\theta) \equiv \varphi(\theta).$$

If the point ζ is exterior to γ, and if the equality (41.1) is true for all values of ζ, then

$$f(\theta) = \varphi(\theta) + \text{const}.$$

We consider first the case when the point ζ is inside γ. It follows from equality (41.1) that

$$\frac{1}{2\pi i} \int_\gamma \frac{f(\theta) - \varphi(\theta)}{\sigma - \zeta} d\sigma \equiv \frac{1}{2\pi i} \int_\gamma \frac{F(\theta)}{\sigma - \zeta} d\sigma \equiv 0,$$

where $F(\theta) \equiv f(\theta) - \varphi(\theta)$, and we shall prove that $F(\theta) \equiv 0$.

Now since $|\zeta| < 1$, we have

$$\frac{1}{\sigma - \zeta} = \frac{1}{\sigma} + \frac{\zeta}{\sigma^2} + \frac{\zeta^2}{\sigma^3} + \cdots,$$

and

$$(41.2) \quad \frac{1}{2\pi i} \int_\gamma \frac{F(\theta)}{\sigma - \zeta} d\sigma = \frac{1}{2\pi i} \int_\gamma \sum_{n=0}^{\infty} \zeta^n \frac{F(\theta)\, d\sigma}{\sigma^{n+1}}$$

$$= \sum_{n=0}^{\infty} (a_n - ib_n)\zeta^n,$$

where [see (39.4) and (39.5)]

$$a_n - ib_n = \frac{1}{2\pi i} \int_\gamma F(\theta) \sigma^{-n-1}\, d\sigma = \frac{1}{2\pi} \int_0^{2\pi} F(\theta) e^{-in\theta}\, d\theta.$$

But (41.2) vanishes for all values of ζ; hence $a_n = b_n = 0$, ($n = 0, 1, 2, \cdots$). A reference to formula (39.4) shows that all Fourier coefficients of the function $F(\theta)$ vanish, and hence $F(\theta) \equiv 0$.

Consider now the case when $|\zeta| > 1$; then

$$\frac{1}{\sigma - \zeta} = -\frac{1}{\zeta} - \frac{\sigma}{\zeta^2} - \frac{\sigma^2}{\zeta^3} - \cdots ,$$

and

(41.3) $$\frac{1}{2\pi i} \int_\gamma \frac{F(\theta)}{\sigma - \zeta} d\sigma = -\frac{1}{2\pi i} \int \sum_{n=1}^{\infty} \frac{\sigma^{n-1} F(\theta) d\sigma}{\zeta^n}$$

$$= -\sum_{n=1}^{\infty} \frac{a_n + ib_n}{\zeta^n},$$

where

$$a_n + ib_n = \frac{1}{2\pi i} \int_\gamma F(\theta) \sigma^{n-1} d\sigma$$

$$= \frac{1}{2\pi} \int_0^{2\pi} F(\theta) e^{in\theta} d\theta, \qquad (n = 1, 2, 3, \cdots).$$

Since (41.3) vanishes for all values of $|\zeta| > 1$, $a_n = b_n = 0$, $(n = 1, 2, 3, \cdots)$. Thus, all Fourier coefficients of $F(\theta)$, with the possible exception of a_0, vanish, and hence

$$\varphi(\theta) = f(\theta) + \text{const.}$$

It follows from this proof that if the point ζ is outside γ, and if *in addition to the equality* (41.1) we have the equality

$$\frac{1}{2\pi i} \int_\gamma f(\theta) \frac{d\sigma}{\sigma} = \frac{1}{2\pi i} \int_\gamma \varphi(\theta) \frac{d\sigma}{\sigma},$$

then $f(\theta) = \varphi(\theta)$.

An important corollary follows from this theorem.

COROLLARY: *If we have four real continuous functions f_1, f_2, φ_1, φ_2 and the following simultaneous equalities for all values of ζ:*

$$\frac{1}{2\pi i} \int_\gamma \frac{f_1 + if_2}{\sigma - \zeta} d\sigma = \frac{1}{2\pi i} \int_\gamma \frac{\varphi_1 + i\varphi_2}{\sigma - \zeta} d\sigma,$$

$$\frac{1}{2\pi i} \int_\gamma \frac{f_1 - if_2}{\sigma - \zeta} d\sigma = \frac{1}{2\pi i} \int_\gamma \frac{\varphi_1 - i\varphi_2}{\sigma - \zeta} d\sigma,$$

then

$$\varphi_1 = f_1, \qquad \varphi_2 = f_2, \qquad \text{if } |\zeta| < 1,$$

and

$$\varphi_1 = f_1 + \text{const.}, \qquad \varphi_2 = f_2 + \text{const.}, \qquad \text{if } |\zeta| > 1.$$

This corollary follows at once from Harnack's Theorem when we consider the results of adding and subtracting the equalities in question.

References for Collateral Reading

E. C. TITCHMARSH: The Theory of Functions, Oxford University Press, New York, 2nd ed., pp. 64–101, pp. 399–428.

W. F. OSGOOD: Lehrbuch der Funktionentheorie, B. G. Teubner, Berlin, vol. 1.

É. GOURSAT: Cours d'analyse, Gauthiers-Villars et Cie, Paris, vol. 2.

E. PICARD: Leçons sur quelques types simples d'équations aux dérivées partielles, Gauthiers-Villars et Cie, Paris.

42. Formulas of Schwarz and Poisson. We have already seen that the determination of the torsion function $\varphi(x, y)$ and its conjugate function $\psi(x, y)$ are special cases of the fundamental boundary-value problems of Potential Theory—the so-called problems of Dirichlet and Neumann. These problems occur also in other branches of applied mathematics, notably in hydrodynamics and in electrodynamics. While the solution of the two-dimensional problems of Dirichlet and Neumann for special types of boundaries is likely to present serious calculational difficulties, it is possible to write down general formulas for the case when the boundary of the region is a circle. We shall give a derivation of formulas associated with the names of Schwarz and Poisson that solve the problem of Dirichlet for a circular region.

Consider a region bounded by a circle, which we can take, without loss of generality, to be a unit circle with center at the origin. As in the preceding section, we denote the boundary of the circle $|\zeta| = 1$ by γ, and any point on the boundary by $\sigma = e^{i\theta}$.

Let it be required to determine a harmonic function $u(\xi, \eta)$, which on the boundary of the circle γ assumes the values

$$(42.1) \qquad u|_\gamma = f(\theta),$$

where $f(\theta)$ is a continuous, real function of θ. Denote the conjugate harmonic function by $v(\xi, \eta)$; the function $v(\xi, \eta)$ is determined to within an arbitrary constant from the knowledge of the function $u(\xi, \eta)$ [see (35.2)]. Then the function

$$F(\zeta) = u(\xi, \eta) + iv(\xi, \eta)$$

is an analytic function of the complex variable $\zeta = \xi + i\eta$ for all values of ζ interior to $|\zeta| = 1$. If we assume that $F(\zeta)$ is continuous in the closed region $|\zeta| \leq 1$, then we can rewrite the boundary condition (42.1) in the form[1]

(42.2) $\qquad\qquad F(\sigma) + \bar{F}(\bar{\sigma}) = 2f(\theta) \qquad\qquad$ on γ.

If we multiply both members of Eq. (42.2) by $\dfrac{1}{2\pi i}\dfrac{d\sigma}{\sigma - \zeta}$, where ζ is any point interior to γ, and integrate over the circle γ, we obtain the formula

(42.3) $\qquad \dfrac{1}{2\pi i}\int_\gamma \dfrac{F(\sigma)}{\sigma - \zeta}\,d\sigma + \dfrac{1}{2\pi i}\int_\gamma \dfrac{\bar{F}(\bar{\sigma})}{\sigma - \zeta}\,d\sigma = \dfrac{1}{\pi i}\int_\gamma \dfrac{f(\theta)}{\sigma - \zeta}\,d\sigma,$

which by the Theorem of Harnack is entirely equivalent to (42.2).

The first of the integrals in the left-hand member of (42.3), by Cauchy's Integral Formula, is equal to $F(\zeta)$, while the second is equal[2] to $\bar{F}(0)$. Let $\bar{F}(0) = a_0 - ib_0$; then (42.3) becomes

(42.4) $\qquad\qquad F(\zeta) = \dfrac{1}{\pi i}\int_\gamma \dfrac{f(\theta)}{\sigma - \zeta}\,d\sigma - a_0 + ib_0.$

If we set $\zeta = 0$ in (42.4), we obtain

$$a_0 + ib_0 = \dfrac{1}{\pi i}\int_\gamma \dfrac{f(\theta)\,d\sigma}{\sigma} - a_0 + ib_0,$$

---

[1] It is possible to prove that if $f(\theta)$ satisfies Hölder's condition, then the function $F(\zeta)$ given by (42.6) will be continuous in the closed region $|\zeta| \leq 1$. We recall that a function $f(\theta)$ is said to satisfy Hölder's condition (or a Lipschitz condition) if for any pair of values θ' and θ'' in the interval in question

$$|f(\theta'') - f(\theta')| \leq M|\theta'' - \theta'|^\alpha,$$

where M and α are positive constants. This condition is less restrictive than the requirement of the existence of a bounded derivative.

[2] Since $F(\zeta) = F(0) + F'(0)\zeta + \dfrac{1}{2!}F''(0)\zeta^2 + \cdots$, and since on $|\zeta| = 1$ $\zeta = 1/\sigma$, $\bar{F}(\bar{\sigma}) = \bar{F}(0) + \bar{F}'(0)\dfrac{1}{\sigma} + \dfrac{1}{2!}\bar{F}''(0)\dfrac{1}{\sigma^2} + \cdots$, and term-by-term integration gives the desired result upon noting that

$$\dfrac{1}{2\pi i}\int_\gamma \dfrac{d\sigma}{\sigma^n(\sigma - \zeta)} = 0, \qquad \text{if } n > 0,$$
$$= 1, \qquad \text{if } n = 0.$$

and hence

$$(42.5) \qquad 2a_0 = \frac{1}{\pi i} \int_\gamma \frac{f(\theta)\, d\sigma}{\sigma} = \frac{1}{\pi} \int_0^{2\pi} f(\theta)\, d\theta.$$

The quantity b_0 is left undetermined, as one would expect, since the function $v(\xi, \eta)$ is determined to within an arbitrary real constant.

Inserting the value of a_0 from (42.5) in (42.4), we have

$$(42.6) \qquad F(\zeta) = \frac{1}{\pi i} \int_\gamma \frac{f(\theta)\, d\sigma}{\sigma - \zeta} - \frac{1}{2\pi i} \int_\gamma \frac{f(\theta)\, d\sigma}{\sigma} + ib_0$$
$$= \frac{1}{2\pi i} \int_\gamma f(\theta) \frac{\sigma + \zeta}{\sigma - \zeta} \frac{d\sigma}{\sigma} + ib_0,$$

which is the desired formula of Schwarz.

Now if we substitute $\zeta = \rho e^{i\psi}$ and $\sigma = e^{i\theta}$ in (42.6) and separate the real and imaginary parts, we find

$$(42.7) \quad \Re F(\zeta) \equiv u(\xi, \eta) = \frac{1}{2\pi} \int_0^{2\pi} \frac{(1 - \rho^2) f(\theta)\, d\theta}{1 - 2\rho \cos(\theta - \psi) + \rho^2}.$$

This is the integral of Poisson, which gives the solution of the problem of Dirichlet. It is possible to prove that (42.7) represents the solution of the problem of Dirichlet under the assumption that $f(\theta)$ is merely a piecewise continuous function.[1]

The discussion of this section was confined to the general solution of the first boundary-value problems of Potential Theory, when the boundary curve is a circle. It is possible to generalize the formulas obtained above so as to make them apply to any simply connected region. This is done by introducing a mapping function, and we proceed next to an outline of some basic notions that underlie the idea of conformal mapping of simply connected domains.

43. Conformal Mapping. Let the functional relationship $\mathfrak{z} = \omega(\zeta)$ set up a correspondence between the points $\zeta = \xi + i\eta$ of the complex ζ-plane and $\mathfrak{z} = x + iy$ of the complex $\mathfrak{z}$-plane. If $\mathfrak{z} = \omega(\zeta)$ is analytic in some region R of the ζ-plane, then the

---

[1] See O. D. Kellogg, Foundations of Potential Theory, and G. C. Evans, The Logarithmic Potential, Chap. IV, for a discussion of the Problem of Neumann.

totality of values $\mathfrak{z}$ belongs to some region R' of the $\mathfrak{z}$-plane, and it is said that the region R is mapped into the region R' by the mapping function $\omega(\zeta)$. If C is some curve drawn in the region R, and the point ζ is allowed to move along C, then the corresponding point $\mathfrak{z}$ will trace a curve C' in the $\mathfrak{z}$-plane, and C' is called the map of C (Fig. 29).

The relationship between the curves C and C' is interesting. Consider a pair of points ζ and $\zeta + \Delta\zeta$ on C, and let the arc length between them be $\Delta s = \widehat{PQ}$. The corresponding points in the region R' are denoted by $\mathfrak{z}$ and $\mathfrak{z} + \Delta\mathfrak{z}$, and the distance between them, measured along the curve C', is $\Delta s' = \widehat{P'Q'}$.

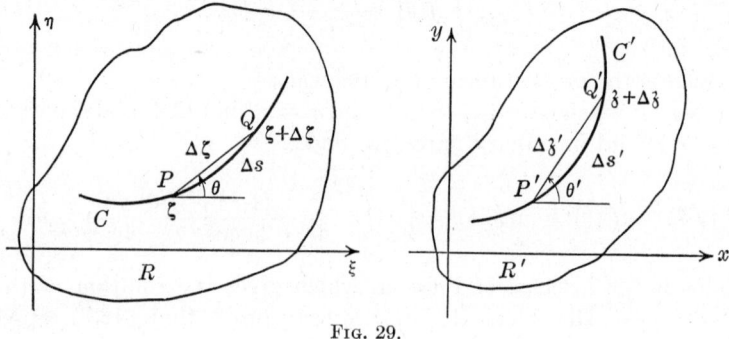

Fig. 29.

Since the ratio of the lengths of arc elements has the same limit as the ratio of the lengths of the corresponding chords,

$$\lim_{\Delta\zeta \to 0} \frac{\Delta s'}{\Delta s} = \lim_{\Delta\zeta \to 0} \left|\frac{\Delta\mathfrak{z}}{\Delta\zeta}\right| = \left|\frac{d\mathfrak{z}}{d\zeta}\right|.$$

Since $\mathfrak{z} = \omega(\zeta)$ is assumed to be analytic, $\dfrac{d\mathfrak{z}}{d\zeta}$ has a unique value independent of the manner in which $\Delta\zeta \to 0$, and it follows that the transformation causes elements of arc passing through P *in any direction* to experience a change in length whose magnitude is determined by the modulus of $\dfrac{d\mathfrak{z}}{d\zeta}$ calculated at P.

It will be shown next that the argument of $\dfrac{d\mathfrak{z}}{d\zeta}$ determines the orientation of the element of arc $\Delta s'$ relative to Δs. The argument of the complex number $\Delta\zeta$ is measured by the angle θ made by the chord PQ with the ξ-axis, while the argument of

§43 EXTENSION, TORSION, AND FLEXURE OF BEAMS

$\Delta\mathfrak{z}$ is measured by the corresponding angle θ' between the x-axis and the chord $P'Q'$. Hence the difference between the angles θ' and θ is equal to

$$\arg \Delta\mathfrak{z} - \arg \Delta\zeta = \arg \frac{\Delta\mathfrak{z}}{\Delta\zeta}.$$

As $\Delta\zeta \to 0$, the vectors $\Delta\zeta$ and $\Delta\mathfrak{z}$ tend to coincide with the tangents to C at P and to C' at P', respectively, and hence[1] $\arg \dfrac{d\mathfrak{z}}{d\zeta}$ is the angle of rotation of the element of arc $\Delta s'$ relative to Δs. It follows immediately from this statement that if C_1 and C_2 are two curves in the ζ-plane that intersect at an angle τ, then the corresponding curves C_1' and C_2' in the $\mathfrak{z}$-plane also intersect at an angle τ, since the tangents to these curves are rotated through the same angle. A transformation that preserves angles is called *conformal*, and thus one can state the following theorem.

THEOREM: *The mapping performed by an analytic function $\omega(\zeta)$ is conformal at all points of the ζ-plane where $\omega'(\zeta) \neq 0$.*

We shall be concerned, for the most part, with the mapping of simply connected regions, where the mapping is one to one and hence $\omega'(\zeta) \neq 0$. The regions R and R' may, however, be finite or infinite. It should be noted that if the region R is finite and R' is infinite, then the function $\omega(\zeta)$ must become infinite at some point a of the region R; otherwise we could not have a point in the region R that corresponds to the point at infinity in the region R'. It is possible to show that at such points the function $\omega(\zeta)$ has a simple pole, so that its structure in the neighborhood of the point is

$$\omega(\zeta) = \frac{c}{\zeta - a} + f(\zeta),$$

where c is a constant and $f(\zeta)$ is analytic at all points of the region R. Other types of singularities cannot be present, since the mapping is assumed to be one to one.

If both regions R and R' are infinite, and if the points at infinity correspond, then the function $\omega(\zeta)$ has the form

$$\omega(\zeta) = c\zeta + f(\zeta),$$

---

[1] Note that this statement assumes that $\dfrac{d\mathfrak{z}}{d\zeta} \neq 0$ at P.

where c is a constant and $f(\zeta)$ is analytic in the infinite region R. We recall that a function is said to be analytic in an infinite region R if, for sufficiently large ζ, it has the structure

$$a_0 + \frac{a_1}{\zeta} + \frac{a_2}{\zeta^2} + \cdots + \frac{a_n}{\zeta^n} + \cdots.$$

Let there be given two arbitrary simply connected regions R and R', each of which is bounded by a simple closed contour that can be represented parametrically by[1]

$$\xi = \xi(t), \qquad \eta = \eta(t), \qquad (0 \leq t \leq t_1).$$

Is it possible to find the mapping function $\mathfrak{z} = \omega(\zeta)$ that will map the region R on R' conformally and in such a way that the mapping is continuous up to and including the boundaries C and C' of the regions R and R'? The answer to this question was given in the affirmative by C. Caratheodory[2] in 1913, and it is known that the mapping function $\omega(\zeta)$ is determined uniquely if we specify the correspondence of two arbitrary chosen points ζ_0 and $\mathfrak{z}_0$ of the regions R and R' and the directions of arbitrarily chosen linear elements passing through these points.[3]

We can assume without loss of generality[4] that the region R in the ζ-plane is bounded by a unit circle $|\zeta| = 1$, and it is clear that if the region R' is mapped by the function $\mathfrak{z} = \omega(\zeta)$ on the unit circle $|\zeta| \leq 1$, then the function

$$\mathfrak{z} = \omega\left(\frac{1}{\zeta}\right)$$

maps the region R' on an infinite plane with a circular hole. In general, it will be found convenient to map finite, simply connected regions on the unit circle $|\zeta| \leq 1$, and infinite regions on the portion of the ζ-plane defined by the equation $|\zeta| \geq 1$.

---

[1] Since the curves are closed, we must have $\xi(0) = \xi(t_1)$, $\eta(0) = \eta(t_1)$. The term *simple contour* is used to mean rectifiable contour.

[2] *Mathematische Annalen*, vol. 73, pp. 305–320.

[3] This theorem is associated with the name of Riemann, who proved the existence of $\omega(\zeta)$ under conditions that are less general than those enunciated above.

[4] If the regions R_1 and R_2 in the planes $\mathfrak{z}_1$ and $\mathfrak{z}_2$, respectively, are mapped on the unit circle in the ζ-plane by functions $\mathfrak{z}_1 = \omega_1(\zeta)$ and $\mathfrak{z}_2 = \omega_2(\zeta)$, then the region R_1 is mapped on the region R_2 by a transformation $\mathfrak{z}_1 = \Omega(\mathfrak{z}_2)$, obtained by eliminating ζ from $\mathfrak{z}_1 = \omega_1(\zeta)$ and $\mathfrak{z}_2 = \omega_2(\zeta)$.

In regard to the mapping of multiply connected regions, we shall make a few general remarks. It can be shown that a doubly connected region R' can be mapped on a circular ring, but that the radii of the circles making up the ring cannot be chosen arbitrarily. It is obvious that in general one can map in one-to-one manner only regions of like connectivity. The condition of like connectivity, however, is not sufficient for the existence of a mapping function.

The affirmative answer to the question of the existence of a function $\mathfrak{z} = \omega(\zeta)$ that maps conformally a given region in the $\mathfrak{z}$-plane on another given region in the ζ-plane helps little in the matter of the actual construction of mapping functions for specified regions.[1] However, there are some explicit formulas that permit one to construct mapping functions for certain classes of regions. If, for example, the region R' is that bounded by a rectilinear polygon of n sides, then the function $\omega(\zeta)$ that maps the interior of the polygon on the unit circle $|\zeta| \leq 1$ has the form

$$(43.1) \quad \mathfrak{z} = A \int_0^\zeta (\zeta - \zeta_1)^{\alpha_1-1}(\zeta - \zeta_2)^{\alpha_2-1} \cdots (\zeta - \zeta_n)^{\alpha_n-1} d\zeta + B,$$

where ζ_i are the points on the boundary γ of the unit circle that correspond to the vertices of the polygon in the $\mathfrak{z}$-plane, and the numbers $\alpha_i \pi$ are the interior angles at the vertices of the polygon.[2]

The formula (43.1) was derived by Schwarz and Christoffel,[3] and is known as the Schwarz-Christoffel transformation.

[1] Recently, some methods of constructing polynomials and rational functions that map a given region conformally on a unit circle with sufficiently high degree of accuracy have been developed by L. Kantorovich, V. Kryloff, G. Golusin, P. Melentieff, M. Mooratoff, N. Stenin, and others. The work of these mathematicians promises to be of great usefulness in applications. See V. I. Smirnov, Konformnoe Otobrazhenie (Conformal Representation), which contains a number of papers by these mathematicians. This book (in Russian) is in the Brown University Library. See also papers by L. V. Kantorovich, *Matematicheski Sbornik*, vol. 10 (1933), no. 3, pp. 294–325, and vol. 41 (1934), no. 1, pp. 179–182. This latter paper contains several corrections of errors that crept into the earlier paper.

[2] The formula is usually phrased in terms of mapping of the polygon on the half plane, but the transformation that maps the unit circle on the half plane does not alter the form of (43.1).

[3] For derivation of this formula see H. A. Schwarz, Gesammelte Abhandlungen, vol. 2, pp. 65–83; E. B. Christoffel, Annali di matematica pura ed

References for Collateral Reading

L. Bieberbach: Einführung in die konforme Abbildung, Walter de Gruyter & Company, Berlin.

C. Caratheodory: Conformal Representation, Cambridge University Press, London.

A. R. Forsyth: Theory of Functions of a Complex Variable, Cambridge University Press, London.

W. F. Osgood: Lehrbuch der Funktionentheorie, vol. 1, B. G. Teubner, Berlin.

R. Rothe, F. Ollendorff, and K. Pohlhausen: Theory of Functions Applied to Engineering Problems, Technology Press, Massachusetts Institute of Technology, Cambridge, Mass.

M. Walker: Conjugate Functions for Engineers, Oxford University Press, London.

Burkhardt-Rasor: Theory of Functions of a Complex Variable, D. C. Heath and Company, Boston.

44. Solution of the Torsion Problem by Means of Conformal Mapping. Let the torsion function $\varphi(x, y)$ (Secs. 31, 35) be combined with its conjugate function $\psi(x, y)$ to form the function $F(\mathfrak{z}) = \varphi(x, y) + i\psi(x, y)$, where $\mathfrak{z} = x + iy$. The function $F(\mathfrak{z})$ is analytic throughout the region R representing the cross section of the beam. If the region R is simply connected, we can map it conformally on a unit circle in the ζ-plane. Let

$$(44.1) \qquad \mathfrak{z} = \omega(\zeta)$$

be the function that maps the region R on the unit circle $|\zeta| \leq 1$. The function $F(\mathfrak{z})$ can be expressed in terms of the variable ζ, so that

$$(44.2) \qquad \varphi + i\psi = F[\omega(\zeta)] = f(\zeta),$$

where $f(\zeta)$ is analytic in the interior of the circle $|\zeta| = 1$.

It will be recalled [see (35.4)] that the function ψ satisfies on the boundary C of the region R the condition

$$\psi = \tfrac{1}{2}(x^2 + y^2) = \tfrac{1}{2}\mathfrak{z}\bar{\mathfrak{z}};$$

---

applicata, vol. 1 (1867), pp. 95–103, and vol. 4 (1871), pp. 1–9. For a detailed discussion of the Schwarz-Christoffel transformation and of the Schwarz reflection principle, see, for example, Forsyth's Theory of Functions of a Complex Variable, Secs. 36, 267, 268, and Burkhardt-Rasor's Theory of Functions of a Complex Variable, Secs. 71, 73. S. L. Green, The Theory and Use of the Complex Variable, Chap. VI, works out in detail several elementary examples.

hence the imaginary part of the function $f(\zeta)$ defined by (44.2) must satisfy the condition

(44.3) $$\psi = \tfrac{1}{2}\omega(\zeta)\bar{\omega}(\bar{\zeta}) \qquad \text{on } \gamma,$$

where γ is the boundary of the circle $|\zeta| = 1$.

Thus, the torsion problem will be solved if we succeed in determining the real part ψ of the analytic function

$$\frac{1}{i}f(\zeta) = \psi - i\varphi,$$

which on the boundary γ of the unit circle $|\zeta| = 1$ assumes the values

$$\psi = \tfrac{1}{2}\omega(\zeta)\bar{\omega}(\bar{\zeta}).$$

But this is a special case of the problem treated in Sec. 42, and a reference to (42.4) shows that

$$\frac{1}{i}f(\zeta) = \frac{1}{\pi i}\int_\gamma \frac{1}{2}\frac{\omega(\sigma)\bar{\omega}(\bar{\sigma})}{\sigma - \zeta}\,d\sigma - a_0 + ib_0$$

or

(44.4) $$f(\zeta) = \frac{1}{2\pi}\int_\gamma \frac{\omega(\sigma)\bar{\omega}(\bar{\sigma})}{\sigma - \zeta}\,d\sigma + \text{const.}$$

Noting that on the boundary γ of the unit circle $|\zeta| = 1$, $\sigma = e^{i\theta}$ and hence $\bar{\sigma} = e^{-i\theta} = 1/\sigma$, one sees that the integral (44.4) can be written as

(44.5) $$f(\zeta) = \varphi + i\psi = \frac{1}{2\pi}\int_\gamma \frac{\omega(\sigma)\bar{\omega}(1/\sigma)}{\sigma - \zeta}\,d\sigma + \text{const.}$$

The formula (44.5) gives us at once the torsion function φ and its conjugate ψ, so that the solution of the torsion problem is reduced to quadratures. If the numerator $\omega(\sigma)\bar{\omega}(1/\sigma)$, of the integrand, happens to be a rational function of σ, then the integral can be evaluated with the aid of the theorems on residues.

It is not difficult to express[1] the torsional rigidity D directly in terms of the function $f(\zeta)$. From (34.10)

---

[1] The calculations leading to formulas (44.7), (44.8), and (44.10) are due to N. I. Muschelišvili. See, for example, his paper "Sur le problème de torsion des cylindres élastiques isotropes," *Atti della reale accademia nazionale dei lincei*, ser. 6, vol. 9 (1929), pp. 295–300.

(44.6) $$D = \mu \iint_R (x^2 + y^2) \, dx \, dy + \mu \iint_R \left(x \frac{\partial \varphi}{\partial y} - y \frac{\partial \varphi}{\partial x}\right) dx \, dy$$
$$\equiv \mu I_0 + \mu D_0,$$

where I_0 is the polar moment of inertia of the area bounded by C, and

$$D_0 = \iint_R \left(x \frac{\partial \varphi}{\partial y} - y \frac{\partial \varphi}{\partial x}\right) dx \, dy$$
$$= \iint_R \left[\frac{\partial}{\partial y}(x\varphi) - \frac{\partial}{\partial x}(y\varphi)\right] dx \, dy.$$

An application of Green's formula to this integral gives

$$D_0 = -\int_C \varphi(x \, dx + y \, dy) = -\int_C \varphi \, d\tfrac{1}{2} r^2,$$

where $r^2 = x^2 + y^2$. But on the contour C

$$r = \mathfrak{z}\bar{\mathfrak{z}} = \omega(\sigma)\bar{\omega}(\bar{\sigma}) \quad \text{and} \quad \varphi = \tfrac{1}{2}[f(\sigma) + \bar{f}(\bar{\sigma})];$$

hence

(44.7) $$D_0 = -\tfrac{1}{4} \int_\gamma [f(\sigma) + \bar{f}(\bar{\sigma})] \, d[\omega(\sigma)\bar{\omega}(\bar{\sigma})].$$

Also

$$I_0 = \iint_{\tilde{R}} (x^2 + y^2) \, dx \, dy = \iint_R \left[\frac{\partial}{\partial x}(x^2 y) + \frac{\partial}{\partial y}(xy^2)\right] dx \, dy$$
$$= -\int_C xy(x \, dx - y \, dy).$$

But

$$x = \frac{\mathfrak{z} + \bar{\mathfrak{z}}}{2}, \qquad y = \frac{\mathfrak{z} - \bar{\mathfrak{z}}}{2i},$$

and we find that

$$I_0 = -\frac{1}{8i} \int_C (\mathfrak{z}^2 - \bar{\mathfrak{z}}^2)(\mathfrak{z} \, d\mathfrak{z} - \bar{\mathfrak{z}} \, d\bar{\mathfrak{z}}).$$

But

$$\int_C \mathfrak{z}^3 \, d\mathfrak{z} = 0, \qquad \int_C \bar{\mathfrak{z}}^3 \, d\bar{\mathfrak{z}} = 0,$$

and

$$\int_C \mathfrak{z}^2 \bar{\mathfrak{z}} \, d\bar{\mathfrak{z}} = \int_C \mathfrak{z}^2 \, d(\tfrac{1}{2}\bar{\mathfrak{z}}^2) = -\int_C \bar{\mathfrak{z}}^2 \mathfrak{z} \, d\mathfrak{z},$$

where we make use of integration by parts. Hence we can write the polar moment of inertia I_0 in the form

$$(44.8) \quad I_0 = -\frac{i}{4}\int_C \bar{\mathfrak{z}}\,\mathfrak{z}\, d\mathfrak{z} = -\frac{i}{4}\int_\gamma [\bar{\omega}(\sigma)]^2 \omega(\sigma)\, d\omega(\sigma).$$

If $\omega(\sigma)$ is a rational function, then the integrands of (44.7) and (44.8) can be easily evaluated with the aid of theorems on residues, and the expression for the torsional rigidity D can be obtained in closed form.

We may note that the shear components τ_{zx} and τ_{zy} of the stress tensor can likewise be expressed directly in terms of the functions $F(\mathfrak{z})$ and $f(\zeta)$. It follows from the formulas (34.4) and (35.1) that

$$\tau_{zx} - i\tau_{zy} = \mu\alpha\left(\frac{\partial \varphi}{\partial x} - i\frac{\partial \varphi}{\partial y} - y - ix\right)$$

$$= \mu\alpha\left[\frac{\partial \varphi}{\partial x} + i\frac{\partial \psi}{\partial x} - i(x - iy)\right].$$

But

$$\frac{\partial \varphi}{\partial x} + i\frac{\partial \psi}{\partial x} = \frac{dF}{d\mathfrak{z}},$$

and we get

$$(44.9) \quad \tau_{zx} - i\tau_{zy} = \mu\alpha[F'(\mathfrak{z}) - i\bar{\mathfrak{z}}].$$

Since $F(\mathfrak{z}) = F[\omega(\zeta)] = f(\zeta)$, we have

$$F'(\mathfrak{z}) = f'(\zeta)\frac{d\zeta}{d\mathfrak{z}} = f'(\zeta)\frac{1}{\omega'(\zeta)}.$$

Hence (44.9) becomes

$$(44.10) \quad \tau_{zx} - i\tau_{zy} = \mu\alpha\left[\frac{f'(\zeta)}{\omega'(\zeta)} - i\bar{\omega}(\zeta)\right].$$

This formula is extremely useful in calculating the components of shear.

If the mapping function is written in the form

$$\mathfrak{z} = \omega(\zeta) = \sum_{n=0}^{\infty} a_n \zeta^n,$$

then it is not difficult to give a formal solution of the torsion problem in terms of the coefficients a_n. We have

(44.11) $$\omega(\sigma)\bar{\omega}\left(\frac{1}{\sigma}\right) = \sum_{m=0}^{\infty} a_m \sigma^m \sum_{n=0}^{\infty} \bar{a}_n \sigma^{-n}$$
$$= \sum_{n=0}^{\infty} b_n \sigma^n + \sum_{n=1}^{\infty} \bar{b}_n \sigma^{-n},$$

where

(44.12) $$b_n = \sum_{r=0}^{\infty} a_{n+r} \bar{a}_r.$$

Upon inserting the expression (44.11) in (44.5), it is seen that

$$f(\zeta) = \frac{1}{2\pi} \int_{\gamma} \left[\sum_{n=0}^{\infty} b_n \sigma^n + \sum_{n=1}^{\infty} \bar{b}_n \sigma^{-n} \right] \frac{d\sigma}{(\sigma - \zeta)}$$

or

(44.13) $$f(\zeta) = \varphi + i\psi = i \sum_{n=0}^{\infty} b_n \zeta^n,$$

where Cauchy's Integral Formula and Eq. (41.8) have been used. The expression (44.13) for the complex stress function $\varphi + i\psi$ was derived by R. M. Morris[1] by a different method, and was used to obtain formal solutions of the problem of torsion for those cases in which the complex constants a_n are known.

A formal expression involving the constants a_n can be given for the torsional rigidity $D = \mu(I_0 + D_0)$ [see (44.6)]. Eq. (44.8) for the moment of inertia I_0 can be written as

$$I_0 = -\frac{i}{4} \int_{\gamma} \left[\omega(\sigma)\bar{\omega}\left(\frac{1}{\sigma}\right) \right] \bar{\omega}\left(\frac{1}{\sigma}\right) d\omega(\sigma).$$

Now from $\omega(\sigma) = \sum_{n=0}^{\infty} a_n \sigma^n$, it follows that

$$\bar{\omega}\left(\frac{1}{\sigma}\right) d\omega(\sigma) = i \sum_{n=-\infty}^{\infty} c_n \sigma^n \, d\theta,$$

---

[1] R. M. Morris, "The Internal Problems of Two-dimensional Potential Theory," *Mathematische Annalen*, vol. 116 (1939), pp. 374–400; vol. 117 (1939), pp. 31–38.

where

(44.14) $$\begin{cases} c_n = \sum_{r=0}^{\infty} (n+r)a_{n+r}\bar{a}_r, \\ c_{-n} = \sum_{r=0}^{\infty} r\bar{a}_{n+r}a_r, \quad (n = 0, 1, 2, \cdots). \end{cases}$$

From this relation and from (44.11), we get

$$I_0 = \tfrac{1}{4}\int_0^{2\pi} \left(\sum_{m=0}^{\infty} b_m\sigma^m + \sum_{m=1}^{\infty} \bar{b}_m\sigma^{-m}\right)\left(\sum_{n=0}^{\infty} c_n\sigma^n + \sum_{n=1}^{\infty} c_{-n}\sigma^{-n}\right) d\theta.$$

Since $\sigma = e^{i\theta}$, we see that the integral of every term involving σ^n, $(n \neq 0)$ vanishes, and we are left with

$$I_0 = \frac{\pi}{2}\left[b_0 c_0 + \sum_{n=1}^{\infty} (b_n c_{-n} + \bar{b}_n c_n)\right].$$

Similarly we can write

$$D_0 = \tfrac{1}{4}\int_0^{2\pi}\left(\sum_{m=0}^{\infty} b_m\sigma^m - \sum_{m=0}^{\infty} b_m\sigma^{-m}\right)\left(\sum_{n=1}^{\infty} nb_n\sigma^n - \sum_{n=1}^{\infty} n\bar{b}_n\sigma^{-n}\right) d\theta$$

$$= -\pi \sum_{n=1}^{\infty} nb_n\bar{b}_n.$$

Combining the expressions for I_0 and D_0, we get finally

(44.15) $$D = \frac{\mu\pi}{2}\left[b_0 c_0 + \sum_{n=1}^{\infty} (c_n\bar{b}_n + c_{-n}b_n - 2nb_n\bar{b}_n)\right].$$

As an illustration of the application of the foregoing procedure, we consider a beam whose cross section is the cardioid

$$r = 2c(1 + \cos \alpha)$$

$(r^2 = x^2 + y^2, \tan \alpha = y/x)$. It is readily verified that in this case a suitable mapping function is

$$\mathfrak{z} = c(1 - \zeta)^2,$$

so that the only nonvanishing coefficients a_n are

$$a_0 = c, \quad a_1 = -2c, \quad a_2 = c.$$

The nonvanishing constants b_n and c_n are easily found to be

$$b_0 = 6c^2, \quad b_1 = -4c^2, \quad b_2 = c^2,$$
$$c_{-1} = -2c^2, \quad c_0 = 6c^2, \quad c_1 = -6c^2, \quad c_2 = 2c^2.$$

The complex stress function is

(44.16) $\quad f(\zeta) = \varphi + i\psi = i \sum_{n=0}^{\infty} b_n \zeta^n = ic^2(6 - 4\zeta + \zeta^2),$

while

(44.17) $\qquad\qquad D = 17\mu\pi c^4.$

It should be noted that the method outlined above is readily applicable whenever the mapping function $\omega(\zeta)$ for the region R is a polynomial, or whenever the mapping can be approximated with sufficient accuracy by a polynomial. If the mapping function is known as a power series in ζ, then a formal solution can be given in terms of the coefficients a_n of the mapping function. If the mapping function is known in a closed form, then it may be easier to proceed directly from Eq. (44.5) rather than expand $\omega(\zeta)$ in a power series and then deal with the resulting infinite series (44.12), (44.13), (44.14), and (44.15). The reader will verify that formula (44.4) can be used in this case to obtain the result (44.16) with no calculational effort.

The problem of the cardioid is of some interest inasmuch as it indicates an approximate behavior of a checked beam.

PROBLEM

Obtain the solution of the torsion problem for a cardioid by utilizing formula (44.4), and thus verify (44.14).

45. Applications of Conformal Mapping. This section contains several illustrations of the application of the foregoing theory to the solution of the torsion problem.

Consider first the mapping function

(45.1) $\qquad\qquad z = \omega(\zeta) = \dfrac{k\zeta}{\zeta^2 + a^2}, \quad (a > 1, k > 0).$

A study of the map of the unit circle $|\zeta| \leq 1$ obtained with the aid of (45.1) shows that when a is near unity, the resulting figure in the z-plane differs little from the figure consisting of a

§45 EXTENSION, TORSION, AND FLEXURE OF BEAMS

pair of tangent circles (Fig. 30). The function $f(\zeta)$ that solves the torsion problem for a beam whose cross section is shown in Fig. 30 is thus [see (44.5)]

(45.2) $\quad f(\zeta) = \varphi + i\psi = \dfrac{1}{2\pi} \displaystyle\int_\gamma \dfrac{k^2 \sigma^2\, d\sigma}{(\sigma^2 + a^2)(1 + a^2\sigma^2)(\sigma - \zeta)}.$

The integrand of (45.2), viewed as a function of σ, has two simple poles $\sigma_1 = ia$ and $\sigma_2 = -ia$ that lie outside the unit circle. For large values of $|\sigma|$ the integrand is of the order $1/|\sigma|^3$, and we have[1]

$$f(\zeta) = -i(R_1 + R_2),$$

where R_1 and R_2 are the residues of the integrand at $\sigma_1 = ia$ and $\sigma_2 = -ia$. Since

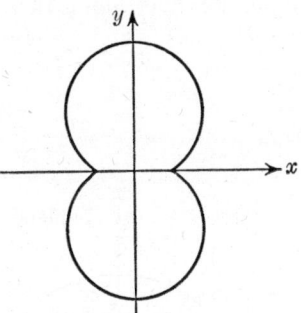

Fig. 30.

$$R_1 = \left[\dfrac{(\sigma - ia)k^2\sigma^2}{(\sigma^2 + a^2)(1 + a^2\sigma^2)(\sigma - \zeta)}\right]_{\sigma=ia} = \dfrac{-k^2 a}{2i(1 - a^4)(ia - \zeta)},$$

and

$$R_2 = \left[\dfrac{(\sigma + ia)k^2\sigma^2}{(\sigma^2 + a^2)(1 + a^2\sigma^2)(\sigma - \zeta)}\right]_{\sigma=-ia} = \dfrac{-k^2 a}{2i(1 - a^4)(ia + \zeta)},$$

we have

(45.3) $\quad f(\zeta) = \dfrac{ia^2 k^2}{(a^4 - 1)(\zeta^2 + a^2)}.$

The components of stress τ_{zx} and τ_{zy} can be calculated with the aid of (44.10), and it is not difficult to show with the aid of (44.7) and (44.8) that the torsional rigidity D of the section is

$$D = \dfrac{\mu \pi k^4 (a^8 + 1)}{2(a^4 - 1)^4}.$$

As an illustration of the type of calculations required when the mapping function is not a rational function, consider the map of the unit circle obtained with the aid of

(45.4) $\quad z = \omega(\zeta) = a\sqrt{1 + \zeta}, \quad$ where $a > 0$.

---

[1] See the discussion of residues in Sec. 40.

We shall deal with that branch of the multiple-valued function $\sqrt{1+\zeta}$ that gives $+1$ for $\zeta = 0$.

Then from Fig. 31

$$\mathfrak{z} = a\sqrt{r}\, e^{i\varphi/2}, \qquad \left(-\frac{\pi}{2} \leq \varphi \leq \frac{\pi}{2}\right).$$

When ζ moves along the circle γ,

$$\varphi = \tfrac{1}{2}\theta, \qquad (-\pi \leq \theta \leq \pi).$$

and

$$r = 2\cos \tfrac{1}{2}\theta.$$

Hence

$$\mathfrak{z} = a\sqrt{2\cos \tfrac{1}{2}\theta}\, e^{i\theta/4}, \qquad (-\pi \leq \theta \leq \pi).$$

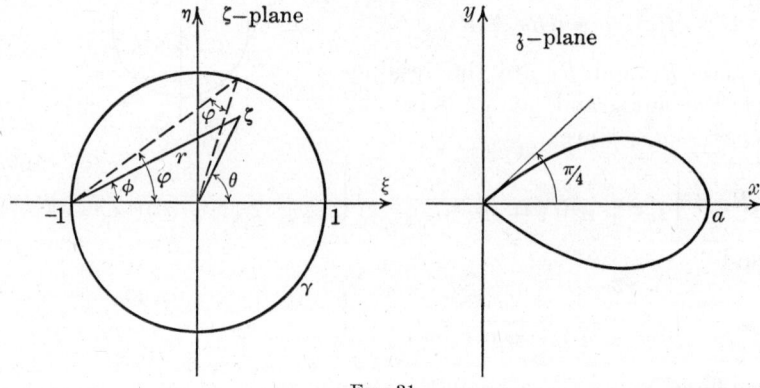

Fig. 31.

Let R and ψ denote the modulus and the argument of $\mathfrak{z}$; then

$$R e^{i\psi} = a\sqrt{2\cos \tfrac{1}{2}\theta}\, e^{i\theta/4}.$$

Hence

$$R = a\sqrt{2\cos \tfrac{1}{2}\theta}, \qquad \psi = \frac{\theta}{4},$$

and it follows that

$$R = a\sqrt{2\cos 2\psi}.$$

Thus, the map of the unit circle is one loop of the lemniscate shown in Fig. 31.

Substituting (45.4) in the formula (44.5), we have

$$(45.5) \quad f(\zeta) = \frac{1}{2\pi}\int_\gamma \frac{a^2\sqrt{1+\sigma}\sqrt{1+\dfrac{1}{\sigma}}}{\sigma - \zeta}\, d\sigma = \frac{a^2}{2\pi}\int_\gamma \frac{1+\sigma}{\sqrt{\sigma}}\frac{d\sigma}{\sigma - \zeta}.$$

Since the sign of the square root must be chosen positive, we can write $\sqrt{\sigma} = e^{i\theta/2}$.

If we cut the negative axis as shown in Fig. 32, then the integrand of (45.5) will be a single-valued function in the simply connected region indicated in the figure, and the only singularity of the integrand is the pole at $\sigma = \zeta$. Hence

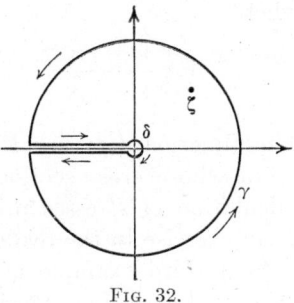

Fig. 32.

$$\frac{1}{2\pi i}\left[\int_\gamma g(\sigma, \zeta)\, d\sigma + \int_{-1}^{0} g(\sigma, \zeta)\, d\sigma + \int_\delta g(\sigma, \zeta)\, d\sigma \right.$$
$$\left. + \int_0^{-1} g(\sigma, \zeta)\, d\sigma\right] = R,$$

where

$$g(\sigma, \zeta) \equiv \frac{1+\sigma}{\sqrt{\sigma}\,(\sigma-\zeta)}.$$

R is the residue of $g(\sigma, \zeta)$ at $\sigma = \zeta$, and δ is a small circle about the origin. But the residue R at $\sigma = \zeta$ is obviously

$$R = \frac{1+\zeta}{\sqrt{\zeta}}.$$

Thus,

$$\frac{1}{2\pi i}\int_\gamma g(\sigma, \zeta)\, d\sigma = -\frac{1}{2\pi i}\left[\int_{-1}^{0} g(\sigma, \zeta)\, d\sigma + \int_0^{-1} g(\sigma, \zeta)\, d\sigma \right.$$
$$\left. + \int_\delta g(\sigma, \zeta)\, d\sigma\right] + R$$
$$= -\frac{1}{\pi}\int_0^1 \frac{1-t}{\sqrt{t}}\frac{dt}{t+\zeta} + \frac{1+\zeta}{\sqrt{\zeta}},$$

where we have dropped the integral over the small circle δ, since it vanishes when the radius of δ tends to zero, and where the integrals over the portion of the real axis between 0 and -1 are combined by making an obvious change of variable and by noting the difference in sign of the function $\sqrt{\sigma}$ on the upper and lower banks of the cut. Integrating and dropping the nonessential additive constant, we have finally

(45.6) $$f(\zeta) = \frac{a^2}{\pi}\frac{1+\zeta}{\sqrt{\zeta}}\log\frac{1+i\sqrt{\zeta}}{1-i\sqrt{\zeta}},$$

where

$$\log \frac{1 + i\sqrt{\zeta}}{1 - i\sqrt{\zeta}} = 2i\sqrt{\zeta}\left(1 - \frac{\zeta}{3} + \frac{\zeta^2}{5} - \cdots\right).$$

The function $f(\zeta)$ in (45.6) solves the torsion problem for a beam whose cross section is one loop of the lemniscate. The calculation of stresses presents no serious difficulty and is left as an exercise to the reader.

As a third example of this general method of attack upon problems of torsion, consider the case of a cylinder whose cross section is bounded by two circular arcs.[1]

Consider a region R of the complex $\mathfrak{z}$-plane bounded by two circular arcs C_1 and C_2 making an angle $\alpha \neq 0$ at their points

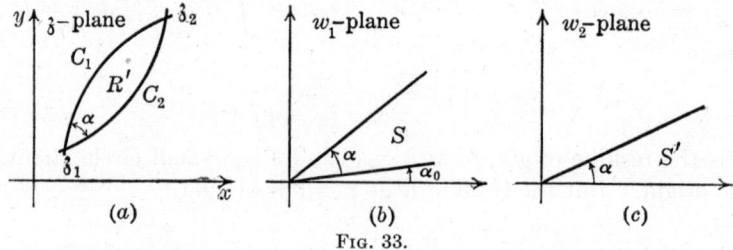

Fig. 33.

of intersection $\mathfrak{z} = \mathfrak{z}_1$ and $\mathfrak{z} = \mathfrak{z}_2$ (Fig. 33a). It is obvious that the transformation $w_1 = (\mathfrak{z} - \mathfrak{z}_1)/(\mathfrak{z} - \mathfrak{z}_2)$ maps the point $\mathfrak{z}_1$ on the origin of the w_1-plane, the point $\mathfrak{z}_2$ on the point at infinity, and the region R on the infinite sector S shown in Fig. 33b. The sector S is rotated through an angle α_0 to bring one radius into coincidence with the real axis of the w_2-plane by the transformation $w_2 = e^{-i\alpha_0}w_1$. If the transformation $w_3 = w_2^{\pi/\alpha}$ is applied, then the domain S' is mapped on the upper half plane of the complex w_3-plane. Finally, this upper half plane is carried into the unit circle in the ζ-plane by the mapping function $\zeta = (i - w_3)/(1 - iw_3)$. If these successive transformations are combined, one obtains the mapping function

$$\zeta = \frac{Ci(\mathfrak{z} - \mathfrak{z}_1)^{\pi/\alpha} + (\mathfrak{z} - \mathfrak{z}_2)^{\pi/\alpha}}{Ci(\mathfrak{z} - \mathfrak{z}_1)^{\pi/\alpha} - (\mathfrak{z} - \mathfrak{z}_2)^{\pi/\alpha}},$$

[1] The discussion that follows is taken from the paper "Torsion of Regions Bounded by Circular Arcs," by I. S. and E. S. Sokolnikoff, *Bulletin of the American Mathematical Society*, vol. 44 (1938), pp. 384–387.

§45 EXTENSION, TORSION, AND FLEXURE OF BEAMS

which effects conformal mapping of the region R on the unit circle in the complex ζ-plane. The choice of the constants C, $\mathfrak{z}_1$, and $\mathfrak{z}_2$ is uniquely determined by the geometrical configuration and the scale used in mapping the region R on the unit circle. For $\alpha = \pi/n$, where n is an integer, the mapping function becomes rational, a fact that greatly simplifies the evaluation of (44.5). To illustrate the procedure, it will suffice to consider two important special cases when the region R is one of the following:

1. A lune formed by two circular arcs of equal radius and intersecting at right angles;
2. A semicircle.

The first of these regions suggests the cross section of a propeller blade. It is easily verified that the mapping functions appropriate to the regions defined in (1) and (2) are, respectively,

$$\zeta = \frac{2\mathfrak{z}}{\mathfrak{z}^2 + 1}$$

and

$$\zeta = \frac{(\mathfrak{z} + 1)^2 - i(\mathfrak{z} - 1)^2}{(\mathfrak{z} + 1)^2 + i(\mathfrak{z} - 1)^2}.$$

The solution of the torsion problem for the region described in (1) above is given next.

From

$$\zeta = \frac{2\mathfrak{z}}{\mathfrak{z}^2 + 1},$$

it follows that on the boundary of R, $\mathfrak{z} = [1 - (1 - \sigma^2)^{1/2}]/\sigma$, where the appropriate branch of the square root is determined from the observation that the imaginary part of $\mathfrak{z}$ is positive whenever θ in $\sigma = e^{i\theta}$ lies between 0 and π. Then the numerator of the integrand in (44.5) is

$$\omega(\sigma)\omega\left(\frac{1}{\sigma}\right) = [1 - (1 - \sigma^2)^{1/2}]\left[1 - \left(1 - \frac{1}{\sigma^2}\right)^{1/2}\right].$$

Substituting this expression in (44.5) and evaluating the resulting integrals, we get

$$f(\zeta) = -i(1 - \zeta^2)^{1/2} - \frac{i(1 - \zeta^2)}{\pi\zeta} \log\frac{1 - \zeta}{1 + \zeta} + \text{const.},$$

or if we return to the $\mathfrak{z}$-plane with the aid of the mapping function,

$$F(\mathfrak{z}) = i\frac{\mathfrak{z}^2 - 1}{\mathfrak{z}^2 + 1} - \frac{i(1 - \mathfrak{z}^2)^2}{\pi\mathfrak{z}(1 + \mathfrak{z}^2)} \log \frac{1 - \mathfrak{z}}{1 + \mathfrak{z}} + \text{const.}$$

The imaginary part of $F(\mathfrak{z})$, which is the desired solution, is

$$\psi(x, y) = \{\pi(x^2 + y^2)[(x^2 + y^2 + 1)^2 - 4y^2]\}^{-1}\{\pi(x^2 + y^2)$$
$$\cdot [(x^2 + y^2)^2 - 1]$$
$$+ x(x^2 + y^2 + 1)[4y^2 - (x^2 + y^2 - 1)^2] \log S$$
$$+ y(x^2 + y^2 - 1)[(x^2 + y^2 + 1)^2 + 4x^2]\beta\},$$

where

$$S = \frac{[(1 - x^2 - y^2)^2 + 4y^2]^{\frac{1}{2}}}{(1 + x)^2 + y^2} \quad \text{and} \quad \beta = \tan^{-1}\frac{-2y}{1 - x^2 - y^2}.$$

A simple calculation shows that, on the boundary of the lune formed by $x^2 + (y \pm 1)^2 = 2$, $\psi(x, y)$ reduces to

$$\psi = \tfrac{1}{2}(x^2 + y^2) + \text{const.},$$

as it should.

An analogous calculation gives the solution of the torsion problem for the semicircular region. We obtain

$$f(\zeta) = \frac{i 2^{\frac{1}{2}}(1 + i\zeta)(1 - \zeta^2)^{\frac{1}{2}}}{(1 + \zeta)^2} + \frac{4}{\pi(i + \zeta)} + \frac{2}{\pi}\frac{(1 - \zeta^2)}{(i + \zeta)^2} \log \frac{i(1 - \zeta)}{1 + \zeta}$$
$$+ \text{const.},$$

or, if we return to the $\mathfrak{z}$-plane with the aid of the mapping function,

$$F(\mathfrak{z}) = \frac{1}{2\pi}\left[\pi i \mathfrak{z}^2 + \frac{2(\mathfrak{z}^2 + 1)}{\mathfrak{z}} + \frac{(\mathfrak{z}^2 - 1)^2}{\mathfrak{z}^3} \log \frac{1 - \mathfrak{z}}{1 + \mathfrak{z}} + \text{const.}\right].$$

This result agrees with that obtained by Greenhill by an entirely different method.[1]

---

[1] A. G. GREENHILL, "Fluid Motion in a Rotating Quadrantal Cylinder," *Messenger of Mathematics*, vol. 8 (1879), p. 89; "On the Motion of a Frictionless Liquid in a Rotating Sector," *Messenger of Mathematics*, vol. 10 (1881), p. 83.
See also A. E. H. Love, A Treatise on the Mathematical Theory of Elasticity, p. 319.

§45   EXTENSION, TORSION, AND FLEXURE OF BEAMS

As a final example of the solution of the torsion problem in terms of the mapping function $z = \omega(\zeta)$, we consider a cylinder whose cross section R is bounded by the inverse of an ellipse with respect to its center.

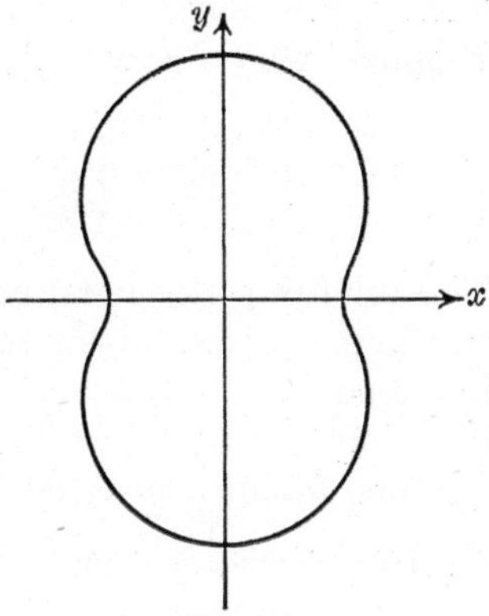

If the ellipse

$$\frac{x^2}{a^2} + \frac{y^2}{b^2} = 1$$

is inverted with respect to its center, then the point (x, y) is carried to the point (x', y'), which is such that

FIG. 34.

$$r^2(r')^2 \equiv [x^2 + y^2][(x')^2 + (y')^2] = 1.$$

The resulting curve C (Fig. 34) is given in terms of the parameter u by the equations

$$\frac{x}{x^2 + y^2} = \frac{1}{c} \cosh k \cos u,$$

$$\frac{y}{x^2 + y^2} = \frac{1}{c} \sinh k \sin u,$$

or

(45.7) $$x + iy = c \sec (u + ik)$$

with $c = 1/\sqrt{a^2 - b^2}$, $\tanh k = b/a$.

Equation (45.7) can evidently be written

(45.8) $$z = c \sec (w + ik), \quad v = 0,$$

with $z = x + iy$, $w = u + iv$.

If we put $\zeta = e^{iw}$ in (45.8), we see that the resulting function

(45.9) $$z = \omega(\zeta) \equiv \frac{2ce^k \zeta}{\zeta^2 + e^{2k}}$$

maps the cross section R of the cylinder upon the interior of the unit circle $|\zeta| \leq 1$.

In the preceding section, expressions were derived that give a formal solution of the torsion problem when the mapping function is expanded in an infinite series

$$z = \omega(\zeta) = \sum_{n=0}^{\infty} a_n \zeta^n = \sum_{n=0}^{\infty} a_n e^{inw}.$$

Such expressions, given by R. M. Morris, were used by T. J. Higgins,[1] who observed that in this case

$$a_n = \begin{cases} 0, & \text{if } n = 0, 2, 4, \cdots, \\ 2c(-1)^{\frac{(n-1)}{2}} e^{-nk}, & \text{if } n = 1, 3, 5, \cdots. \end{cases}$$

The infinite series entering into (44.13) and (44.15) were then summed, and the torsion function and twisting moment obtained in closed form. However, since $\omega(\zeta)$ is a rational function of ζ, it is simpler to proceed directly[2] from (44.5), (44.7), and (44.8).

When (45.9) is inserted in (44.5), it is seen that

$$(45.10) \quad f(\zeta) = \varphi + i\psi$$
$$= \frac{2c^2}{\pi} \int_\gamma \frac{\sigma^2 \, d\sigma}{(\sigma^2 + e^{2k})(\sigma^2 + e^{-2k})(\sigma - \zeta)} + \text{const.}$$
$$= -4c^2 i (R_1 + R_2) + \text{const.},$$

where R_1 and R_2 are the residues of the integrand at $\sigma = ie^k$ and $\sigma = -ie^k$, respectively. We have

$$R_1 = \left[\frac{\sigma^2}{(\sigma + ie^k)(\sigma^2 + e^{-2k})(\sigma - \zeta)} \right]_{\sigma = ie^k} = \frac{ie^k}{4(\zeta - ie^k) \sinh 2k},$$
$$R_2 = \frac{-ie^k}{4(\zeta + ie^k) \sinh 2k},$$

and hence

$$f(\zeta) = \varphi + i\psi = c^2 \operatorname{csch} 2k \, \tan (w + ik),$$

where the constant in (45.10) has been taken equal to $-ic^2 \operatorname{csch} 2k$.

The shearing stresses may be found either from the relations

$$\tau_{zx} = \mu\alpha \frac{\partial \Psi}{\partial y}, \quad \tau_{yz} = -\mu\alpha \frac{\partial \Psi}{\partial x}, \quad \Psi = \psi - \frac{1}{2}(x^2 + y^2),$$

or from Eq. (44.10), to be

$$\tau_{zy} = -2\mu\alpha c \sin u \left[\frac{\operatorname{csch} 2k \, \cosh (v + k)}{\cos 2u - \cosh 2(v + k)} + \frac{\sinh (v + k)}{\cos 2u + \cosh 2(v + k)} \right],$$
$$\tau_{zy} = -2\mu\alpha c \cos u \left[\frac{\operatorname{csch} 2k \, \sinh (v + k)}{\cos 2u - \cosh 2(v + k)} - \frac{\cosh (v + k)}{\cos 2u + \cosh 2(v + k)} \right].$$

[1] "The Torsion of a Prism with Cross Section the Inverse of an Ellipse," *Journal of Applied Physics*, vol. 13 (1942), pp. 457–459.

[2] See I. S. Sokolnikoff and R. D. Specht, "Two Dimensional Boundary Value Problems in Potential Theory," *Journal of Applied Physics*, vol. 14 (1943), pp. 91–95.

§45 EXTENSION, TORSION, AND FLEXURE OF BEAMS

Equation (44.8) for the moment of inertia I_0 takes the form

$$I_0 = 4c^4 i \int_\gamma \frac{\sigma^3(\sigma^2 - e^{2k})}{(\sigma^2 + e^{2k})^3(\sigma^2 + e^{-2k})^2} \, d\sigma$$
$$= -8\pi c^4 (R_3 + R_4),$$

where R_3 and R_4 are the residues of the integrand at $\sigma = ie^{-k}$ and $\sigma = -ie^{-k}$, respectively. The residues are

$$R_3 = \frac{d}{d\sigma}\left[\frac{\sigma^3(\sigma^2 - e^{2k})}{(\sigma^2 + e^{2k})^3(\sigma + ie^{-k})^2}\right]_{\sigma=ie^{-k}}$$
$$= \frac{-\operatorname{csch}^4 2k(2 + \cosh 4k)}{16} = R_4,$$

and therefore

$$I_0 = \pi c^4 (2 + \cosh 4k) \operatorname{csch}^4 2k.$$

Similarly from (44.7) we get

$$D_0 = -i4c^4 \operatorname{csch} 2k \int_\gamma \frac{\sigma(1 - \sigma^4)^2}{(\sigma^2 + e^{2k})^3(\sigma^2 + e^{-2k})^3} \, d\sigma$$
$$= 8\pi c^4 \operatorname{csch} 2k (R_5 + R_6),$$

in which R_5 and R_6 are the residues at $\sigma = ie^{-k}$ and $\sigma = -ie^{-k}$. We find that

$$R_5 = \frac{1}{2}\frac{d^2}{d\sigma^2}\left[\frac{\sigma(1 - \sigma^4)^2}{(\sigma^2 + e^{2k})^3(\sigma + ie^{-k})^3}\right]_{\sigma=ie^{-k}}$$
$$= -\tfrac{1}{8} \operatorname{csch}^3 2k = R_6,$$

and hence

$$D_0 = -2\pi c^4 \operatorname{csch}^4 2k.$$

The twisting moment is given by

$$M = \mu\alpha(I_0 + D_0)$$
$$= \mu\alpha\pi c^4 (2 \operatorname{csch}^2 2k + \operatorname{csch}^4 2k).$$

The examples discussed in this section illustrate the remarkable ease with which we can find displacements, stresses, and torsional rigidity with the help of this general formulation of the problem of torsion, whenever the mapping function is of simple form. In the case of some polygonal cross sections, the solution can be obtained with the help of the Schwarz-Christoffel transformation.¹ Approximate mapping functions may be used, as mentioned in Sec. 43, in problems that are otherwise difficult to handle. Even when the mapping function is known, it may be advantageous to use some other method, as was done in the case of the rectangular cross section (Sec. 38), where series of orthogonal functions were employed. The torsion problem for multiply connected regions will be considered briefly in Sec. 47.

If beams were to be designed to withstand torsional loads alone, then one need consider only circular cylinders, for, as noted in Sec. 37, such cylinders have the greatest torsional rigidity for a given cross-sectional area. In practice, however, structural members not designed primarily for torsion are yet subjected to twisting loads, and it becomes necessary to consider the torsion problem for a beam with arbitrary cross section. If the particular cross sections exhibited seem of little technical importance, it should be noted that the first example gives us information about the distribution of stress in the neighborhood of a reentrant corner, while the second and third display the elastic behavior at a rectangular corner.

References for Collateral Reading

N. W. McLachlan: Complex Variable and Operational Calculus, Cambridge University Press, London, Chap. II, Sec. 2.62, and Chap. III.
A readable account of the theory of residues.

PROBLEM

Discuss the derivation of formula (45.3) with the aid of formula (44.13).

[1] E. Trefftz, "Über die Torsion prismatischer Stäbe von polygonalen Querschnitt," *Mathematische Annalen*, vol. 82 (1921), pp. 97–112.

S. Bergmann, "Über die Berechnung des magnetischen Feldes in einem Einphasen-Transformator," *Zeitschrift für angewandte Mathematik und Mechanik*, vol. 5 (1925), pp. 319–331.

I. S. Sokolnikoff, "On a Solution of Laplace's Equation with an Application to the Torsion Problem for a Polygon with Re-entrant Angles," *Transactions of the American Mathematical Society*, vol. 33 (1931), pp. 719–732.

P. F. Kufarev, "Torsion and Bending of Members of Polygonal Sections," *Applied Mathematics and Mechanics*, new ser., vol. 1 (1937), pp. 43–76.

Bruno Bernstein, "Zum Torsion des Winkelquerschnittes," private publication (available at the Brown University Library).

46. Membrane and Other Analogies. It is clear from the discussion given in Secs. 38 and 45 that a rigorous solution of the torsion problem for beams whose cross sections are in the shape of the letters I, U, L, T, etc. is likely to prove extremely vexing. While there are some rigorous solutions of the torsion problem for beams of polygonal cross section,[1] the resultant formulas are too involved to be of much value to a practical designing engineer, who requires some simple, reasonably accurate formulas for drafting-room use. Within the last two decades a variety of approximate formulas have been developed for the torsion constants of sections whose components are rectangles. Many such formulas are based on the mathematical analogy between the torsion problem and the behavior of a stretched elastic membrane subjected to a uniform excess of pressure on one side.[2] A reference to this analogy was made in Sec. 35, and we proceed to discuss it here in detail.

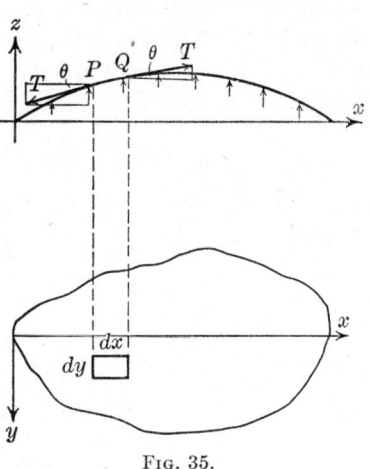

FIG. 35.

Let a very thin homogeneous membrane, such as a soap film, be stretched under a uniform tension T per unit length over an opening made in a rigid plate. The opening in the plate

---

[1] See the references in the footnote on p. 186.

[2] A detailed discussion of the procedure employed in deriving some approximate formulas for sections whose components are rectangles (as well as for some tubular sections) is given in the National Advisory Committee for Aeronautics, *Report* 334, by G. W. Trayer and H. W. March, entitled "The Torsion of Members Having Sections Common in Aircraft Construction." This report contains an extensive bibliography and a comparison of their formulas with those obtained by other investigators. A description of the experimental procedure used in studying torsion of beams with the aid of soap films is given by Trayer and March and also by A. A. Griffith and G. I. Taylor in the Technical Report of the Advisory Committee for Aeronautics (British), 1917–1918. A brief account of the procedure employed by Griffith and Taylor is found in S. Timoshenko, Theory of Elasticity, Sec. 82.

is assumed to have the same shape as the cross section of the beam subjected to torsion, and the membrane is supposed to be fixed at the edge of the opening. If p is the pressure per unit area of the membrane, and if the membrane is in equilibrium, then the force $p\,dx\,dy$, acting on an element of area $dx\,dy$ (Fig. 35), must be balanced by the resultant of the vertical components of the tensile stresses acting on the boundary of the element of area. Now the resultant of the vertical components of the tensile forces acting on the edges dy is

$$(T\,dy\,\sin\theta)_Q - (T\,dy\,\sin\theta)_P \doteq \left(T\,dy\,\frac{\partial z}{\partial x}\right)_Q - \left(T\,dy\,\frac{\partial z}{\partial x}\right)_P$$
$$\doteq \left(T\,dy\,\frac{\partial z}{\partial x}\right)_P + \frac{\partial}{\partial x}\left(T\,dy\,\frac{\partial z}{\partial x}\right)_P dx - \left(T\,dy\,\frac{\partial z}{\partial x}\right)_P$$
$$= T\,\frac{\partial^2 z}{\partial x^2}\,dx\,dy,$$

where it is assumed that the deflection z is small. Similarly the resultant of the vertical components acting on the edges dx is

$$T\,\frac{\partial^2 z}{\partial y^2}\,dx\,dy.$$

Hence the equation of equilibrium of the element is

$$p\,dx\,dy + T\,\frac{\partial^2 z}{\partial x^2}\,dx\,dy + T\,\frac{\partial^2 z}{\partial y^2}\,dx\,dy = 0,$$

and we have

(46.1) $$\frac{\partial^2 z}{\partial x^2} + \frac{\partial^2 z}{\partial y^2} = \frac{-p}{T}.$$

This equation must be solved subject to the condition $z = 0$ on the edge of the opening. If we substitute in (46.1)

(46.2) $$z = \frac{p}{2T}\,\Psi,$$

the equation becomes

(46.3) $$\frac{\partial^2 \Psi}{\partial x^2} + \frac{\partial^2 \Psi}{\partial y^2} = -2,$$

subject to the condition that

(46.4) $$\Psi = 0 \quad \text{on the boundary.}$$

§46 EXTENSION, TORSION, AND FLEXURE OF BEAMS

Equation (46.3) is identical with that obtained in Sec. 35 for the stress function Ψ, and it is clear from the discussion there that the slope of the membrane at any point is proportional to the magnitude of the shearing stress

$$\tau = \mu\alpha \sqrt{\left(\frac{\partial \Psi}{\partial x}\right)^2 + \left(\frac{\partial \Psi}{\partial y}\right)^2} = \mu\alpha \frac{d\Psi}{d\nu}$$

at the corresponding point of the section subjected to torsion. The contour lines $z =$ constant of the membrane correspond to the lines of shearing stress $\Psi(x, y) =$ constant. Recalling that the torsional rigidity of the beam is given by Eq. (35.10),

$$D = 2\mu \int \int_R \Psi \, dx \, dy,$$

it becomes clear that the volume between the plane of the opening $z = 0$ and the surface of the membrane is proportional to the torsional rigidity D of the section. Since the contour lines of the membrane can be mapped out, and the slope at each point and the volume under the membrane can be measured, one can secure the desired information concerning the lines of shearing stress and the torsional rigidity of the beam from experimental measurements.

A consideration of the equation of the unloaded membrane,

$$\frac{\partial^2 z}{\partial x^2} + \frac{\partial^2 z}{\partial y^2} = 0,$$

which is so supported at the edges that

(46.5) $\qquad\qquad z = \frac{1}{2}(x^2 + y^2) \qquad$ on the boundary,

shows that one can also determine experimentally the function $\psi(x, y)$ [see (35.3)] from a study of an unloaded soap film stretched so that the heights of the membrane over the contour of the section have the values given by (46.5).

The membrane analogy has been used in an interesting way by Timoshenko to discuss an approximate behavior of a beam of narrow, rectangular cross section, and in analyzing the stress concentration near fillets in channel sections and I beams. It is interesting to note that the maximum shearing stress in a narrow beam of thickness c is twice as great as in a circular shaft of diameter c and subjected to the same twist. The details of the calculations and further discussion will be found in Timoshenko's Theory of Elasticity, Secs. 77 and 81.

In addition to the references given in Timoshenko's Theory of Elasticity (Secs. 76, 82, 84), the following papers may be consulted. The technique of measuring the ordinates of the membrane has been discussed by Thiel,[1] who used stereoscopic photography, while Reichenbächer[2] has described an optical device for the automatic plotting of the contour lines of the membrane. The soap film has been replaced by a paraffin surface by Kopf and Weber,[3] and by the interface between two immiscible liquids by Piccard and Baes[4] and by Sunatani, Matuyama, and Hatamura.[5] L. Föppl[6] and Deutler[7] have discussed the form of the membrane analogy in which the film is under zero resultant pressure and its boundary has variable height.

The boundary-value problems of torsion can also be interpreted in terms of various hydrodynamical analogies. These are discussed briefly in Timoshenko's Theory of Elasticity, Sec. 83, where several references are given. To these may be added a paper by Den Hartog and McGivern[8] in which experimental technique is described.

[1] A. Thiel, "Photogrammetrisches Verfahren zur versuchsmässigen Lösung von Torsionsaufgaben (nach einem Seifenhautgleichnis von L. Föppl)," *Ingenieur Archiv*, vol. 5 (1934), pp. 417–429.

[2] H. Reichenbächer, "Selbsttätige Ausmessung von Seifenhautmodellen (Anwendung auf das Torsionsproblem)," *Ingenieur Archiv*, vol. 7 (1936), pp. 257–272.

[3] E. Kopf and E. Weber, "Verfahren zur Ermittlung der Torsionsbeanspruchung mittels Membranmodell," *Zeitschrift des Vereines deutscher Ingenieure*, vol. 78 (1934), pp. 913–914.

[4] A. Piccard and L. Baes, "Mode expérimental nouveau relatif à l'application des surfaces à courbure constante à la solution du problème de la torsion des barres prismatiques," *Proceedings of the Second International Congress for Applied Mechanics, Zürich*, 1927, pp. 195–199.

[5] Chidô Sunatani, Tokuzo Matuyama, and Motomune Hatamura, "The Solution of Torsion Problems by Means of a Liquid Surface," *Technical Reports of the Tôhoku Imperial University*, vol. 12 (1937), pp. 374–396.

[6] L. Föppl, "Eine Ergänzung des Prandtlschen Seifenhaut-Gleichnisses zur Torsion," *Zeitschrift für angewandte Mathematik und Mechanik*, vol. 15 (1935), pp. 37–40.

[7] H. Deutler, "Zur versuchsmässigen Lösung von Torsionsaufgaben mit Hilfe des Seifenhautgleichnisses," *Ingenieur Archiv.*, vol. 9 (1938), pp. 280–282.

[8] J. P. Den Hartog, and J. G. McGivern, "On the Hydrodynamic Analogy of Torsion," *Journal of Applied Mechanics*, vol. 2 (1935), pp. A46–A48.

§47 EXTENSION, TORSION, AND FLEXURE OF BEAMS

The analogy between the torsion of a cylinder and the potential of a plane electric field affords another way of obtaining experimental solutions of the torsion problem. This is described in Sec. 7, Chap. III, of Technische Dynamik by C. B. Biezeno and R. Grammel and in a paper by H. Cranz.[1]

The equation for current flow in a conductor of variable thickness is identical with that describing the torsion of a shaft of varying circular section.[2] This analogy, which yields a practical method for studying stress concentration in the neighborhood of fillets or grooves in shafts under torsion, is described in Sec. 87 of Timoshenko's Theory of Elasticity and in papers by Thum and Bautz,[3] Jacobsen,[4] and Salet.[5]

References for Collateral Reading

A. E. H. LOVE: A Treatise on the Mathematical Theory of Elasticity, Cambridge University Press, London, Sec. 224.

S. TIMOSHENKO: Theory of Elasticity, McGraw-Hill Book Company, Inc., New York, Secs. 76, 82.

R. V. SOUTHWELL: Theory of Elasticity for Engineers and Physicists, Oxford University Press, New York, Secs. 386–389.

W. TRAYER and H. W. MARCH: National Advisory Committee for Aeronautics, *Report* 334.

PROBLEM

Show from the membrane analogy that upper and lower bounds can be put on the torsional stiffness of an irregular section by considering the largest inscribed and smallest circumscribed circular section.

47. Torsion of Hollow Beams. The discussion of the torsion problem has been confined thus far to solid beams, so that the region of the cross section has been simply connected. Hollow or tubular beams are of considerable technical importance, and

---

[1] H. CRANZ, "Experimentelle Lösung von Torsionsaufgaben," *Ingenieur Archiv*, vol. 4, pp. 506–509, (1933).

[2] See Sec. 49.

[3] A. THUM and W. BAUTZ, "Die Ermittlung von Spannungsspitzen in verdrehbeanspruchten Wellen durch ein elektrisches Modell," *Zeitschrift des Vereines deutscher Ingenieure*, vol. 78 (1934), pp. 17–19.

[4] L. S. JACOBSEN, "Torsional Stresses in Shafts Having Grooves or Fillets," *Journal of Applied Mechanics*, vol. 2 (1935), pp. A154–A155.

[5] G. SALET, "Détermination des pointes de tension dans les arbres de révolution soumis à torsion au moyen d'un modèle électrique," *Bulletin de l'association technique maritime et aéronautique* (1936), 40, pp. 341–350, 351.

it is necessary to extend the formulation of the torsion problem so as to include multiply connected regions.

Let it be assumed that a beam has several longitudinal cavities so that the boundary of the cross section of the beam is made up of several simple closed contours. Denote the exterior contour by C_0, and let $C_1, C_2, \cdots, C_n$ be the simple closed contours lying entirely within the contour C_0 (Fig. 36). The contours $C_1, C_2, \cdots, C_n$ correspond to the cavities of the beam. The discussion that led to the formulation of the differential equation (34.5) is valid in this case, and we have the differential equation

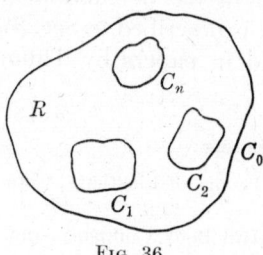

Fig. 36.

$$(47.1) \quad \frac{\partial^2 \varphi}{\partial x^2} + \frac{\partial^2 \varphi}{\partial y^2} = 0 \quad \text{in } R,$$

where R is the multiply connected region interior to C_0 and exterior to $C_1, C_2, \cdots, C_n$. Since the longitudinal cavities and the outer surface are free from external loads, we have, as shown in Sec. 34, the boundary conditions

$$(47.2) \quad \frac{d\varphi}{d\nu} = y \cos(x, \nu) - x \cos(y, \nu) \text{ on } C_i, (i = 0, 1, 2, \cdots, n).$$

It follows from the discussion in Sec. 35 that the function $\psi(x, y)$ conjugate to the torsion function $\varphi(x, y)$ satisfies on each contour C_i the condition

$$\frac{d\psi}{ds} = \frac{d}{ds}\left[\frac{1}{2}(x^2 + y^2)\right],$$

or

$$(47.3) \quad \psi = \tfrac{1}{2}(x^2 + y^2) + k_i \quad \text{on } C_i, \quad (i = 0, 1, 2, \cdots, n),$$

where the k_i are constants that will, in general, have different values on the contours C_i.

Ordinarily the function $\psi(x, y)$ determined with the aid of (35.2) will be multiple-valued, but in our case ψ must satisfy Eq. (47.3), and hence ψ must return to its original value after the point $\mathfrak{z} = x + iy$ traverses each contour C_i. Moreover, the function $\varphi(x, y)$ is determined by the boundary conditions (47.2) to within an arbitrary constant, and it follows that its conjugate function $\psi(x, y)$ must likewise be determined to within an

§47   EXTENSION, TORSION, AND FLEXURE OF BEAMS   193

arbitrary constant. Accordingly, the value of only one of the arbitrary constants k_i entering into (47.3) can be assigned at will, and the remaining ones must be so determined as to make the function ψ single-valued in the region R.

The stress function $\Psi(x, y)$ can be defined, as in Sec. 35, by the formula

(47.4) $\quad\quad \Psi(x, y) = \psi(x, y) - \tfrac{1}{2}(x^2 + y^2),$

so that $\Psi(x, y)$ satisfies in the region R the equation

(47.5) $\quad\quad \nabla^2 \Psi = -2.$

As a consequence of the definition (47.4), it follows that on the contours, C_i, $(i = 0, 1, 2, \cdots, n)$, the function Ψ satisfies the conditions

(47.6) $\quad\quad \Psi = k_i, \quad (i = 0, 1, 2, \cdots, n),$

where the value of only one of the constants k_i may be assigned arbitrarily.

The formula (34.10) for the calculation of the torsional rigidity D is still available, and we have

(47.7) $\quad D = \mu \iint_R \left(x^2 + y^2 + x \frac{\partial \varphi}{\partial y} - y \frac{\partial \varphi}{\partial x} \right) dx\, dy$

$\quad\quad\quad = \mu \iint_R - \left(x \frac{\partial \Psi}{\partial x} + y \frac{\partial \Psi}{\partial y} \right) dx\, dy,$

where we make use of the relations

$$\frac{\partial \varphi}{\partial x} = y + \frac{\partial \Psi}{\partial y} \quad \text{and} \quad \frac{\partial \varphi}{\partial y} = -\left(x + \frac{\partial \Psi}{\partial x} \right),$$

obtained in Sec. 35. The integration in (47.7) is performed over the multiply connected region R. The right-hand member of (47.7) can be rewritten as

$$D = \mu \iint_R \left[2\Psi - \frac{\partial}{\partial x}(x\Psi) - \frac{\partial}{\partial y}(y\Psi) \right] dx\, dy$$

$$= 2\mu \iint_R \Psi\, dx\, dy + \mu \int_C \Psi(y\, dx - x\, dy),$$

where we make use of Green's Theorem, and the subscript C on the line integral means that the integration is to be performed in appropriate directions over all the contours C_i, $(i = 0, 1, 2,$

..., n). Now if we choose the value of Ψ over the contour C_0 to be zero (that is, $k_0 = 0$) and note the boundary conditions (47.6), we have

$$D = 2\mu \int\int_R \Psi\, dx\, dy + \mu \sum_{i=1}^{n} k_i \int_{C_i} (y\, dx - x\, dy).$$

But

$$\int_{C_i} (y\, dx - x\, dy) = 2\int\int_{A_i} dx\, dy = 2A_i,$$

where A_i is the area enclosed by the contour C_i, and we have

(47.8) $$D = 2\mu \int\int_R \Psi\, dx\, dy + \sum_{i=1}^{n} 2\mu k_i A_i.$$

The expression for the twisting moment M is

(47.9) $$M = 2\mu\alpha \left(\int\int_R \Psi\, dx\, dy + \sum_{i=1}^{n} k_i A_i \right).$$

Some important approximate formulas that are applicable to thin tubes follow at once from these formulas. While it is not the purpose of this volume to deal with approximate engineering formulas, a brief reference to their mode of development is made in the remainder of this section.

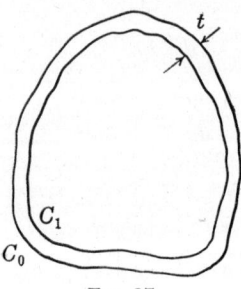

Fig. 37.

Let a thin tubular section of thickness t be bounded by an exterior contour C_0 and an interior contour C_1 (Fig. 37). If the section is thin, then the value of Ψ under the integral sign can be replaced by its average value taken over the contours C_0 and C_1, but since the value of Ψ over the contour C_0 is taken as zero, we have

$$\overline{\Psi} = \tfrac{1}{2}\Psi_1 = \tfrac{1}{2}k_1,$$

where $\overline{\Psi}$ denotes the average value of Ψ. Hence the approximate value of the torque M from (47.9) is

$$M \doteq 2\mu\alpha k_1(\tfrac{1}{2}A + A_1),$$

where A is the area of the ring and A_1 is the area of the hole.

§47 EXTENSION, TORSION, AND FLEXURE OF BEAMS

The resultant shearing stress at any point in the cross section is given by the formula (see Sec. 35)

$$\tau = \mu\alpha \sqrt{\left(\frac{\partial\Psi}{\partial x}\right)^2 + \left(\frac{\partial\Psi}{\partial y}\right)^2} = \mu\alpha \frac{d\Psi}{d\nu}.$$

Hence the approximate average shearing stress $\bar{\tau}$ in the tube is

$$\bar{\tau} = \mu\alpha \frac{\delta\Psi}{t} = \mu\alpha \frac{k_1}{t},$$

where $\delta\Psi$ denotes the difference in the values of the stress function Ψ on the outer and the inner boundaries. If the perimeter of the inner boundary is l, then one can deduce the following approximate relations for very thin tubes:[1]

$$\bar{\tau} \doteq 2\mu\alpha \frac{A_1}{l},$$

$$M \doteq 4\mu\alpha t \frac{A_1^2}{l}.$$

The membrane analogy is also used in an approximate discussion of the torsion problem for thin slit tubes.[2]

The application of the membrane analogy to obtain experimental solutions of the torsion problem for hollow beams is described by Timoshenko[3] and by Biezeno and Grammel.[4] This method was applied to the torsion problem of an eccentric circular annulus by Engelmann.[5] Neményi[6] has discussed the use of numerical and experimental (membrane) methods in the torsion problem for beams of multiply connected cross section.

[1] For a discussion of approximate formulas see R. V. Southwell, Theory of Elasticity for Engineers and Physicists, Sec. 393; S. Timoshenko, Theory of Elasticity, Sec. 85; G. W. Trayer and H. W. March, National Advisory Committee for Aeronautics, *Report* 334, pp. 17–20.

[2] See S. Timoshenko, Theory of Elasticity, Secs. 77, 81, and G. W. Trayer and H. W. March, National Advisory Committee for Aeronautics, *Report* 334, p. 20.

[3] Theory of Elasticity, Sec. 84.

[4] Technische Dynamik, Chap. III, Sec. 26, pp. 199–201.

[5] Fritz Engelmann, "Verdrehung von Stäben mit einseitig-ring-förmigem Querschnitt," *Forschung auf dem Gebiete des Ingenieurwesens*, vol. 6 (1935), pp. 146–154.

[6] P. Neményi, "Lösung des Torsionsproblems für Stäbe mit mehrfach zusammenhängendem Quershnitt," *Zeitschrift für angewandte Mathematik und Mechanik*, vol. 1 (1921), pp. 364–367.

It will be recalled that the curves $\Psi(x, y) = $ constant determine the lines of shearing stress (see Sec. 35), and it follows from the boundary conditions (47.6) that one can obtain a rigorous solution for the torsion problem of a hollow shaft from the solution of the torsion problem of a solid shaft by deleting the portion of material contained within the curve $\Psi(x, y) = $ constant. Thus, in the discussion of the problem of torsion for an elliptic cylinder in Sec. 36, it was shown that the lines of shearing stress are similar ellipses, concentric with the ellipse

$$\frac{x^2}{a^2} + \frac{y^2}{b^2} = 1,$$

representing the cross section of the cylinder. Accordingly, if we delete the portion of material contained within the elliptical cylinder

(47.10) $$\frac{x^2}{a^2} + \frac{y^2}{b^2} = (1 - k)^2, \qquad (0 < k < 1),$$

then the stress function Ψ for an elliptical beam of semi-axes a and b will have a constant value over the curve (47.10), and the same function Ψ will thus solve the torsion problem for a hollow beam bounded by similar elliptical cylinders.

The lines of shearing stress for a beam of circular cross section are circles concentric with the outer boundary, and it follows at once that the formulas contained in Sec. 33 are applicable to hollow circular shafts.[1] In particular, the torsional rigidity D is

(47.11) $$D = \frac{\mu\pi}{2}(a^4 - a_0^4),$$

where a_0 is the radius of the inner circle and a is that of the outer one.

References for Collateral Reading

S. Timoshenko: Theory of Elasticity, McGraw-Hill Book Company, Inc., New York, Secs. 84, 85.

R. V. Southwell: Theory of Elasticity for Engineers and Physicists, Oxford University Press, New York, Secs. 391–393.

PROBLEM

Find the conjugate torsion functions Φ, ψ, the stress function Ψ, and the constant k entering into (47.6) for a hollow circular shaft. Derive

[1] See problem at the end of this section.

the expression (47.11) for the torsional rigidity of the shaft both from (34.10) and from (47.8).

48. Curvilinear Coordinates. In many problems it is more convenient to deal with a curvilinear coordinate system rather than with a rectilinear one. For instance, in discussing the torsion problem for a body having axial symmetry, it is convenient to make use of a cylindrical coordinate system. One such problem will be discussed in the next section, and we proceed to develop the expressions for stresses and strains in curvilinear coordinates. We shall limit ourselves to a treatment of orthogonal coordinate systems, since only such systems occur commonly in practice.

Consider a set of three independent functions of the Cartesian variables x, y, z,

(48.1) $$\alpha_i = \alpha_i(x, y, z), \qquad (i = 1, 2, 3).$$

The intersections of the surfaces

$$\alpha_i(x, y, z) = \text{const.}, \qquad (i = 1, 2, 3)$$

pair by pair determine the coordinate lines of our curvilinear coordinate system, and the intersection of coordinate lines determines a point that will be labeled $(\alpha_1, \alpha_2, \alpha_3)$. It is assumed that the coordinate system (α_i) is orthogonal, so that the element of arc ds has the form

(48.2) $$ds^2 = \sum_{i=1}^{3} g_{ii} \, d\alpha_i^2,$$

where g_{ii} are the metric coefficients that can be calculated[1] from (48.1).

Let P_0 and P be two neighboring points in an unstrained medium, and let these points take the positions P_0' and P' after deformation. We shall confine our discussion to infinitesimal deformations[2] and shall represent the displacements in the directions normal to the coordinate surfaces α_1, α_2, α_3 by u_1, u_2, u_3 respectively. The curvilinear coordinates of the points P_0 and

---

[1] See Prob. 1 at the end of this section. All summations in this section will be indicated by a summation sign.

[2] See Sec. 7.

P are α_i and $\alpha_i + d\alpha_i$ respectively. The coordinates of the points P'_0 and P' will be denoted by $\alpha_i + \xi_i$ and

$$\alpha_i + \xi_i + d\alpha_i + d\xi_i.$$

Then it follows from (48.2) that

$$u_1 = \sqrt{g_{11}}\,\xi_1, \qquad u_2 = \sqrt{g_{22}}\,\xi_2, \qquad u_3 = \sqrt{g_{33}}\,\xi_3.$$

Now the length of the element of arc ds joining the points P_0 and P is given by

$$(48.3) \qquad ds^2 = \sum_{i=1}^{3} g_{ii}(\alpha_1, \alpha_2, \alpha_3)\, d\alpha_i^2,$$

while the length of the same element in the deformed state is given by

$$(48.4) \quad (ds')^2 = \sum_{i=1}^{3} g_{ii}(\alpha_1 + \xi_1, \alpha_2 + \xi_2, \alpha_3 + \xi_3)(d\alpha_i + d\xi_i)^2.$$

But

$$g_{ii}(\alpha_1 + \xi_1, \alpha_2 + \xi_2, \alpha_3 + \xi_3) = g_{ii}(\alpha_1, \alpha_2, \alpha_3) + \sum_{j=1}^{3} \frac{\partial g_{ii}}{\partial \alpha_j}\xi_j$$

to the order of approximation contemplated by the linear theory, and

$$(d\alpha_i + d\xi_i)^2 = (d\alpha_i)^2 + 2\, d\alpha_i\, d\xi_i + d\xi_i^2 \doteq (d\alpha_i)^2 + 2\sum_{j=1}^{3} \frac{\partial \xi_i}{\partial \alpha_j}\, d\alpha_j\, d\alpha_i.$$

Hence (48.4) can be written as

$$(48.5) \qquad (ds')^2 = \sum_{i=1}^{3}\sum_{j=1}^{3} G_{ij}\, d\alpha_i\, d\alpha_j,$$

where

$$(48.6) \qquad G_{ij} = \delta_{ij}\left(g_{ii} + \sum_{k=1}^{3} \frac{\partial g_{ii}}{\partial \alpha_k}\xi_k\right) + g_{ii}\frac{\partial \xi_i}{\partial \alpha_j} + g_{jj}\frac{\partial \xi_j}{\partial \alpha_i},$$

and where we neglect the terms involving the products of ξ_j and $\dfrac{\partial \xi_i}{\partial \alpha_j}$. The symbol δ_{ij}, as usual, denotes the Kroenecker delta.

The expression for G_{ij} has been symmetrized by replacing G_{ij} by $\frac{1}{2}(G_{ij} + G_{ji})$.

It is clear from (48.2) and (48.5) that the elongations of linear elements and shears are characterized by the coefficients g_{ii} and G_{ij}. Thus, consider a linear element ds_i directed along one of the coordinate lines α_i. From (48.2), its length is

$$ds_i = \sqrt{g_{ii}}\, d\alpha_i,$$

while the length of the same element after deformation is

$$ds'_i = \sqrt{G_{ii}}\, d\alpha_i.$$

Accordingly, the extension e_{ii} of this element is

$$e_{ii} = \frac{\sqrt{G_{ii}}\, d\alpha_i - \sqrt{g_{ii}}\, d\alpha_i}{\sqrt{g_{ii}}\, d\alpha_i}$$
$$= \sqrt{1 + \frac{G_{ii} - g_{ii}}{g_{ii}}} - 1 \doteq \frac{1}{2}\frac{G_{ii} - g_{ii}}{g_{ii}},$$

if we neglect the nonlinear terms in the ξ_i and their derivatives. Noting the definitions (48.6), we have

$$(48.7) \qquad e_{ii} = \frac{1}{2}\frac{G_{ii} - g_{ii}}{g_{ii}} = \frac{\partial \xi_i}{\partial \alpha_i} + \frac{1}{2g_{ii}} \sum_{k=1}^{3} \frac{\partial g_{ii}}{\partial \alpha_k} \xi_k$$
$$= \frac{\partial}{\partial \alpha_i}\left(\frac{u_i}{\sqrt{g_{ii}}}\right) + \frac{1}{2g_{ii}} \sum_{k=1}^{3} \frac{\partial g_{ii}}{\partial \alpha_k} \frac{u_k}{\sqrt{g_{kk}}}.$$

The cosine of the angle θ_{ij} between the directions of linear elements in the deformed state that were originally directed parallel to the coordinate lines α_i and α_j is given by[1]

$$(48.8) \qquad \cos \theta_{ij} = \frac{G_{ij}}{\sqrt{G_{ii}G_{jj}}}.$$

Just as in Sec. 4, we define the angle α_{ij} by the formula

$$\theta_{ij} = \frac{\pi}{2} - \alpha_{ij};$$

then

$$\cos \theta_{ij} = \sin \alpha_{ij} \doteq \alpha_{ij}.$$

[1] See Prob. 3 at the end of this section.

The shear components of the strain tensor are defined by the relation $\alpha_{ij} = 2e_{ij}$. Substituting in (48.8), we get[1]

$$e_{ij} = \frac{1}{2} \frac{G_{ij}}{\sqrt{G_{ii}G_{jj}}} \doteq \frac{1}{2} \frac{G_{ij}}{\sqrt{g_{ii}g_{jj}}}.$$

Finally, substituting in the foregoing formula the definitions (48.6), we obtain the shear components of the strain tensor in the form

$$(48.9) \quad e_{ij} = \frac{1}{2\sqrt{g_{ii}g_{jj}}} \left(g_{ii} \frac{\partial \xi_i}{\partial \alpha_j} + g_{jj} \frac{\partial \xi_j}{\partial \alpha_i} \right)$$
$$= \frac{1}{2\sqrt{g_{ii}g_{jj}}} \left[g_{ii} \frac{\partial}{\partial \alpha_j} \left(\frac{u_i}{\sqrt{g_{ii}}} \right) + g_{jj} \frac{\partial}{\partial \alpha_i} \left(\frac{u_j}{\sqrt{g_{jj}}} \right) \right], \text{ if } i \neq j.$$

The components τ_{ij} of the stress tensor in curvilinear coordinates are defined in precisely the same way as they were in the cartesian system. Thus, the component of stress normal to the element of area perpendicular to the coordinate line α_i is denoted by τ_{ii}, and the component of shear associated with the coordinate lines α_i and α_j is written as τ_{ij}.

In this notation,[2] Hooke's law for a homogeneous isotropic medium assumes the form

$$(48.10) \quad \begin{cases} \tau_{ii} = \lambda \vartheta + 2\mu e_{ii} \text{ or } \tau_{ii} = \dfrac{E\sigma}{(1+\sigma)(1-2\sigma)} \vartheta + \dfrac{E}{1+\sigma} e_{ii} \\ \tau_{ij} = 2\mu e_{ij}, \quad \text{or} \quad \tau_{ij} = \dfrac{E}{1+\sigma} e_{ij}, \quad \text{if } i \neq j, \end{cases}$$

where the invariant $\vartheta \equiv e_{11} + e_{22} + e_{33}$. Solving the system (48.10) for the components of strain yields

$$(48.11) \quad e_{ij} = \frac{1+\sigma}{E} \tau_{ij} - \frac{\sigma}{E} \delta_{ij} \Theta,$$

where the invariant $\Theta = \tau_{11} + \tau_{22} + \tau_{33}$.

[1] Note that

$$G_{ii}G_{jj} = \left(g_{ii} + \sum_{k=1}^{3} \frac{\partial g_{ii}}{\partial \alpha_k} \xi_k + 2g_{ii} \frac{\partial \xi_i}{\partial \alpha_i} \right) \left(g_{jj} + \sum_{k=1}^{3} \frac{\partial g_{jj}}{\partial \alpha_k} \xi_k + 2g_{jj} \frac{\partial \xi_j}{\partial \alpha_j} \right)$$
$$= g_{ii}g_{jj} + \sum_{k=1}^{3} \left(\frac{\partial g_{jj}}{\partial \alpha_k} g_{ii} + \frac{\partial g_{ii}}{\partial \alpha_k} g_{jj} \right) \xi_k + 2g_{ii}g_{jj} \left(\frac{\partial \xi_i}{\partial \alpha_i} + \frac{\partial \xi_j}{\partial \alpha_j} \right)$$

+ terms involving products of ξ_k and its derivatives, terms that were neglected previously.

[2] Cf. Sec. 22 and 23.

§48 EXTENSION, TORSION, AND FLEXURE OF BEAMS

A somewhat lengthy calculation, although in essence similar to that outlined in Sec. 15, leads to the following equations of equilibrium in curvilinear coordinates:[1]

$$(48.12) \quad \frac{\partial(g\tau_{ii})}{\partial \alpha_i} - \frac{1}{2} \sum_{j=1}^{3} \frac{g\tau_{jj}}{g_{jj}} \frac{\partial g_{jj}}{\partial \alpha_i} + \sum_{j \neq i} \frac{\partial}{\partial \alpha_j}\left(\frac{g g_{ii}\tau_{ij}}{\sqrt{g_{ii}g_{jj}}}\right) + F_i g \sqrt{g_{ii}} = 0, \quad (i = 1, 2, 3),$$

where $g \equiv \sqrt{g_{11}g_{22}g_{33}}$, and the F_i are the components, in the directions of the coordinate axes, of the body force $\mathbf{F}$.

We shall write out the expressions for the strain components (48.7) and (48.9) and the equations of equilibrium (48.12) for three important special cases of curvilinear coordinates.

a. Plane Polar Coordinates. In this case, the index i assumes the values 1, 2, and according to the usual notation

$$\alpha_1 = r, \quad \alpha_2 = \theta.$$

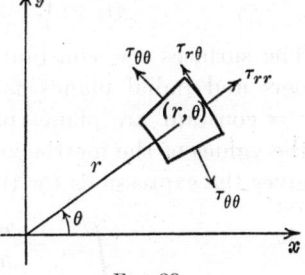

Fig. 38.

The coordinate surfaces in this case are circular cylinders perpendicular to the xy-plane ($r =$ constant) and radial planes through the origin ($\theta =$ constant). The element of arc is given by

$$ds^2 = dr^2 + r^2 \, d\theta^2,$$

so that

$$g_{11} = 1, \quad g_{12} = 0, \quad g_{22} = r^2.$$

Noting the formulas (48.7) and (48.9), we see that the strain components in this case are

$$(48.13) \quad \begin{cases} e_{rr} = \dfrac{\partial u_r}{\partial r}, \\[4pt] e_{\theta\theta} = \dfrac{1}{r}\dfrac{\partial u_\theta}{\partial \theta} + \dfrac{u_r}{r}, \\[4pt] e_{r\theta} = \dfrac{1}{2r}\left(\dfrac{\partial u_r}{\partial \theta} - u_\theta + r\dfrac{\partial u_\theta}{\partial r}\right), \end{cases}$$

---

[1] See A. E. H. Love, A Treatise on the Mathematical Theory of Elasticity, Secs. 19, 20, 21, 22c, 58, 59, 96, 97, 99; E. Trefftz, Handbuch der Physik, vol. 6, Secs. 24, 25.

while the equations of equilibrium (48.12) become

(48.14)
$$\begin{cases} \dfrac{\partial \tau_{rr}}{\partial r} + \dfrac{1}{r}\dfrac{\partial \tau_{r\theta}}{\partial \theta} + \dfrac{\tau_{rr} - \tau_{\theta\theta}}{r} + F_r = 0, \\ \dfrac{\partial \tau_{r\theta}}{\partial r} + \dfrac{1}{r}\dfrac{\partial \tau_{\theta\theta}}{\partial \theta} + \dfrac{2}{r}\tau_{r\theta} + F_\theta = 0. \end{cases}$$

b. Cylindrical Coordinates. The variables involved here are

$$\alpha_1 = r, \qquad \alpha_2 = \theta, \qquad \alpha_3 = z,$$

and the element of arc in cylindrical coordinates is given by

$$ds^2 = dr^2 + r^2\, d\theta^2 + dz^2,$$

so that

$$g_{11} = 1, \qquad g_{22} = r^2, \qquad g_{33} = 1.$$

The surfaces $r = $ constant and $\theta = $ constant are circular cylinders and radial planes as in case a above, while the surfaces $z = $ constant are planes parallel to the xy-plane. Substituting the values of the metric coefficients in (48.7), (48.9), and (48.12) gives the expressions for the strain components

(48.15)
$$\begin{cases} e_{rr} = \dfrac{\partial u_r}{\partial r}, \\ e_{\theta\theta} = \dfrac{1}{r}\dfrac{\partial u_\theta}{\partial \theta} + \dfrac{u_r}{r}, \\ e_{zz} = \dfrac{\partial u_z}{\partial z}, \\ e_{r\theta} = \dfrac{1}{2}\left(\dfrac{1}{r}\dfrac{\partial u_r}{\partial \theta} + \dfrac{\partial u_\theta}{\partial r} - \dfrac{u_\theta}{r}\right), \\ e_{rz} = \dfrac{1}{2}\left(\dfrac{\partial u_z}{\partial r} + \dfrac{\partial u_r}{\partial z}\right), \\ e_{\theta z} = \dfrac{1}{2}\left(\dfrac{\partial u_\theta}{\partial z} + \dfrac{1}{r}\dfrac{\partial u_z}{\partial \theta}\right), \end{cases}$$

and the equations of equilibrium

(48.16)
$$\begin{cases} \dfrac{\partial \tau_{rr}}{\partial r} + \dfrac{1}{r}\dfrac{\partial \tau_{r\theta}}{\partial \theta} + \dfrac{\partial \tau_{rz}}{\partial z} + \dfrac{\tau_{rr} - \tau_{\theta\theta}}{r} + F_r = 0, \\ \dfrac{\partial \tau_{r\theta}}{\partial r} + \dfrac{1}{r}\dfrac{\partial \tau_{\theta\theta}}{\partial \theta} + \dfrac{\partial \tau_{\theta z}}{\partial z} + \dfrac{2}{r}\tau_{r\theta} + F_\theta = 0, \\ \dfrac{\partial \tau_{rz}}{\partial r} + \dfrac{1}{r}\dfrac{\partial \tau_{\theta z}}{\partial \theta} + \dfrac{\partial \tau_{zz}}{\partial z} + \dfrac{1}{r}\tau_{rz} + F_z = 0. \end{cases}$$

§48 EXTENSION, TORSION, AND FLEXURE OF BEAMS

c. Spherical Coordinates. For this coordinate system, we have $\alpha_1 = r$, $\alpha_2 = \alpha$, $\alpha_3 = \theta$. The surfaces are the right circular cylinders, $r = $ constant, the radial planes perpendicular to the xy-plane, $\alpha = $ constant, and the right circular cones with vertices at the origin, $\theta = $ constant. Since the element of arc is given by

$$ds^2 = dr^2 + r^2 \sin^2 \theta \, d\alpha^2 + r^2 \, d\theta^2$$

we have

$$g_{11} = 1, \qquad g_{22} = r^2 \sin^2 \theta, \qquad g_{33} = r^2.$$

The strain components, in this case, are

(48.17) $$\begin{cases} e_{rr} = \dfrac{\partial u_r}{\partial r}, \\[4pt] e_{\theta\theta} = \dfrac{1}{r}\dfrac{\partial u_\theta}{\partial \theta} + \dfrac{u_r}{r}, \\[4pt] e_{\alpha\alpha} = \dfrac{1}{r \sin \theta}\dfrac{\partial u_\alpha}{\partial \alpha} + \dfrac{u_r}{r} + u_\theta \dfrac{\cot \theta}{r}, \\[4pt] e_{r\alpha} = \dfrac{1}{2}\left(\dfrac{1}{r \sin \theta}\dfrac{\partial u_r}{\partial \alpha} - \dfrac{u_\alpha}{r} + \dfrac{\partial u_\alpha}{\partial r} \right), \\[4pt] e_{r\theta} = \dfrac{1}{2}\left(\dfrac{1}{r}\dfrac{\partial u_r}{\partial \theta} - \dfrac{u_\theta}{r} + \dfrac{\partial u_\theta}{\partial r} \right), \\[4pt] e_{\alpha\theta} = \dfrac{1}{2}\left(\dfrac{1}{r}\dfrac{\partial u_\alpha}{\partial \theta} - \dfrac{u_\alpha \cot \theta}{r} + \dfrac{1}{r \sin \theta}\dfrac{\partial u_\theta}{\partial \alpha} \right), \end{cases}$$

and the equations of equilibrium are

(48.18) $$\begin{cases} \dfrac{\partial \tau_{rr}}{\partial r} + \dfrac{1}{r \sin \theta}\dfrac{\partial \tau_{r\alpha}}{\partial \alpha} + \dfrac{1}{r}\dfrac{\partial \tau_{r\theta}}{\partial \theta} + \dfrac{2\tau_{rr} - \tau_{\alpha\alpha} - \tau_{\theta\theta} + \tau_{r\theta}\cot\theta}{r} + F_r = 0, \\[6pt] \dfrac{\partial \tau_{r\alpha}}{\partial r} + \dfrac{1}{r \sin \theta}\dfrac{\partial \tau_{\alpha\alpha}}{\partial \alpha} + \dfrac{1}{r}\dfrac{\partial \tau_{\alpha\theta}}{\partial \theta} + \dfrac{3\tau_{r\alpha} + 2\tau_{\alpha\theta}\cot\theta}{r} + F_\alpha = 0, \\[6pt] \dfrac{\partial \tau_{r\theta}}{\partial r} + \dfrac{1}{r \sin \theta}\dfrac{\partial \tau_{\alpha\theta}}{\partial \alpha} + \dfrac{1}{r}\dfrac{\partial \tau_{\theta\theta}}{\partial \theta} + \dfrac{3\tau_{r\theta} + (\tau_{\theta\theta} - \tau_{\alpha\alpha})\cot\theta}{r} + F_\theta = 0. \end{cases}$$

References for Collateral Reading

A. E. H. Love: A Treatise on the Mathematical Theory of Elasticity, Cambridge University Press, London, Secs. 19–22c, 58, 59, 96, 97, 99.

E. Trefftz: Handbuch der Physik, Verlag von Julius Springer, Berlin, vol. 6, Secs. 24, 25.

A. J. McConnell: Applications of the Absolute Differential Calculus, Blackie & Son, Ltd., Glasgow, pp. 307–311.

F. Odqvist: "Equations de compatibilité pour un système de coordonnées triplex orthogonaux quelconques," *Comptes rendus de l'académie des sciences* (Paris), vol. 205 (1937), pp. 202–204.

F. K. G. Odqvist: "Kompatibilitätsgleichungen bei Zylinderkoordinater," *Zeitschrift für angewandte Mathematik und Mechanik*, vol. 14 (1934), pp. 123–124.

L. Brillouin: "Les lois de l'élasticité en coordonnées quelconques," Congrès international de mathématique, Toronto, 1924, *Annales physique*, vol. 3 (1925), pp. 251–298.

T. N. Blinchikov: "Differential Equations of the Equilibrium of the Theory of Elasticity in the Curvilinear Coordinate System," *Applied Mathematics and Mechanics*, new ser., vol. 2 (1939), pp. 407–413 (in Russian with an English summary).

E. Volterra: "Questioni di elasticitá vincolata I. Componenti di deformazione e potenziale elastico in coordinate qualsivogliano," *Atti della reale accademia nazionale dei lincei*, ser. 6, vol. 20 (1934), pp. 424–428.

H. Thirring: "On the Tensor Analytical Representation of the Theory of Elasticity," *Physikalische Zeitschrift*, vol. 26 (1925), pp. 518–522.

PROBLEMS

1. Show that the metric coefficients g_{ij} can be calculated by observing that

$$dx_k = \sum_{i=1}^{3} \frac{\partial x_k}{\partial \alpha_i} d\alpha_i, \qquad (dx_k)^2 = \sum_{i,j=1}^{3} \frac{\partial x_k}{\partial \alpha_i} \frac{\partial x_k}{\partial \alpha_j} d\alpha_i \, d\alpha_j.$$

It follows that

$$ds^2 = \sum_{k=1}^{3} (dx_k)^2 = \sum_{k,i,j} \frac{\partial x_k}{\partial \alpha_i} \frac{\partial x_k}{\partial \alpha_j} d\alpha_i \, d\alpha_j,$$

or

$$ds^2 = \sum_{i,j=1}^{3} g_{ij} \, d\alpha_i \, d\alpha_j,$$

where

$$g_{ij} = \sum_{k=1}^{3} \frac{\partial x_k}{\partial \alpha_i} \frac{\partial x_k}{\partial \alpha_j}.$$

2. Calculate the metric coefficient g_{ij} for plane polar coordinates $\alpha_1 \equiv r$, $\alpha_2 \equiv \theta$ from the relations

$$x_1 = \alpha_1 \cos \alpha_2, \qquad x_2 = \alpha_1 \sin \alpha_2$$

and

$$g_{ij} = \sum_{k=1}^{3} \frac{\partial x_k}{\partial \alpha_i} \frac{\partial x_k}{\partial \alpha_j}.$$

3. Consider a curvilinear triangle in the undeformed state, with sides directed parallel to the coordinate lines α_1 and α_2. The increments in the coordinates α_j along the sides can be written as $(d\alpha_1, 0, 0)$ and $(0, d\alpha_2, 0)$, while the "hypotenuse" corresponds to coordinate changes $(-d\alpha_1, d\alpha_2, 0)$. Show from (48.5) that after deformation, the sides have lengths $\sqrt{G_{11}}\, d\alpha_1$, $\sqrt{G_{22}}\, d\alpha_2$ and include an angle θ_{12}, while the length of the hypotenuse is $\sqrt{G_{11}\, d\alpha_1^2 + G_{22}\, d\alpha_2^2 - 2G_{12}\, d\alpha_1\, d\alpha_2}$. Use the law of cosines to show that

$$\cos \theta_{12} = \frac{G_{12}}{\sqrt{G_{11} G_{22}}}.$$

49. Torsion of Shafts of Varying Circular Cross Section.

In discussing the torsion by terminal couples of a circular shaft of varying diameter, it is convenient to make use of the cylindrical coordinates r, θ, z introduced in the preceding section. We shall direct the axis of the shaft along the z-axis, and in order to avoid using subscripts, we shall denote the displacements in the radial and tangential directions by u and v respectively. The displacement in the direction of the axis of the shaft will be called w.

It will be recalled that in the case of a uniform circular shaft twisted by terminal couples, the displacement of points in any cross section is in the tangential direction,[1] and that the displacement in the direction of the axis of the shaft vanishes. We shall attempt to solve the torsion problem for a shaft of varying diameter by assuming that in this case, we also have

$$u = w = 0,$$

and then prove that the solution based on this assumption fulfills all conditions of the problem, and hence is the desired one.

On account of the circular symmetry, the tangential displacement v cannot depend on the angle θ, and thus will be function of the variables r and z.

Since the displacements u and w vanish, the formulas (48.15) (with $u_r \equiv u$, $u_\theta \equiv v$, $u_z \equiv w$) give

(49.1) $\quad e_{rr} = e_{\theta\theta} = e_{zz} = e_{rz} = 0, \quad e_{r\theta} = \frac{1}{2}\left(\frac{\partial v}{\partial r} - \frac{v}{r}\right),$

$$e_{\theta z} = \frac{1}{2}\frac{\partial v}{\partial z},$$

[1] See Fig. 20.

and it follows from (48.10) that the corresponding stresses are

(49.2) $\quad \tau_{rr} = \tau_{\theta\theta} = \tau_{zz} = \tau_{rz} = 0, \quad \tau_{r\theta} = \mu\left(\dfrac{\partial v}{\partial r} - \dfrac{v}{r}\right),$

$$\tau_{\theta z} = \mu \dfrac{\partial v}{\partial z}.$$

Inserting these expressions in the three equilibrium equations (48.16) shows that two of them are satisfied identically, and the remaining one requires that

$$\frac{\partial^2 v}{\partial r^2} + \frac{1}{r}\frac{\partial v}{\partial r} - \frac{v}{r^2} + \frac{\partial^2 v}{\partial z^2} = 0.$$

This equation can be rewritten in the form

$$\frac{\partial}{\partial r}\left[r^3 \frac{\partial}{\partial r}\left(\frac{v}{r}\right)\right] + \frac{\partial}{\partial z}\left[r^3 \frac{\partial}{\partial z}\left(\frac{v}{r}\right)\right] = 0,$$

and it follows that there exists a function $F(r, z)$ such that

(49.3) $\quad \dfrac{\partial F}{\partial r} = r^3 \dfrac{\partial}{\partial z}\left(\dfrac{v}{r}\right), \quad \dfrac{\partial F}{\partial z} = -r^3 \dfrac{\partial}{\partial r}\left(\dfrac{v}{r}\right),$

so that

$$\frac{\partial}{\partial z}\left(\frac{v}{r}\right) = \frac{1}{r^3}\frac{\partial F}{\partial r}, \quad \text{and} \quad -\frac{\partial}{\partial r}\left(\frac{v}{r}\right) = \frac{1}{r^3}\frac{\partial F}{\partial z}.$$

Differentiating the first of these equations with respect to r, the second with respect to z, and adding gives the equation on the function $F(r, z)$ in the form

(49.4) $\quad \dfrac{\partial^2 F}{\partial r^2} - \dfrac{3}{r}\dfrac{\partial F}{\partial r} + \dfrac{\partial^2 F}{\partial z^2} = 0.$

The stress components will now be expressed in terms of the function $F(r, z)$. It follows from formulas (49.3) that

$$\frac{1}{r^2}\frac{\partial F}{\partial r} = r\frac{\partial}{\partial z}\left(\frac{v}{r}\right) = \frac{\partial v}{\partial z},$$

$$-\frac{1}{r^2}\frac{\partial F}{\partial z} = r\frac{\partial}{\partial r}\left(\frac{v}{r}\right) = \frac{\partial v}{\partial r} - \frac{v}{r},$$

and comparison of these expressions with the last two of the formulas (49.2) shows that the nonvanishing components of stress are given in terms of the function F by the formulas

(49.5) $$\tau_{\theta z} = \frac{\mu}{r^2} \frac{\partial F}{\partial r}, \qquad \tau_{r\theta} = -\frac{\mu}{r^2} \frac{\partial F}{\partial z}.$$

Since the lateral surface of the shaft is free from external loads, it follows that the resultant shearing stress must be directed along the tangent to the boundary of the axial section. Accordingly, the component of the resultant stress in the direction ν

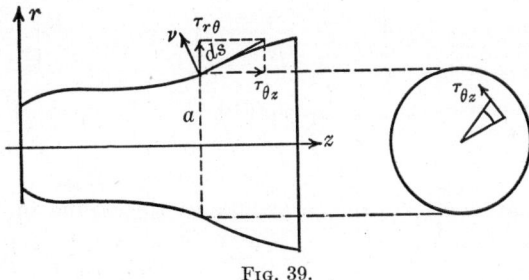

Fig. 39.

normal to this boundary must vanish, and we have the boundary condition

$$\tau_{\theta z} \cos(z, \nu) + \tau_{r\theta} \cos(r, \nu) = 0.$$

But $\cos(z, \nu) = -\dfrac{dr}{ds}$, $\cos(r, \nu) = \dfrac{dz}{ds}$, where ds is the element of arc along the boundary of the axial section (Fig. 39), and we have

$$-\tau_{\theta z} \frac{dr}{ds} + \tau_{r\theta} \frac{dz}{ds} = 0 \quad \text{on the boundary.}$$

Substituting in this expression from (49.5), we get

$$\frac{\partial F}{\partial r} \frac{dr}{ds} + \frac{\partial F}{\partial z} \frac{dz}{ds} = 0,$$

or

$$\frac{dF}{ds} = 0 \quad \text{on the boundary.}$$

Thus, the condition that the lateral surface be free from external loads demands that the function $F(r, z)$ assume a constant value on the boundary of the axial section.

The twisting moment on any cross section whose radius is a is easily computed. Thus,

$$(49.6) \quad M = \int_0^{2\pi}\int_0^a \tau_{\theta z} r^2\, dr\, d\theta = 2\pi \int_0^a r^2 \tau_{\theta z}\, dr = 2\pi\mu \int_0^a \frac{\partial F}{\partial r}\, dr$$
$$= 2\pi\mu[F(a, z) - F(0, z)].$$

The solution of Eq. (49.4) is quite simple for the case of a conical shaft, shown in Fig. 40. It is easily checked that the function

$$(49.7) \quad F(r, z) = c\left\{\frac{z}{(r^2 + z^2)^{\frac{1}{2}}} - \frac{1}{3}\left[\frac{z}{(r^2 + z^2)^{\frac{1}{2}}}\right]^3\right\},$$

where c is a constant, satisfies Eq. (49.4). Moreover, the expression $z/(r^2 + z^2)^{\frac{1}{2}}$ is constant on the lateral surface, since it is equal to the cosine of one-half the vertical angle of the cone, and hence the function $F(r, z)$ in (49.7) assumes a constant value on the lateral surface of the cone.

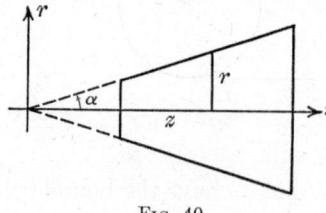

Fig. 40.

The magnitude of the shearing stresses $\tau_{\theta z}$ and $\tau_{r\theta}$ is given at once by the formulas (49.5), and a simple calculation shows that

$$\tau_{\theta z} = -\frac{\mu c r z}{(r^2 + z^2)^{5/2}}, \qquad \tau_{r\theta} = -\frac{\mu c r^2}{(r^2 + z^2)^{5/2}}.$$

The value of the constant c can be determined from (49.6) when the twisting couple in the terminal section is known.

It is possible to solve by this method torsion problems for shafts of elliptical, parabolic, and hyperbolic sections.[1]

---

[1] A review of the literature on the subject will be found in an article by Th. Pöschl in *Zeitschrift für angewandte Mathematik und Mechanik*, vol. 2 (1922), p. 137. In a paper entitled "Some Practically Important Stress-systems in Solids of Revolution," *Proceedings of the Royal Society* (London), ser. A, vol. 180 (1942), pp. 367–396, R. V. Southwell uses the semi-inverse method of Saint-Venant to discuss the torsion of a shaft of varying circular cross section, the torsion and flexure of an incomplete tore, shearing stresses in a toroidal hook, and symmetrical strain in a solid of revolution. Some references to approximate methods of investigating the concentration of stress near fillets of shafts of varying diameters are given in S. Timoshenko,

§50   EXTENSION, TORSION, AND FLEXURE OF BEAMS   209

We shall make use of some particular solutions of Eq. (49.4) in the next section, which is concerned with a study of local effects near the ends of a twisted circular cylinder in which the distribution of stress at one end differs from that demanded by Saint-Venant's theory.

50. Local Effects. It was already noted that Saint-Venant's theory of torsion, discussed in Secs. 33 and 34, imposes a requirement that the distribution of stress over the ends of the cylinder be the same as in every other cross section of the cylinder. The theory developed in these sections yields stresses that at the end sections are statically equivalent to the applied twisting couples. The distribution of these end stresses cannot be arbitrarily specified, since the end couples must be applied in a way demanded by the solution of the torsion problem for the particular section under discussion. If the distribution of stresses over the ends of the cylinder differs from that demanded by the theory, there will be some local irregularities in the neighborhood of the ends, and it is to be expected from Saint-Venant's principle that the effect of local perturbations will not be felt far from the ends. We proceed to investigate the character of local disturbances in a circular cylinder of radius a that is twisted by some prescribed distribution of stresses $\tau_{\theta z}$ over the end $z = 0$.

If we assume a particular solution of Eq. (49.4) in the form $e^{-kz}R(r)$, where $R(r)$ is a function of r alone, then it follows from

---

Theory of Elasticity, p. 281. In addition, the following papers may be consulted:

H. NEUBER, "Elastisch-strenge Lösungen zur Kerbwirkung bei Scheiben und Umdrehungskörpern," *Zeitschrift für angewandte Mathematik und Mechanik*, vol. 13 (1933–1934), pp. 439–442.

A. THUM and W. BAUTZ, "Die Ermittlung von Spannungspitzen in verdrehbeaupruchten Wellen dürch ein electrisches Modell," *Zeitschrift des Vereines deutscher Ingenieure*, vol. 78 (1934), pp. 17–19.

L. S. JACOBSEN, "Torsional Stresses in Shafts having Grooves or Fillets," *Journal of Applied Mechanics*, vol. 2 (1935), pp. A151–A155.

G. SALET, "Détermination des pointes de tension dans les arbres de révolution soumis à torsion au moyen d'un modèle électrique," *Bulletin de l'association technique maritime et aéronautique*, no. 40 (1936), pp. 341–351.

G. SALET, "Amélioration de la méthode de détermination des pointes de tension dans les arbres de révolution soumis à torsion au moyen d'un modèle électrique," *Bulletin de l'association technique maritime et aéronautique*, no. 41 (1937), pp. 295–303.

(49.4) that the function $R(r)$ must satisfy the equation

$$\frac{d^2R}{dr^2} - \frac{3}{r}\frac{dR}{dr} + k^2R = 0.$$

This is a well-known differential equation, and it is easy to show[1] that it is satisfied by the function $R = r^2 J_2(kr)$, where $J_2(kr)$ is the Bessel function of the first kind and of second order. Since the differential equation (49.4) is linear, a linear combination of solutions of the type $A_n r^2 e^{-k_n z} J_2(k_n r)$ will satisfy the equation, and we take the function $F(r, z)$ in the form

$$(50.1) \qquad F(r, z) = \tfrac{1}{4}\alpha r^4 + \sum_{n=1}^{\infty} A_n r^2 e^{-k_n z} J_2(k_n r),$$

where α and A_i are constants. If the constants k_i are chosen to be the successive roots of the equation $J_2(ka) = 0$, then on the boundary of the axial section we have $F(a, z) = \tfrac{1}{4}\alpha a^4 =$ constant, which is the required boundary condition. It is obvious from (49.5) that the distribution of stress in the cylinder corresponding to the choice $A_n = 0$, $(n = 1, 2, \cdots)$ is precisely that required by Saint-Venant's theory.

The expression for the tangential stress $\tau_{\theta z}$ is given by the first of formulas (49.5), and we obtain formally

$$(50.2) \qquad \tau_{\theta z} = \mu \left\{ \alpha r + \sum_{n=1}^{\infty} A_n e^{-k_n z} \left[\frac{2}{r} J_2(k_n r) + J_2'(k_n r) k_n \right] \right\},$$

where the prime denotes the derivative with respect to the argument $k_n r$. But[2]

$$(50.3) \qquad k\left[\frac{2}{kr} J_2(kr) + J_2'(kr) \right] = k J_1(kr),$$

where $J_1(kr)$ stands for the Bessel function of order one, which is known to satisfy the equation

$$(50.4) \qquad \left(\frac{d^2}{dr^2} + \frac{1}{r}\frac{d}{dr} + k^2 - \frac{1}{r^2} \right) J_1(kr) = 0.$$

---

[1] See, for example, I. S. and E. S. Sokolnikoff, Higher Mathematics for Engineers and Physicists, 2d ed., p. 339.

[2] See G. N. Watson's Theory of Bessel Functions or J. M. MacRobert's Treatise on Bessel Functions.

Substituting (50.3) in (50.2) and setting $z = 0$ gives the expression for the distribution of stresses $\tau_{\theta z}$ over the end $z = 0$ of the cylinder,

(50.5) $\qquad (\tau_{\theta z})_{z=0} = \mu \left\{ \alpha r + \sum_{n=1}^{\infty} A_n k_n J_1(k_n r) \right\}.$

We proceed to calculate the torque M acting on the end $z = 0$ of the cylinder. Now

(50.6) $\quad M = \int_0^a (2\pi r^2 (\tau_{\theta z})_{z=0}\, dr$

$\qquad = 2\pi \mu \left[\int_0^a \alpha r^3\, dr + \sum_{n=1}^{\infty} A_n k_n \int_0^a r^2 J_1(k_n r)\, dr \right].$

and it is easy to show that $\int_0^a r^2 J_1(k_n r)\, dr = 0$. We note from (50.4) that

$$J_1(kr) = -\frac{1}{k^2} \left(\frac{d^2}{dr^2} + \frac{1}{r}\frac{d}{dr} - \frac{1}{r^2} \right) J_1(kr),$$

so that

$$\int_0^a r^2 J_1(k_n r)\, dr = -\frac{1}{k_n^2} \int_0^a \left(r^2 \frac{d^2}{dr^2} + r\frac{d}{dr} - 1 \right) J_1(k_n r)\, dr$$

$$= -\frac{1}{k_n^2} \int_0^a \frac{d}{dr}\left[r^2 \frac{dJ_1(k_n r)}{dr} - r J_1(k_n r) \right] dr$$

$$= -\frac{1}{k_n^2} [a^2 J_1'(k_n a) k_n - a J_1(k_n a)]$$

$$= -\frac{a^2}{k_n} \left[J_1'(k_n a) - \frac{1}{k_n a} J_1(k_n a) \right].$$

But it is known that

$$J_2(kr) = -J_1'(kr) + \frac{1}{kr} J_1(kr),$$

and if the numbers k_n are the roots of the equation $J_2(ka) = 0$, then $J_1'(k_n a) - \frac{1}{k_n a} J_1(k_n a) = 0$. Thus, the integrals in (50.6) involving Bessel's functions vanish, and we get

$$M = \frac{\pi a^4}{2} \mu \alpha,$$

which is the same expression for the moment M as previously obtained in Sec. 33.

Since a suitably restricted function of r defined in the interval $(0, a)$ can be expanded in a series of the form $\sum_{n=0}^{\infty} a_n J_1(k_n r)$, we see from (50.5) that we can obtain the solution of the torsion problem that corresponds to distribution of stress $(\tau_{\theta z})_0$ over the end $z = 0$, where $(\tau_{\theta z})_0$ is a prescribed function of r. It is obvious from (50.2) that the effect of the terms involving the factors $e^{-k_n z}$ diminishes with an increase in z.

Instead of prescribing the distribution of stress over one end of the cylinder, one may impose a requirement that one of the sections of the twisted cylinder remain plane.[1]

References for Collateral Reading

A. E. H. LOVE: A Treatise on the Mathematical Theory of Elasticity, Cambridge University Press, London, Secs. 226A, 226B.

S. TIMOSHENKO: Theory of Elasticity, McGraw-Hill Book Company, Inc., New York, Secs. 86, 87.

51. Torsion of Nonisotropic Beams. Many important structural materials (wood, for example) are definitely nonisotropic, and the formulation of the torsion problem given above is not applicable to beams made of such materials. Fortunately, the

---

[1] A solution by energy methods of such a torsion problem for a beam of elliptical section was found by A. Föppl (1920) and is given in A. and L. Föppl, Drang and Zwang, vol. 2, Sec. 77. The beam of rectangular section was considered by S. Timoshenko, *Proceedings of the London Mathematical Society*, vol. 20 (1922), p. 389. Energy methods are also used by N. V. Zvolinskij in "Angenäherte Lösung der Torsionsaufgabe für einen elastischen zylindrischen Stab mit einem nicht verwolbten Querschnitt," *Bulletin de l'académie des sciences de l'U.R.S.S., classe des sciences mathématiques et naturelles*, no. 8 (1939), pp. 91–100 (in Russian). The problem of flexure of such a beam has been treated by R. Sonntag in "Über Biegung bei verhinderter Querschnittskrümmung," *Ingenieur Archiv*, vol. 4 (1944), pp. 415–420.

The effect of local stresses corresponding to different modes of applying torsional couples to a circular cylinder has been discussed in detail by Wolf and by Deimel [K. WOLF, *Sitzungsberichte der Akademie der Wissenschaften in Wien*, vol. 125 (1916), p. 1149; R. F. DEIMEL, "The Torsion of a Circular Cylinder," *Proceedings of the National Academy of Sciences of the United States of America*, vol. 21 (1935), pp. 637–642].

§51   EXTENSION, TORSION, AND FLEXURE OF BEAMS   213

extension of Saint-Venant's theory of torsion to nonisotropic media is not a difficult matter when the normal section of the beam is a plane of elastic symmetry of the medium. We shall confine our discussion of the torsion theory to beams made of nonisotropic materials that have three mutually orthogonal planes of elastic symmetry, and we shall assume that the axis of the beam is perpendicular to one of these planes.

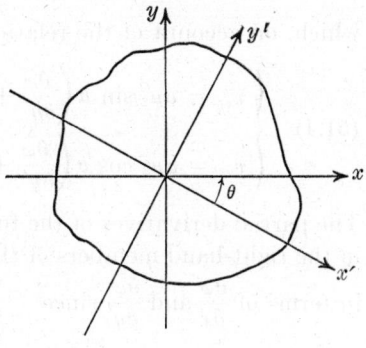

Fig. 41.

As in the isotropic case, we let the axis of the beam coincide with the z-axis, and we choose the x and y axes in one end of the beam. The longitudinal planes of elastic symmetry will be denoted by $x'z$ and $y'z$, and the shear moduli associated with the axes z and x' and z and y' are denoted by μ_1 and μ_2 respectively (Fig. 41).

The components of shear $\tau_{x'z}$ and $\tau_{y'z}$ are connected with the shearing strains $e_{x'z}$ and $e_{y'z}$ by the formulas

(51.1) $$\tau_{x'z} = 2\mu_1 e_{x'z}, \qquad \tau_{y'z} = 2\mu_2 e_{y'z},$$

where

(51.2) $$e_{x'z} = \frac{1}{2}\left(\frac{\partial u}{\partial z} + \frac{\partial w}{\partial x'}\right), \qquad e_{y'z} = \frac{1}{2}\left(\frac{\partial w}{\partial y'} + \frac{\partial v}{\partial z}\right),$$

and u, v, and w are the components of displacement in the directions of the x'-, y'-, and z-axes, respectively.

We assume, as in the isotropic case,[1] that the displacements u, v and w are given by the formulas

(51.3) $$u = -\alpha z y', \qquad v = \alpha z x', \qquad w = \alpha \varphi(x', y'),$$

where α is the angle of twist per unit length of the bar, and $\varphi(x', y')$ is the torsion function associated with this problem.

If we let θ denote the angle between the axes x and x', then the expressions for the nonvanishing components of stress, τ_{xz}

---

[1] See Sec. 34.

and τ_{yz}, are related to the components $\tau_{x'z}$ and $\tau_{y'z}$ by the formulas[1]

$$\tau_{xz} = \tau_{y'z} \sin \theta + \tau_{x'z} \cos \theta,$$
$$\tau_{yz} = \tau_{y'z} \cos \theta - \tau_{x'z} \sin \theta,$$

which, on account of the relations (51.1) and (51.3), become

(51.4)
$$\begin{cases} \tau_{xz} = \alpha\mu_2 \sin \theta \left(\dfrac{\partial \varphi}{\partial y'} + x'\right) + \alpha\mu_1 \cos \theta \left(\dfrac{\partial \varphi}{\partial x'} - y'\right), \\ \tau_{yz} = \alpha\mu_2 \cos \theta \left(\dfrac{\partial \varphi}{\partial y'} + x'\right) - \alpha\mu_1 \sin \theta \left(\dfrac{\partial \varphi}{\partial x'} - y'\right). \end{cases}$$

The partial derivatives of the torsion function $\varphi(x', y')$ appearing in the right-hand members of these expressions can be calculated in terms of $\dfrac{\partial \varphi}{\partial x}$ and $\dfrac{\partial \varphi}{\partial y}$, since

$$x = x' \cos \theta + y' \sin \theta,$$
$$y = -x' \sin \theta + y' \cos \theta,$$

We have

(51.5)
$$\begin{cases} \dfrac{\partial \varphi}{\partial x'} = \dfrac{\partial \varphi}{\partial x} \cos \theta - \dfrac{\partial \varphi}{\partial y} \sin \theta, \\ \dfrac{\partial \varphi}{\partial y'} = \dfrac{\partial \varphi}{\partial x} \sin \theta + \dfrac{\partial \varphi}{\partial y} \cos \theta. \end{cases}$$

Inserting the values from (51.5) in (51.4) and introducing the abbreviations

$$A = \mu_2 \sin^2 \theta + \mu_1 \cos^2 \theta,$$
$$B = (\mu_2 - \mu_1) \sin \theta \cos \theta,$$
$$C = \mu_2 \cos^2 \theta + \mu_1 \sin^2 \theta,$$

we get

(51.6)
$$\begin{cases} \tau_{xz} = \alpha \left(A \dfrac{\partial \varphi}{\partial x} + B \dfrac{\partial \varphi}{\partial y} + Bx - Ay\right), \\ \tau_{yz} = \alpha \left(B \dfrac{\partial \varphi}{\partial x} + C \dfrac{\partial \varphi}{\partial y} + Cx - By\right). \end{cases}$$

Since $\tau_{xx} = \tau_{yy} = \tau_{zz} = \tau_{xy} = 0$, and τ_{xz} and τ_{yz} are independent of z, the first two of the equilibrium equations (15.3) are identically satisfied, and the third one gives the equation

(51.7) $$A \frac{\partial^2 \varphi}{\partial x^2} + 2B \frac{\partial^2 \varphi}{\partial x \partial y} + C \frac{\partial^2 \varphi}{\partial y^2} = 0.$$

[1] Note formulas (16.5).

Thus, in this case, the torsion function φ no longer satisfies Laplace's equation.

Let the boundary C of the cross section have the equation $f(x, y) = 0$; then the components $\cos(x, \nu)$ and $\cos(y, \nu)$ of the normal ν to the boundary C are proportional to $\dfrac{\partial f}{\partial x}$ and $\dfrac{\partial f}{\partial y}$, respectively, and we can write the boundary condition

$$\tau_{xz} \cos(x, \nu) + \tau_{yz} \cos(y, \nu) = 0 \qquad \text{on } C$$

in the form

(51.8) $\left(A\dfrac{\partial \varphi}{\partial x} + B\dfrac{\partial \varphi}{\partial y}\right)\dfrac{\partial f}{\partial x} + \left(B\dfrac{\partial \varphi}{\partial x} + C\dfrac{\partial \varphi}{\partial y}\right)\dfrac{\partial f}{\partial y}$
$= (Ay - Bx)\dfrac{\partial f}{\partial x} + (By - Cx)\dfrac{\partial f}{\partial y} \qquad \text{on } C.$

Equation (51.7) and the boundary condition (51.8) can be simplified by introducing new independent variables ξ and η, defined by the formulas

(51.9) $\qquad \xi = \dfrac{\sqrt{\mu_1\mu_2}}{A} x, \qquad \eta = y - \dfrac{B}{A} x.$

A simple calculation shows that Eq. (51.7) becomes

(51.10) $\qquad \dfrac{\partial^2 \varphi}{\partial \xi^2} + \dfrac{\partial^2 \varphi}{\partial \eta^2} = 0,$

while the equation of the boundary $f(x, y) = 0$ is changed into

(51.11) $\qquad F(\xi, \eta) = 0.$

Making the corresponding change of variables in the boundary condition (51.8) gives

(51.12) $\qquad \dfrac{\sqrt{\mu_1\mu_2}}{A}\left(\dfrac{\partial \varphi}{\partial \xi}\dfrac{\partial F}{\partial \xi} + \dfrac{\partial \varphi}{\partial \eta}\dfrac{\partial F}{\partial \eta}\right) = \eta\dfrac{\partial F}{\partial \xi} - \xi\dfrac{\partial F}{\partial \eta} \qquad \text{on } C',$

where C' is the transformed boundary defined by (51.11).

Finally, if we set

$$\varphi'(\xi, \eta) = \dfrac{\sqrt{\mu_1\mu_2}}{A} \varphi(\xi, \eta),$$

then

(51.13)
$$\frac{\partial^2 \varphi'}{\partial \xi^2} + \frac{\partial^2 \varphi'}{\partial \eta^2} = 0,$$

and (51.12) becomes

$$\frac{\partial \varphi'}{\partial \xi}\frac{\partial F}{\partial \xi} + \frac{\partial \varphi'}{\partial \eta}\frac{\partial F}{\partial \eta} = \eta \frac{\partial F}{\partial \xi} - \xi \frac{\partial F}{\partial \eta} \qquad \text{on } C'$$

or

(51.14) $$\frac{\partial \varphi'}{\partial \nu} = \eta \cos(\xi, \nu) - \xi \cos(\eta, \nu) \qquad \text{on } C',$$

where ν is the normal to the boundary C'.

The boundary condition (51.14) is precisely of the same form as that appearing in a study of torsion of isotropic cylinders; hence the solution of the torsion problem for a cylinder of nonisotropic material (having three orthogonal planes of elastic symmetry), whose cross section is C, is reduced to the solution of the torsion problem for an isotropic bar whose cross section has a different boundary C', defined by Eq. (51.11).

It is not difficult to calculate the torsional rigidity of a nonisotropic cylinder in terms of the torsional rigidity of the corresponding isotropic cylinder. Substituting from (51.6) in the expression for the couple M, we obtain

$$M = \int\int_R (\tau_{yz} x - \tau_{xz} y)\, dx\, dy$$
$$= \frac{\alpha A}{\sqrt{\mu_1 \mu_2}} \int\int_{R'} \left[\sqrt{\mu_1 \mu_2} \left(\xi \frac{\partial \varphi}{\partial \eta} - \eta \frac{\partial \varphi}{\partial \xi} \right) + A(\xi^2 + \eta^2) \right] d\xi\, d\eta,$$

where the integration now extends over the region R' bounded by the curve C'. Recalling that $\varphi' \equiv \dfrac{\sqrt{\mu_1 \mu_2}}{A} \varphi$, we have

$$M = \frac{\alpha A^2}{\sqrt{\mu_1 \mu_2}} \int\int_{R'} \left[\xi \frac{\partial \varphi'}{\partial \eta} - \eta \frac{\partial \varphi'}{\partial \xi} + \xi^2 + \eta^2 \right] d\xi\, d\eta,$$

and since

$$M = \alpha D,$$

where D is the torsional rigidity, we see that the torsional rigidity of a nonisotropic cylinder can be deduced from the torsional rigidity of the isotropic cylinder obtained from the nonisotropic one by a homogeneous deformation (51.9).

We conclude this formulation of the torsion problem for a nonisotropic prism by remarking that the transformation (51.9) changes the boundary of an ellipse

$$\frac{x^2}{a^2} + \frac{y^2}{b^2} = 1$$

into another ellipse, and since the solution of the torsion problem for an isotropic elliptical cylinder is known, we can write down at once the solution of the corresponding problem for a nonisotropic elliptical cylinder.

The transformation (51.9), in general, carries a rectangle into a parallelogram, and hence the solution of the torsion problem for a nonisotropic rectangular beam is not covered by the discussion contained in Sec. 38, unless the x'-axis coincides with the x-axis. If these axes coincide, then $B = 0$, and the rectangle will be transformed into another rectangle of different length. The solution corresponding to this case is written out in Love's Treatise, on page 325.

References for Collateral Reading

G. W. TRAYER and H. W. MARCH: National Advisory Committee for Aeronautics, *Report* 334, pp. 37–44.

A. E. H. LOVE: Treatise on the Mathematical Theory of Elasticity, Cambridge University Press, London, Sec. 226.

52. Flexure of Beams by Terminal Loads. Let a cantilever beam of uniform cross section have one end ($z = 0$) fixed and the other end ($z = l$) loaded by some distribution of forces that is statically equivalent to a single force (W_x, W_y, 0) lying in the plane $z = l$ and acting at the *load point* (x_0, y_0, l). The z-axis is taken along the central line of the beam, while the x- and y-axes are any orthogonal axes intersecting at the centroid of the end $z = 0$ (Fig. 42). The lateral surface of the beam is free from external forces, and the body forces are assumed to vanish.

We shall follow the semi-inverse method of Saint-Venant; that is, we put

(52.1) $$\tau_{xx} = \tau_{xy} = \tau_{yy} = 0.$$

The functions τ_{zx}, τ_{zy}, and τ_{zz} will be so chosen that the equations of equilibrium and compatibility, as well as the boundary conditions, are satisfied.

In writing an expression for τ_{zz}, we shall be guided by an expression for the bending moment M_y that would be produced by the load W_x acting alone. In any cross section z units distant from the fixed end, one would have

(52.2) $$M_y = W_x(l - z),$$

so that the stress distribution (due to W_x alone) in this section would have to be statically equivalent to the moment M_y and to the resultant force W_x. Now in the discussion of the problem

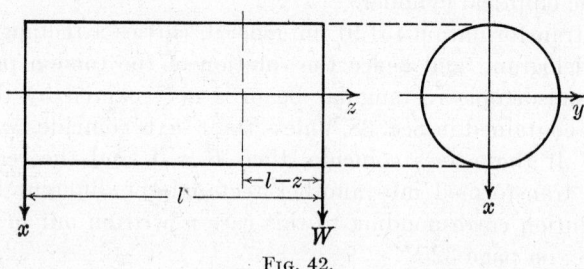

FIG. 42.

of bending of beams by couples applied at the ends,[1] it was found that the stress τ_{zz}, distributed according to the linear relation

(52.3) $$\tau_{zz} = -\frac{M_y}{I_y} x, \quad \text{(bending by couples)},$$

with

$$I_y = \iint_R x^2 \, dx \, dy,$$

is statically equivalent to a couple of moment M_y. Equations (52.2) and (52.3) suggest that we try to satisfy the conditions of the present problem by assuming

(52.4) $$\tau_{zz} = -E(l - z)(K_x x + K_y y),$$

where the constants K_x, K_y are to be determined from the conditions

(52.5) $$\iint_R \tau_{zx} \, dx \, dy = W_x, \quad \iint_R \tau_{zy} \, dx \, dy = W_y.$$

---

[1] See Sec. 32. Note that the coordinate axes there were taken to be the principal axes of the cross section, while the choice of axes here is not restricted.

Substituting from (52.1) and (52.4) in the equations of equilibrium (29.1), we get

(52.6)
$$\begin{cases} \dfrac{\partial \tau_{zx}}{\partial z} = 0, \quad \dfrac{\partial \tau_{zy}}{\partial z} = 0, \\ \dfrac{\partial \tau_{zx}}{\partial x} + \dfrac{\partial \tau_{zy}}{\partial y} + E(K_x x + K_y y) = 0. \end{cases}$$

It follows from the first two of Eqs. (52.6) that the shear components τ_{zx} and τ_{zy} have the same value in all cross sections of the beam, while the third equation can be rearranged to read

$$\frac{\partial}{\partial x}\left(\tau_{zx} + \frac{1}{2} EK_x x^2\right) + \frac{\partial}{\partial y}\left(\tau_{zy} + \frac{1}{2} EK_y y^2\right) = 0.$$

As this equation is of the form

$$\frac{\partial}{\partial x}\left(\frac{\partial F}{\partial y}\right) + \frac{\partial}{\partial y}\left(-\frac{\partial F}{\partial x}\right) = 0,$$

it is evident that there exists a function $F(x, y)$ such that

(52.7)
$$\begin{cases} \tau_{zx} = \dfrac{\partial F}{\partial y} - \dfrac{1}{2} EK_x x^2, \\ \tau_{zy} = -\dfrac{\partial F}{\partial x} - \dfrac{1}{2} EK_y y^2. \end{cases}$$

The conditions to be satisfied by the function $F(x, y)$ can be determined from the Beltrami-Michell compatibility equations (24.15), which reduce in this case to

$$\nabla^2 \tau_{zx} + \frac{EK_x}{1+\sigma} = 0, \quad \nabla^2 \tau_{zy} + \frac{EK_y}{1+\sigma} = 0.$$

Substituting from (52.7) in these equations, we see that the latter will be fulfilled if

$$\frac{\partial}{\partial y}(\nabla^2 F) = 2\mu\sigma K_x, \quad \frac{\partial}{\partial x}(\nabla^2 F) = -2\mu\sigma K_y,$$

from which it follows that

(52.8) $\quad \nabla^2 F(x, y) = -2\mu\sigma K_y x + 2\mu\sigma K_x y - 2\mu\alpha.$

The physical significance of the constant of integration $-2\mu\alpha$ will be discovered presently. It is not difficult to obtain a particular integral of (52.8) in the form of a polynomial, and it is readily verified that the solution of (52.8) is

(52.9) $\quad F(x, y) = f(x, y) - \tfrac{1}{3}\mu\sigma(K_y x^3 - K_x y^3) - \tfrac{1}{2}\mu\alpha(x^2 + y^2),$

where $f(x, y)$ is a harmonic function.

It will prove advantageous to write the stresses τ_{zx}, τ_{zy} not in terms of $f(x, y)$ but in terms of its harmonic conjugate $g(x, y)$, where

(52.10) $\quad \dfrac{\partial f}{\partial x} = \dfrac{\partial g}{\partial y}, \quad \dfrac{\partial f}{\partial y} = -\dfrac{\partial g}{\partial x}.$

Equations (52.7) can now be written, with the help of (52.9) and (52.10), as

(52.11) $\quad \begin{cases} \tau_{zx} = -\dfrac{\partial g}{\partial x} - \mu\alpha y + \mu\sigma K_x y^2 - \dfrac{1}{2} E K_x x^2, \\ \tau_{zy} = -\dfrac{\partial g}{\partial y} + \mu\alpha x + \mu\sigma K_y x^2 - \dfrac{1}{2} E K_y y^2. \end{cases}$

The constant of integration $-2\mu\alpha$ in (52.8) can easily be interpreted physically. Each element of area of a cross section is rotated in its own plane through an angle [see (7.5)]

$$\omega = \frac{1}{2}\left(\frac{\partial v}{\partial x} - \frac{\partial u}{\partial y}\right).$$

The *local twist* at a point (x, y) of a cross section is defined as

$$\frac{\partial \omega}{\partial z} = \frac{1}{2}\left(\frac{\partial^2 v}{\partial z\,\partial x} - \frac{\partial^2 u}{\partial z\,\partial y}\right) = \frac{\partial e_{zy}}{\partial x} - \frac{\partial e_{zx}}{\partial y}$$
$$= \frac{1}{2\mu}\left(\frac{\partial \tau_{zy}}{\partial x} - \frac{\partial \tau_{zx}}{\partial y}\right).$$

Substituting the values of the shear stresses from (52.11), one gets

$$\frac{\partial \omega}{\partial z} = \alpha + \sigma(K_y x - K_x y).$$

The mean value of the local twist over the section (or, equally well, the value of the local twist at the centroid of the section) is just the constant α. Thus, we see that the terms in (52.11) that involve α represent a twist of the beam, and, indeed, the

§52 EXTENSION, TORSION, AND FLEXURE OF BEAMS

terms $-\mu\alpha y$ and $\mu\alpha x$ in these expressions also appear in the solution of the torsion problem (see Sec. 34). In the latter case, one has

$$\tau_{zx} = \mu\alpha\left(\frac{\partial\varphi}{\partial x} - y\right), \qquad \tau_{zy} = \mu\alpha\left(\frac{\partial\varphi}{\partial y} + x\right), \qquad \text{(pure torsion)}.$$

We are thus led to introduce the torsion function $\varphi(x, y)$ into the flexure problem by writing

(52.12) $\quad g(x, y) = -\mu\alpha\varphi(x, y) + \mu[K_x\varphi_1(x, y) + K_y\varphi_2(x, y)],$

where $\varphi(x, y)$, $\varphi_1(x, y)$, and $\varphi_2(x, y)$ are harmonic functions. We can now write

(52.13)
$$\begin{cases} \tau_{zx} = \mu\alpha\left(\dfrac{\partial\varphi}{\partial x} - y\right) + \mu K_x\left[\dfrac{\partial\varphi_1}{\partial x} - x^2 - \sigma(x^2 - y^2)\right] + \mu K_y\dfrac{\partial\varphi_2}{\partial x}, \\ \tau_{zy} = \mu\alpha\left(\dfrac{\partial\varphi}{\partial y} + x\right) + \mu K_y\left[\dfrac{\partial\varphi_2}{\partial y} - y^2 - \sigma(y^2 - x^2)\right] + \mu K_x\dfrac{\partial\varphi_1}{\partial y}. \end{cases}$$

The boundary conditions on the functions φ_1 and φ_2 may be derived from the relation

(52.14) $\quad \tau_{zx}\cos(x, \nu) + \tau_{zy}\cos(y, \nu) = 0,$

which expresses the vanishing of external force on the lateral surface of the cylinder, and from the boundary condition on the torsion function φ [see (34.6)],

(52.15) $\quad \dfrac{d\varphi}{d\nu} = y\cos(x, \nu) - x\cos(y, \nu).$

Inserting Eqs. (52.13) in (52.14) and taking account of (52.15) yields

$$K_x\frac{d\varphi_1}{d\nu} + K_y\frac{d\varphi_2}{d\nu} = K_x[(1+\sigma)x^2 - \sigma y^2]\cos(x, \nu) \\ + K_y[(1+\sigma)y^2 - \sigma x^2]\cos(y, \nu) \quad \text{on } C,$$

and this will be satisfied if the functions φ_1 and φ_2 are subject to the conditions

(52.16)
$$\begin{cases} \dfrac{d\varphi_1}{d\nu} = [(1+\sigma)x^2 - \sigma y^2]\cos(x, \nu) & \text{on } C, \\ \dfrac{d\varphi_2}{d\nu} = [(1+\sigma)y^2 - \sigma x^2]\cos(y, \nu) & \text{on } C. \end{cases}$$

The flexure problem has thus been reduced to the task of finding functions harmonic within the region R of the cross section and whose normal derivatives are prescribed on the boundary C; that is, we have been led to the problem of Neumann. In order to see that the condition of the existence of a solution of this problem is fulfilled, we observe that

$$\int_C \frac{d\varphi_1}{d\nu} ds = \int_C [(1 + \sigma)x^2 - \sigma y^2] dy$$

$$= 2(1 + \sigma) \int\!\!\int_R x \, dx \, dy = 0,$$

$$\int_C \frac{d\varphi_2}{d\nu} ds = -\int_C [(1 + \sigma)y^2 - \sigma x^2] dx$$

$$= 2(1 + \sigma) \int\!\!\int_R y \, dx \, dy = 0,$$

since the origin is at the centroid of the section.

In considering the torsion of a beam by couples, it was seen that the solution could be made to depend upon either a problem of Neumann, that is, the problem of finding a function $\varphi(x, y)$, harmonic in R and such that (in the case of torsion)

$$\frac{d\varphi}{d\nu} = y \cos(x, \nu) - x \cos(y, \nu) \qquad \text{on } C,$$

or upon a problem of Dirichlet, with

$$\psi = \tfrac{1}{2}(x^2 + y^2) + \text{const.} \qquad \text{on } C.$$

The torsion functions φ and ψ are harmonic conjugates; that is, $\varphi + i\psi$ is an analytic function of $x + iy$. The flexure problem may also be reduced to a problem of Dirichlet by introducing the harmonic functions ψ_1, ψ_2, conjugate to φ_1 and φ_2, respectively. Then

$$\frac{\partial \varphi_i}{\partial x} = \frac{\partial \psi_i}{\partial y}, \qquad \frac{\partial \varphi_i}{\partial y} = -\frac{\partial \psi_i}{\partial x}, \qquad \frac{d\varphi_i}{d\nu} = \frac{d\psi_i}{ds}, \qquad (i = 1, 2),$$

and the boundary conditions (52.16) can be written

$$\frac{d\psi_1}{ds} = [(1 + \sigma)x^2 - \sigma y^2]\frac{dy}{ds}, \qquad \frac{d\psi_2}{ds} = -[(1 + \sigma)y^2 - \sigma x^2]\frac{dx}{ds},$$

§52 EXTENSION, TORSION, AND FLEXURE OF BEAMS

or

(52.17)
$$\begin{cases} \psi_1 = -\tfrac{1}{3}\sigma y^3 + (1+\sigma)\int_{(x_0,y_0)}^{(x,y)} x^2\, dy + \text{const. on } C, \\ \psi_2 = \tfrac{1}{3}\sigma x^3 - (1+\sigma)\int_{(x_0,y_0)}^{(x,y)} y^2\, dx + \text{const. on } C, \end{cases}$$

where the line integrals are to be evaluated along the contour C.

We turn now to the determination of the constants K_x and K_y. Since the resultant of the stresses τ_{zx} acting over any cross section must equal the component W_x of the applied load, we have

$$W_x = \int\int_R \tau_{zx}\, dx\, dy,$$

or, substituting from (52.13),

(52.18) $\quad W_x = \mu\alpha \int\int_R \dfrac{\partial\varphi}{\partial x}\, dx\, dy + \mu K_x \int\int_R \dfrac{\partial\varphi_1}{\partial x}\, dx\, dy$

$\qquad\qquad\qquad + \mu K_y \int\int_R \dfrac{\partial\varphi_2}{\partial x}\, dx\, dy$

$\qquad\qquad\qquad + \mu K_x[-(1+\sigma)I_y + \sigma I_x],$

where

$$I_x = \int\int_R y^2\, dx\, dy, \qquad I_y = \int\int_R x^2\, dx\, dy.$$

Now if Φ is any harmonic function, then[1]

$$\int\int_R \dfrac{\partial\Phi}{\partial x}\, dx\, dy = \int_C x\, \dfrac{d\Phi}{d\nu}\, ds.$$

With the aid of this identity and the boundary conditions (52.15) and (52.16), Eq. (52.18) becomes

$W_x = \mu\alpha \int_C (xy\, dy + x^2\, dx) + \mu K_x \int_C [(1+\sigma)x^3 - \sigma xy^2]\, dy$

$\qquad + \mu K_y \int_C [-(1+\sigma)xy^2 + \sigma x^3]\, dx + \mu K_x[-(1+\sigma)I_y + \sigma I_x].$

---

[1] For

$$\int\int_R \dfrac{\partial\Phi}{\partial x}\, dx\, dy = \int\int_R \left[\dfrac{\partial}{\partial x}\left(x\dfrac{\partial\Phi}{\partial x}\right) + \dfrac{\partial}{\partial y}\left(x\dfrac{\partial\Phi}{\partial y}\right)\right] dx\, dy$$

$$= \int_C \left(-x\dfrac{\partial\Phi}{\partial y}\, dx + x\dfrac{\partial\Phi}{\partial y}\, dy\right) = \int_C x\, \dfrac{d\Phi}{d\nu}\, ds.$$

Similarly, it follows that

$$\int\int_R \dfrac{\partial\Phi}{\partial y}\, dx\, dy = \int_C y\, \dfrac{d\Phi}{d\nu}\, ds.$$

Upon applying Green's Theorem and recalling that
$$E = 2\mu(1 + \sigma),$$
the last equation can be written as

(52.19) $\qquad W_x = E(K_x I_y + K_y I_{xy}),$

where
$$I_{xy} = \iint_R xy\, dx\, dy$$
is the product of inertia of the section. Similarly, from
$$W_y = \iint_R \tau_{zy}\, dx\, dy,$$
it follows that

(52.20) $\qquad W_y = E(K_y I_x + K_x I_{xy}).$

Equations (52.19) and (52.20) can be solved for K_x, K_y to give

(52.21) $\qquad \begin{cases} EK_x = \dfrac{I_x W_x - I_{xy} W_y}{I_x I_y - I_{xy}^2}, \\ EK_y = \dfrac{I_y W_y - I_{xy} W_x}{I_x I_y - I_{xy}^2}. \end{cases}$

The form of the expressions (52.13) for the shear stresses τ_{zx} and τ_{zy} suggests that the general flexure problem be resolved into two simpler problems:

1. A simple flexure problem in which α, the mean local twist, is set equal to zero. The position of the load point (x_0, y_0, l) corresponding to this stress distribution is then determined by the condition
$$\iint_R (x\tau_{zy} - y\tau_{zx})\, dx\, dy = x_0 W_y - y_0 W_x,$$
which must hold for arbitrary W_x and W_y. The load point corresponding to $\alpha = 0$ is called the *center of flexure* and is denoted by (x_{cf}, y_{cf}, l);
2. A torsion problem with $K_x = K_y = 0$ and with a twist α due to a couple of moment
$$W_y(x_0 - x_{cf}) - W_x(y_0 - y_{cf}).$$

We can thus think of the load **W**, acting at the load point, as being replaced by an equal load at the center of flexure and a

§52 EXTENSION, TORSION, AND FLEXURE OF BEAMS   225

couple that produces a twist α. The solution of the general flexure problem is then got by superposition of the solutions of these two simpler problems. As the torsion problem has already been considered, we proceed with the simpler flexure problem ($\alpha = 0$).

The solution of the simple flexure problem is given by the harmonic functions φ_1 and φ_2, which satisfy the conditions (52.16) on the boundary. Simpler boundary conditions can be realized by subdividing the problem once more. We define the harmonic functions φ_{11}, φ_{12}, φ_{21}, φ_{22} by the relations

(52.22)
$$\begin{cases} \varphi_1 = (1 + \sigma)\varphi_{11} - \sigma\varphi_{12}, \\ \varphi_2 = (1 + \sigma)\varphi_{22} + \sigma\varphi_{21}. \end{cases}$$

Equations (52.16) now become

(52.23)
$$\begin{cases} (1 + \sigma)\dfrac{d\varphi_{11}}{d\nu} - \sigma\dfrac{d\varphi_{12}}{d\nu} = (1 + \sigma)x^2 \cos(x, \nu) - \sigma y^2 \cos(x, \nu), \\ (1 + \sigma)\dfrac{d\varphi_{22}}{d\nu} + \sigma\dfrac{d\varphi_{21}}{d\nu} = (1 + \sigma)y^2 \cos(y, \nu) - \sigma x^2 \cos(y, \nu). \end{cases}$$

We are at liberty to prescribe arbitrary boundary conditions on the individual functions φ_{ij}, subject only to the restriction that the relations (52.23) be satisfied on C. Boundary conditions that are simple in form and independent of the elastic constants of the material will be realized if it is required that the functions φ_{ij} satisfy conditions

(52.24)
$$\begin{cases} \dfrac{d\varphi_{11}}{d\nu} = x^2 \cos(x, \nu) = \dfrac{d}{d\nu}\left(\dfrac{1}{3}x^3\right), \\ \dfrac{d\varphi_{22}}{d\nu} = y^2 \cos(y, \nu) = \dfrac{d}{d\nu}\left(\dfrac{1}{3}y^3\right), \\ \dfrac{d\varphi_{12}}{d\nu} = y^2 \cos(x, \nu) = \dfrac{d}{ds}\left(\dfrac{1}{3}y^3\right), \\ \dfrac{d\varphi_{21}}{d\nu} = -x^2 \cos(y, \nu) = \dfrac{d}{ds}\left(\dfrac{1}{3}x^3\right). \end{cases}$$

We introduce the conjugate harmonic functions ψ_{12}, ψ_{21} with

$$\dfrac{\partial \varphi_{12}}{\partial x} = \dfrac{\partial \psi_{12}}{\partial y}, \qquad \dfrac{\partial \varphi_{12}}{\partial y} = -\dfrac{\partial \psi_{12}}{\partial x},$$

$$\dfrac{\partial \varphi_{21}}{\partial x} = \dfrac{\partial \psi_{21}}{\partial y}, \qquad \dfrac{\partial \varphi_{21}}{\partial y} = -\dfrac{\partial \psi_{21}}{\partial x},$$

and in terms of these functions the last two boundary conditions can be written as

$$\frac{d\psi_{12}}{ds} = \frac{d}{ds}\left(\frac{1}{3}y^3\right), \qquad \frac{d\psi_{21}}{ds} = \frac{d}{ds}\left(\frac{1}{3}x^3\right),$$

or

(52.25) $\quad \psi_{12} = \frac{1}{3}y^3 + \text{const.}, \qquad \psi_{21} = \frac{1}{3}x^3 + \text{const.} \quad \text{on } C.$

The stress distribution over any cross section is easily seen to be statically equivalent to the load $(W_x, W_y, 0)$; that is,

$$\iint_R \tau_{zx}\,dx\,dy = W_x, \quad \iint_R \tau_{zy}\,dx\,dy = W_y, \quad \iint_R \tau_{zz}\,dx\,dy = 0.$$

The first two equations are satisfied by virtue of our choice of the constants K_x and K_y, while the third follows immediately upon substituting the value of τ_{zz} from (52.4) and recalling that the z-axis passes through the centroids of the cross sections. The torsional moment M_z of the external load is $x_0 W_y - y_0 W_x$, and this is resolved into the two parts

$$\iint_R (x\tau_{zy} - y\tau_{zx})\,dx\,dy = x_{cf} W_y - y_{cf} W_x,$$

(simple flexure, $\alpha = 0$),

$$\iint_R (x\tau_{zy} - y\tau_{zx})\,dx\,dy = (x_0 - x_{cf}) W_y - (y_0 - y_{cf}) W_x,$$

(torsion, $K_x = K_y = 0$).

From (52.4) and (52.21), it follows that

$$M_x = \iint_R y\,\tau_{zz}\,dx\,dy = -(l-z)W_y,$$
$$M_y = \iint_R -x\tau_{zz}\,dx\,dy = (l-z)W_x.$$

The solution of the simple flexure problem in which the applied load $(W_x, W_y, 0)$ acts at the center of flexure (with $\alpha = 0$) is thus given by the stresses

$$\tau_{xx} = \tau_{xy} = \tau_{yy} = 0,$$
$$\tau_{zz} = -E(l-z)(K_x x + K_y y),$$
$$\tau_{zx} = \mu K_x \left[(1+\sigma)\left(\frac{\partial \varphi_{11}}{\partial x} - x^2\right) - \sigma\left(\frac{\partial \psi_{12}}{\partial y} - y^2\right)\right]$$
$$+ \mu K_y \left[(1+\sigma)\frac{\partial \varphi_{22}}{\partial x} + \sigma\frac{\partial \psi_{21}}{\partial y}\right],$$

§52 EXTENSION, TORSION, AND FLEXURE OF BEAMS

$$\tau_{zy} = \mu K_y \left[(1 + \sigma)\left(\frac{\partial \varphi_{22}}{\partial y} - y^2\right) - \sigma\left(\frac{\partial \psi_{12}}{\partial x} - x^2\right) \right]$$
$$+ \mu K_x \left[(1 + \sigma)\frac{\partial \varphi_{11}}{\partial y} + \sigma \frac{\partial \psi_{12}}{\partial x} \right],$$

where

$$EK_x = \frac{I_x W_x - I_{xy} W_y}{I_x I_y - I_{xy}^2}, \qquad EK_y = \frac{I_y W_y - I_{xy} W_x}{I_x I_y - I_{xy}^2},$$

and where the harmonic functions φ_{11}, φ_{22}, ψ_{12}, ψ_{21} satisfy the boundary conditions

$$\frac{d\varphi_{11}}{d\nu} = x^2 \cos(x, \nu), \qquad \frac{d\varphi_{22}}{d\nu} = y^2 \cos(y, \nu),$$
$$\psi_{21} = \tfrac{1}{3}x^3 + \text{const.}, \qquad \psi_{12} = \tfrac{1}{3}y^3 + \text{const.} \quad \text{on } C.$$

The position of the center of flexure is determined from the relation

$$(52.26) \qquad \iint_R (x\tau_{zy} - y\tau_{zx})\, dx\, dy = x_{cf} W_y - y_{cf} W_x,$$

wherein α is to be set equal to zero in Eqs. (52.13) for the stresses τ_{zx}, τ_{zy}. The coordinates x_{cf}, y_{cf} of the center of flexure are readily found to be

$$x_{cf} = J(I_y S_2 - I_{xy} S_1),$$
$$(52.27) \qquad y_{cf} = J(I_{xy} S_2 - I_x S_1),$$

where

$$S_1 = \iint_R \left[x \frac{\partial \varphi_1}{\partial y} - y \frac{\partial \varphi_1}{\partial x} + (1 + \sigma)x^2 y - \sigma y^3 \right] dx\, dy,$$
$$S_2 = \iint_R \left[x \frac{\partial \varphi_2}{\partial y} - y \frac{\partial \varphi_2}{\partial x} - (1 + \sigma)xy^2 + \sigma x^3 \right] dx\, dy,$$
$$\frac{1}{J} = 2(1 + \sigma)(I_x I_y - I_{xy}^2).$$

If the cross section R is symmetrical about the x-axis, then it is evident from the symmetry of the differential equation, the symmetry of the boundary, and the symmetry of the boundary conditions that the function $\varphi_1(x, y)$ is an even function in y, and hence that $x \dfrac{\partial \varphi_1}{\partial y}$ and $y \dfrac{\partial \varphi_1}{\partial x}$ are odd in y. In this case, the formulas given above reduce to

$$S_1 = 0, \qquad \frac{1}{J} = 2(1 + \sigma) I_x I_y,$$

and

(52.28)
$$\begin{cases} x_{cf} = \dfrac{1}{2(1+\sigma)I_x} \int\int_R \left[x\dfrac{\partial \varphi_2}{\partial y} - y\dfrac{\partial \varphi_2}{\partial x} \right. \\ \qquad\qquad \left. - (1+\sigma)xy^2 + \sigma x^3 \right] dx\, dy, \\ y_{cf} = 0; \end{cases}$$

that is, *the center of flexure lies on an axis of symmetry of the section. If the cross section possesses two perpendicular axes of symmetry, then the center of flexure coincides with the centroid of the section.*

The formulation of the flexure problem given in this section is due essentially to A. C. Stevenson,[1] who departed from the classical treatment of Saint-Venant, followed by most writers, by taking the load point not at the centroid of the end section but at any point (x_0, y_0, l) and by abandoning the restriction that the x- and y-axes be the principal axes of the cross section. This freedom in choice of axes is of importance when the section is not symmetrical about two perpendicular axes, for in this case the principal axes do not, in general, afford the most convenient mathematical description of the boundary. The connection between the functions employed in this section and the classical Saint-Venant flexure function will be discussed in the next section.

53. Bending by a Load along a Principal Axis. The general problem considered in the last section will now be specialized to an important particular case, namely, that in which the axes are taken to be the principal axes of the section and the load $(W_x, 0, 0)$ is directed parallel to one of these axes. In this case, one has $I_{xy} = 0$, and Eqs. (52.21) yield

$$K_x = \frac{W_x}{EI_y} = \frac{W_x}{2\mu(1+\sigma)I_y}, \qquad K_y = 0,$$

---

[1] "Flexure with Shear and Associated Torsion in Prisms of Uni-axial and Asymmetric Cross-Sections," *Philosophical Transactions of the Royal Society* (London), ser. A, vol. 237 (1938–39) pp. 161–229. See also Rosa M. Morris, "Some General Solutions of St. Venant's Flexure and Torsion Problem. I," *Proceedings of the London Mathematical Society*, ser. 2, vol. 46 (1940), pp. 81–98.

while Eqs. (52.13) become

$$\tau_{zx} = \mu\alpha\left(\frac{\partial\varphi}{\partial x} - y\right) + \frac{W_x}{2(1+\sigma)I_y}\left[\frac{\partial\varphi_1}{\partial x} - (1+\sigma)x^2 + \sigma y^2\right],$$

$$\tau_{zy} = \mu\alpha\left(\frac{\partial\varphi}{\partial y} + x\right) + \frac{W_x}{2(1+\sigma)I_y}\frac{\partial\varphi_1}{\partial y}.$$

The flexure function $\varphi_1(x, y)$ is not of the same form as the classical Saint-Venant flexure function $\Phi(x, y)$ used by most writers; the two functions (together with their harmonic conjugates ψ_1 and Ψ) are related, in fact, by the expression

$$\varphi_1 + i\psi_1 = -(\Phi + i\Psi) + \tfrac{1}{3}(1 + \tfrac{1}{2}\sigma)(x + iy)^3,$$

or

(53.1) $\quad\begin{cases} \varphi_1 = -\Phi + \tfrac{1}{3}(1 + \tfrac{1}{2}\sigma)(x^3 - 3xy^2), \\ \psi_1 = -\Psi + \tfrac{1}{3}(1 + \tfrac{1}{2}\sigma)(3x^2 - y^3). \end{cases}$

In terms of the harmonic function $\Phi(x, y)$, the stresses can be written as

(53.2) $\quad\begin{cases} \tau_{xx} = \tau_{xy} = \tau_{yy} = 0, \\ \tau_{zz} = -\dfrac{W_x}{I_y}(l - z)x, \\ \tau_{zx} = \mu\alpha\left(\dfrac{\partial\varphi}{\partial x} - y\right) \\ \qquad - \dfrac{W_x}{2(1+\sigma)I_y}\left[\dfrac{\partial\Phi}{\partial x} + \dfrac{1}{2}\sigma x^2 + \left(1 - \dfrac{1}{2}\sigma\right)y^2\right], \\ \tau_{zy} = \mu\alpha\left(\dfrac{\partial\varphi}{\partial y} + x\right) \\ \qquad - \dfrac{W_x}{2(1+\sigma)I_y}\left[\dfrac{\partial\Phi}{\partial y} + (2+\sigma)xy\right]. \end{cases}$

From (53.1) and from the boundary condition (52.16) on the function φ_1, it follows that the harmonic function Φ must satisfy the condition

(53.3) $\quad\dfrac{d\Phi}{d\nu} = -\left[\dfrac{1}{2}\sigma x^2 + \left(1 - \dfrac{1}{2}\sigma\right)y^2\right]\cos(x, \nu)$
$\qquad\qquad - (2+\sigma)xy \cos(y, \nu)$ on C.

This special case of the general flexure problem has been formulated in terms of the torsion function φ and the flexure function Φ as a problem of Neumann. It may be rephrased as a problem of Dirichlet by writing the stresses in terms of the conjugate harmonic functions ψ and Ψ. The appropriate boundary condition on Ψ is seen from (52.17) and (53.1) to be

(53.4) $\quad \Psi = -\tfrac{1}{3}(1 - \tfrac{1}{2}\sigma)y^3 + (1 + \tfrac{1}{2}\sigma)x^2 y$
$\qquad - (1 + \sigma) \int_{(x_0,y_0)}^{(x,y)} x^2\, dy + \text{const.}$
$\quad = -\tfrac{1}{3}(1 - \tfrac{1}{2}\sigma)y^3 - \tfrac{1}{2}\sigma x^2 y$
$\qquad + 2(1 + \sigma) \int_{(x_0,y_0)}^{(x,y)} xy\, dx + \text{const.} \quad \text{on } C.$

where the last step makes use of integration by parts, and where the line integral is to be evaluated along the contour C.

The y-coordinate of the center of flexure is found from (52.22) and (53.2) to be given by

(53.5) $\quad y_{cf} = \dfrac{1}{2(1 + \sigma)I_y} \iint_R \left(x \dfrac{\partial \Phi}{\partial y} - y \dfrac{\partial \Phi}{\partial x} \right) dx\, dy$
$\qquad + \dfrac{1}{2(1 + \sigma)I_y} \iint_R \left[\left(2 + \tfrac{1}{2}\sigma\right) x^2 y - \left(1 - \tfrac{1}{2}\sigma\right) y^3 \right] dx\, dy$
$\quad = \dfrac{1}{2(1 + \sigma)I_y} \iint_R \left(y \dfrac{\partial \varphi_1}{\partial x} - x \dfrac{\partial \varphi_1}{\partial y} \right) dx\, dy$
$\qquad - \dfrac{1}{2(1 + \sigma)I_y} \iint_R [(1 + \sigma)x^2 y + \sigma y^3]\, dx\, dy.$

Since we have set $W_y = 0$, the x-coordinate of the flexural center cannot be determined from (52.26); that is, the mean twist over every section will vanish, provided the load W_x is applied at any point along the line $y = y_{cf}$.

PROBLEM

Consider the problem of determining experimentally the angle of twist of an airplane wing (or any cantilever beam) bent by a load $(W_x, 0, 0)$, direct or parallel to a principal axis. Let the displacements of points (x, y_1, z), (x, y_2, z) on two sections $z = z_1$, $z = z_2$ be observed. Show that the experimental angle of twist per unit length

$$\left[\frac{u(y_2) - u(y_1)}{(y_2 - y_1)(z_2 - z_1)} \right]_{z_1}^{z_2}$$

coincides with the theoretical angle of twist α only if the observed points are symmetrical about the y-axis (which passes through the centroid of the section).

54. The Displacement in a Bent Beam. In this section, expressions for the displacement components u, v, w will be given

§54   EXTENSION, TORSION, AND FLEXURE OF BEAMS   231

in terms of the torsion function φ and the flexure functions φ_1, φ_2. Some conclusions about the state of deformation can then be drawn from these expressions without explicitly determining the functions φ, φ_1, φ_2. The procedure is to substitute the expressions for the stresses found in Sec. 52 in formulas (29.2) and to carry out the integrations in a manner analogous to that used in Secs. 31 and 32. Since the calculation[1] presents no points of interest, we shall merely list the final results, and it is a simple matter to verify that the formulas for the components of displacement lead to the expressions for the stresses found in the preceding section.

In the case of the general flexure problem, discussed in Sec. 52, the expressions for the components of displacement are:

$$
(54.1) \begin{cases}
u = -\alpha yz + K_x[\tfrac{1}{2}\sigma(l-z)(x^2-y^2) - \tfrac{1}{6}z^3 + \tfrac{1}{2}lz^2] \\
\quad + K_y \sigma(l-z)xy, \\
v = \alpha xz + K_y[\tfrac{1}{2}\sigma(l-z)(y^2-x^2) - \tfrac{1}{6}z^3 + \tfrac{1}{2}lz^2] \\
\quad + K_x \sigma(l-z)xy, \\
w = \alpha \varphi(x,y) \\
\quad + K_x[\varphi_1(x,y) - (lz - \tfrac{1}{2}z^2)x - \tfrac{1}{6}(2+\sigma)x^3 + \tfrac{1}{2}\sigma xy^2] \\
\quad + K_y[\varphi_2(x,y) - (lz - \tfrac{1}{2}z^2)y - \tfrac{1}{6}(2+\sigma)y^3 + \tfrac{1}{2}\sigma x^2 y].
\end{cases}
$$

The linear terms that arise in deriving Eqs. (54.1) represent a rigid body displacement and vanish if one imposes the conditions of fixity

$$u = v = w = \frac{\partial u}{\partial z} = \frac{\partial v}{\partial z} = \frac{\partial v}{\partial x} = 0 \quad \text{at } (0,0,0).$$

When the flexure problem is specialized to the case of bending by a load $(W_x, 0, 0)$ along a principal axis (Sec. 53), then Eqs. (54.1) take the form

$$
(54.2) \begin{cases}
u = -\alpha yz + \dfrac{W_x}{EI_y}\left[\dfrac{1}{2}\sigma(l-z)(x^2-y^2) - \dfrac{1}{6}z^3 + \dfrac{1}{2}lz^2\right], \\
v = \alpha xz + \dfrac{W_x}{EI_y}\sigma(l-z)xy, \\
w = \alpha \varphi(x,y) - \dfrac{W_x}{EI_y}\left[\Phi(x,y) + xy^2 + \left(lz - \dfrac{1}{2}z^2\right)x\right],
\end{cases}
$$

where the function Φ is defined in Sec. 53.

[1] The details of such calculation for the special case of Sec. 53 can be found in A. E. H. Love, A Treatise on the Mathematical Theory of Elasticity, Sec. 230.

Some interesting conclusions regarding the state of deformation can be drawn directly from Eqs. (54.1). We note first that points $(0, 0, z)$ lying on the central line of the undeformed beam are carried into points (x', y', z'), with

(54.3) $$\begin{cases} x' = u = K_x(-\tfrac{1}{6}z^3 + \tfrac{1}{2}lz^2), \\ y' = v = K_y(-\tfrac{1}{6}z^3 + \tfrac{1}{2}lz^2); \end{cases}$$

that is, the deformed central line of the beam lies in the *plane of bending*

(54.4) $$y = \frac{K_y}{K_x} x = \frac{I_y W_y - I_{xy} W_x}{I_x W_x - I_{xy} W_y} x.$$

The greatest deflection of the central line of the beam occurs at the loaded end $z = l$, where

$$u = \frac{1}{3} K_x l^3 = \frac{1}{3} \frac{I_x W_x - I_{xy} W_y}{E(I_x I_y - I_{xy}^2)} l^3,$$

$$v = \frac{1}{3} K_y l^3 = \frac{1}{3} \frac{I_y W_y - I_{xy} W_x}{E(I_x I_y - I_{xy}^2)} l^3.$$

If the axes are principal axes of a cross section, then

$$u = \frac{1}{3} \frac{W_x}{EI_y} l^3, \qquad v = \frac{1}{3} \frac{W_y}{EI_x} l^3,$$

while for bending by a load W_x along a principal axis, the end deflection is

$$u = \frac{1}{3} \frac{W_x}{EI_y} l^3.$$

The *plane of the load* (the plane containing the z-axis and the line in the direction of the load) does not, in general, coincide with the plane of bending, since the equation of the former is

$$y = \frac{W_y}{W_x} x.$$

The *neutral plane* is defined as that plane whose filaments are not altered in length; that is, it is characterized by the equation $e_{zz} = 0$. Since

$$e_{zz} = \frac{\partial w}{\partial z} = -K_x(l - z)x - K_y(l - z)y,$$

§54 EXTENSION, TORSION, AND FLEXURE OF BEAMS

we have as the equation of the neutral plane

$$(54.5) \qquad y = -\frac{K_x}{K_y} x.$$

The planes defined by (54.4) and (54.5) are orthogonal, and hence the neutral plane is perpendicular to the plane of bending.

In the case of bending by a load $(W_x, 0, 0)$ along a principal axis (Sec. 53), the xz-plane contains the deformed central line, while the yz-plane is the neutral plane.

Consider now the curvature of the deformed central line of the beam. Taking coordinates $r = \sqrt{x'^2 + y'^2}$ and z in the plane of bending, we have from (54.3)

$$r = \sqrt{K_x^2 + K_y^2}\,(-\tfrac{1}{6}z^3 + \tfrac{1}{2}lz^2).$$

If the displacements and their derivatives are small, then one can write

$$r = \sqrt{K_x^2 + K_y^2}\,(-\tfrac{1}{6}z'^3 + \tfrac{1}{2}lz'^2),$$

from which it follows that the curvature of the central line is given approximately by

$$\frac{1}{R} = \frac{d^2 r}{dz'^2} = \sqrt{K_x^2 + K_y^2}\,(l - z').$$

That the curvature is proportional to the bending moments M_x, M_y is easily seen by referring to Sec. 52, where it was found that

$$M_x = \iint_R y\tau_{zz}\, dx\, dy = -(l-z)W_y,$$
$$M_y = \iint_R -x\tau_{zz}\, dx\, dy = (l-z)W_x,$$

and hence

$$M_x = \frac{-W_y}{R\sqrt{K_x^2 + K_y^2}}, \qquad M_y = \frac{W_x}{R\sqrt{K_x^2 + K_y^2}}.$$

For the case of bending by a load W_x along a principal axis (Sec. 53), these relations become

$$M_x = 0, \qquad M_y = \frac{EI_y}{R}.$$

Thus, the Bernoulli-Euler law is also valid in the case of bending of beams by transverse end loads.

The changes in the cross section of the beam are determined from a study of the terms in u, v, and w that are independent of the twist α, and one can carry out an analysis similar to that given in Sec. 32. The neutral plane is deformed into a saddle-shaped surface, of which the central line is one of the principal lines of curvature. The cross sections $z = c$ of the beam do not remain plane even when the term $\alpha\varphi(x, y)$, which is due to the twisting of the beam by the load, disappears. This can be seen by examining the equation

(54.6) $\quad z' = c + w = c + \alpha\varphi(x, y)$
$\quad\quad\quad + K_x[\varphi_1(x, y) - (lc - \tfrac{1}{2}c^2)x - \tfrac{1}{6}(2 + \sigma)x^3 + \tfrac{1}{2}\sigma xy^2]$
$\quad\quad\quad + K_y[\varphi_2(x, y) - (lc - \tfrac{1}{2}c^2)y - \tfrac{1}{6}(2 + \sigma)y^3 + \tfrac{1}{2}\sigma x^2y].$

For the special case considered in Sec. 53, this takes the form

(54.7) $\quad z' = c + \alpha\varphi(x, y) - \dfrac{W_x}{EI_y}[\Phi(x, y) + xy^2 + (lc - \tfrac{1}{2}c^2)x].$

The nature of the distortion of cross sections and the distribution of stresses can be discussed with more profit after the solutions of the flexure problem for specific cross sections have been deduced. It is not difficult, however, to write down the differential equation of the lines of shearing stress. The directions of these lines are given by the equation

$$\frac{dy}{dx} = \frac{\tau_{zy}}{\tau_{zx}},$$

so that, disregarding the terms in (52.13) that depend on α, we have the differential equation

(54.8) $\quad \left\{K_y\left[\dfrac{\partial \varphi_2}{\partial y} - y^2 - \sigma(y^2 - x^2)\right] + K_x\dfrac{\partial \varphi_1}{\partial y}\right\} dx$
$\quad\quad\quad - \left\{K_x\left[\dfrac{\partial \varphi_1}{\partial x} - x^2 - \sigma(x^2 - y^2)\right] + K_y\dfrac{\partial \varphi_2}{\partial x}\right\} dy = 0.$

For the special case of bending by a load along a principal axis, this becomes

(54.9) $\quad 2\dfrac{\partial \Phi}{\partial y} + (4 + 2\sigma)xy\, dx - \left[2\dfrac{\partial \Phi}{\partial x} + \sigma x^2 + (2 - \sigma)y^2\right] dy = 0.$

This equation will be used to determine the distribution of lines of shearing stress in a bent circular beam.

55. Flexure of Circular Beams.

Let the equation of the boundary of cross section of a beam of length l be

$$x^2 + y^2 = a^2,$$

and let the terminal load W be applied at the centroid of the end section and directed along the x-axis. The form of the boundary suggests the use of polar coordinates (r, θ). In terms of these coordinates, the equation of the boundary assumes the simple form $r = a$, and the boundary condition (53.4) becomes

$$\Psi = -\tfrac{1}{3}(1 - \tfrac{1}{2}\sigma)a^3 \sin^3 \theta - \tfrac{1}{2}\sigma a^3 \cos^2 \theta \sin \theta \\ - 2(1 + \sigma)a^3 \int \sin^2 \theta \cos \theta \, d\theta,$$

or

$$\Psi = -(\tfrac{3}{4} + \tfrac{1}{2}\sigma)a^3 \sin \theta + \tfrac{1}{4}a^3 \sin 3\theta, \quad \text{on } r = a.$$

Since the function Ψ is harmonic in the interior of the circle $r = a$, the appropriate particular solutions of the equation $\nabla^2 \Psi = 0$ in polar coordinates are of the form $r^n \sin n\theta$. Hence we must have

$$\Psi = -(\tfrac{3}{4} + \tfrac{1}{2}\sigma)a^2 r \sin \theta + \tfrac{1}{4}r^3 \sin 3\theta,$$

while the conjugate flexure function is

$$\Phi = -(\tfrac{3}{4} + \tfrac{1}{2}\sigma)a^2 r \cos \theta + \tfrac{1}{4}r^3 \cos 3\theta.$$

Recalling that $x = r \cos \theta$, $y = r \sin \theta$, we get

$$(55.1) \quad \Phi(x, y) = -(\tfrac{3}{4} + \tfrac{1}{2}\sigma)a^2 x + \tfrac{1}{4}(x^3 - 3xy^2).$$

From the symmetry of the cross section, it is seen that the center of flexure coincides with the centroid of the end section, and as the load point has also been taken at the centroid, it follows that in this example $\alpha = 0$. The stress components are found from (53.2) to be

$$(55.2) \quad \begin{cases} \tau_{zx} = \dfrac{(3 + 2\sigma)W}{2\pi a^4 (1 + \sigma)}\left(a^2 - x^2 - \dfrac{1 - 2\sigma}{3 + 2\sigma} y^2\right), \\ \tau_{zy} = -\dfrac{(1 + 2\sigma)W}{\pi a^4 (1 + \sigma)} xy, \\ \tau_{zz} = -\dfrac{4W}{\pi a^4}(l - z)x. \end{cases}$$

Along the diameter $x = 0$,

$$(55.3) \quad \tau_{zy} = 0, \quad \tau_{zx} = \frac{(3 + 2\sigma)W}{2\pi a^4(1 + \sigma)}\left(a^2 - \frac{1 - 2\sigma}{3 + 2\sigma}y^2\right),$$

and it is evident that τ_{zx} takes its maximum value at the center of the circle, where

$$(\tau_{zx})_{\max} = \frac{3 + 2\sigma}{2(1 + \sigma)}\frac{W}{\pi a^2}.$$

The shearing stress at the ends of this diameter is

$$(\tau_{zx})_{y=\pm a} = \frac{1 + 2\sigma}{1 + \sigma}\frac{W}{\pi a^2}.$$

We shall return to the study of the shearing stresses in Sec. 61, where the engineering theory of beams will be discussed briefly.

The distribution of the lines of shearing stress can be determined with the aid of Eq. (54.9) or directly from the defining relation

$$\frac{dy}{dx} = \frac{\tau_{zy}}{\tau_{zx}}.$$

The differential equation of the lines of shearing stress is easily found to be

$$2(1 + 2\sigma)xy\,dx - (3 + 2\sigma)\left(-a^2 + x^2 + \frac{1 - 2\sigma}{3 + 2\sigma}y^2\right)dy = 0,$$

the solution of which is given by

$$x^2 + y^2 = a^2 + cy^{\frac{3+2\sigma}{1+2\sigma}},$$

where c is an arbitrary constant. Several of these lines of shearing stress are indicated in Fig. 43 for $\sigma = 0.3$.

The distortion of cross sections is given by Eq. (54.7), which becomes in this case

$$z' - c = -\frac{W}{\pi a^4 E}[-(3 + 2\sigma)a^2 + 2c(2l - c)]x - \frac{W}{\pi a^4 E}(x^2 + y^2)x.$$

The linear term corresponds to a rigid rotation of the section $z = c$ about the y-axis, whereas the nonlinear terms represent the

§56 EXTENSION, TORSION, AND FLEXURE OF BEAMS

distortion of the section $z = c$ out of a plane. The contour lines of the section are given by

$$-\frac{W}{\pi a^4 E}(x^2 + y^2)x = \text{const.}$$

Some of the contour lines are shown in Fig. 44.

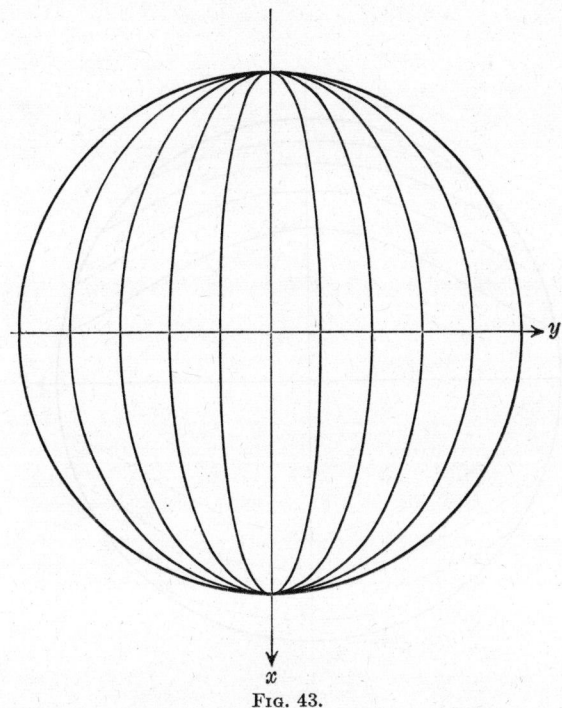

Fig. 43.

56. Bending of a Beam of Elliptical Cross Section. An analysis similar to that used in the preceding section can be applied to determine the flexure function for a beam whose cross section is given by the equation

(56.1) $$f(x, y) \equiv \frac{x^2}{a^2} + \frac{y^2}{b^2} - 1 = 0.$$

We assume, as above, that the load acts in the direction of the x-axis and is applied at the centroid of the end section.

Now the direction cosines $\cos(x, \nu)$ and $\cos(y, \nu)$ of the normal to the boundary of the ellipse are proportional to $\dfrac{\partial f}{\partial x}$ and $\dfrac{\partial f}{\partial y}$, respectively. Hence

$$\frac{\cos(x, \nu)}{\cos(y, \nu)} = \frac{\dfrac{x}{a^2}}{\dfrac{y}{b^2}}.$$

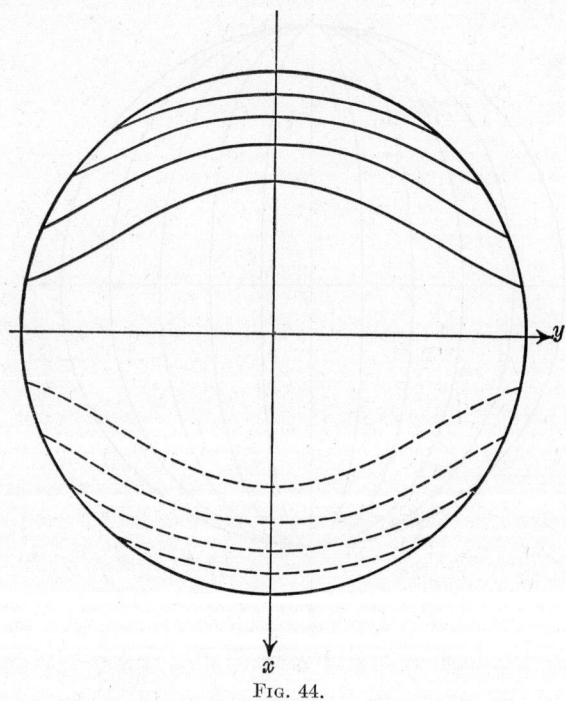

Fig. 44.

On the boundary C of the section, the flexure function Φ satisfies the condition (53.3); hence

$$(56.2) \quad \frac{\partial \Phi}{\partial x} b^2 x + \frac{\partial \Phi}{\partial y} a^2 y = -\left[\frac{1}{2}\sigma x^2 + \left(1 - \frac{1}{2}\sigma\right) y^2\right] b^2 x \\ - (2 + \sigma) a^2 x y^2 \qquad \text{on } C.$$

Since the right-hand member of (56.2) is a homogeneous polynomial in x and y, it is natural to seek a solution in the form of the

§56   EXTENSION, TORSION, AND FLEXURE OF BEAMS   239

sum of integral harmonics. Assuming

$$\Phi + i\Psi = c_1(x + iy) + c_2(x + iy)^3,$$

that is,

(56.3) $$\Phi = c_1 x + c_2(x^3 - 3xy^2),$$

and substituting in the boundary condition (56.2) gives

$$[c_1 + 3c_2(x^2 - y^2)]b^2 - 6c_2 a^2 y^2 =$$
$$- [\tfrac{1}{2}\sigma x^2 + (1 - \tfrac{1}{2}\sigma)y^2]b^2 - (2 + \sigma)a^2 y^2.$$

But on the boundary of the section,

$$x^2 = a^2 - \frac{a^2 y^2}{b^2},$$

and the preceding equation demands that

$$c_1 = \frac{-a^2}{3a^2 + b^2}[2(1 + \sigma)a^2 + b^2],$$

$$c_2 = \frac{(2 + \tfrac{1}{2}\sigma)a^2 + (1 - \tfrac{1}{2}\sigma)b^2}{9a^2 + 3b^2}.$$

The expression (56.3) for the flexure function now becomes

(56.4) $$\Phi = -\frac{a^2[2(1 + \sigma)a^2 + b^2]}{3a^2 + b^2} x$$
$$+ \frac{2a^2 + b^2 + \tfrac{1}{2}\sigma(a^2 - b^2)}{3(3a^2 + b^2)}(x^3 - 3xy^2).$$

Calculating stresses with the aid of formulas (53.2), we find that

(56.5) $$\begin{cases} \tau_{zx} = \dfrac{2W}{\pi a^3 b}\dfrac{2(1+\sigma)a^2 + b^2}{(1+\sigma)(3a^2+b^2)}\left[a^2 - x^2 - \dfrac{(1-2\sigma)a^2}{2(1+\sigma)a^2 + b^2}y^2\right], \\ \tau_{zy} = -\dfrac{4W}{\pi a^3 b}\dfrac{(1+\sigma)a^2 + \sigma b^2}{(1+\sigma)(3a^2+b^2)}xy. \end{cases}$$

It is obvious from these formulas that the y-component, τ_{zy}, of the shearing stress vanishes on the horizontal axis ($x = 0$) of the cross section, and that

$$(\tau_{zx})_{x=0} = \frac{2W}{\pi a^3 b}\frac{2(1+\sigma)a^2 + b^2}{(1+\sigma)(3a^2+b^2)}\left[a^2 - \frac{(1-2\sigma)a^2}{2(1+\sigma)a^2 + b^2}y^2\right].$$

Hence τ_{zx} takes its maximum value at the center of the ellipse where

$$(\tau_{zx})_{\max} = \frac{2W}{A}\frac{2(1+\sigma)a^2 + b^2}{(1+\sigma)(3a^2+b^2)},$$

and $A = \pi ab$ is the area of the cross section. Evidently τ_{zy} reaches its maximum on the boundary, and if we put $x = a \cos \theta$, $y = b \sin \theta$, then $(xy)_{max} = (\tfrac{1}{2} ab \sin 2\theta)_{max} = \tfrac{1}{2} ab$, and it is seen that

$$(\tau_{zy})_{max} = \frac{2W}{A} \frac{b}{a} \frac{(1+\sigma)a^2 + \sigma b^2}{(1+\sigma)(3a^2 + b^2)}.$$

If $b << a$, then the shape of the beam approaches that of a thin rectangular plank loaded parallel to its longer side. In this case, neglecting terms of order b^2/a^2, we get

$$(\tau_{zx})_{max} \doteq \frac{4W}{3A}, \qquad (\tau_{zy})_{max} \doteq \frac{4W}{3A} \frac{b}{2a}, \qquad b << a.$$

On the other hand, if $a << b$, then the load acts along the shorter axis, and

$$(\tau_{zx})_{max} \doteq \frac{2}{1+\sigma} \frac{W}{A}, \qquad (\tau_{zy})_{max} \doteq \frac{2}{1+\sigma} \frac{W}{A} \frac{\sigma b}{a}, \qquad a << b.$$

These results will be discussed further in Sec. 61.

57. Bending of Rectangular Beams. The two preceding sections have illustrated the solution of the boundary-value problems of flexure by forming those combinations of particular solutions of the differential equation

$$\nabla^2 \Phi = 0 \quad \text{or} \quad \nabla^2 \Psi = 0$$

that satisfy the boundary conditions on the function Φ or Ψ. The flexure of circular beams was treated by inspection of the boundary values of Ψ in polar coordinates and by utilizing the particular solutions of the form $r^n \sin n\theta$. Beams of elliptical cross section were handled by observing that the boundary condition on $\dfrac{d\Phi}{d\nu}$ involved only homogeneous polynomials in x and y, and this fact suggested that a solution for the complex flexure function $\Phi + i\Psi$ be sought as a sum of terms of the form $c_n(x + iy)^n$. In this section, the solution of the flexure problem for a beam of rectangular cross section is given as an infinite series of particular solutions $A_n \sinh \alpha x \cos \beta y$, the coefficients A_n being so chosen as to ensure the satisfaction of the boundary conditions. The next two sections will illustrate the use of analytic functions in solving the flexure problems.

§57 EXTENSION, TORSION, AND FLEXURE OF BEAMS

Let the equation of the boundary of the cross section of the beam be
$$(x^2 - a^2)(y^2 - b^2) = 0,$$
and let the terminal load be directed along the positive x-axis and applied at the origin.

A reference to the boundary conditions (53.3) shows that on the sides $x = \pm a$, we must have
$$\frac{\partial \Phi}{\partial x} = -\frac{1}{2}\sigma a^2 - \left(1 - \frac{1}{2}\sigma\right)y^2, \qquad -b < y < b,$$
while on the sides $y = \pm b$, we must satisfy the condition
$$\frac{\partial \Phi}{\partial y} = \mp(2 + \sigma)bx, \qquad -a < x < a.$$

In order to simplify the boundary conditions, we define a harmonic function $f(x, y)$ by the relation
$$f(x, y) = \Phi(x, y) - \tfrac{1}{6}(2 + \sigma)(x^3 - 3xy^2);$$
then the boundary conditions to be satisfied by the function $f(x, y)$ are
$$\frac{\partial f}{\partial x} = -(1 + \sigma)a^2 + \sigma y^2, \qquad \text{on } x = \pm a,$$
$$\frac{\partial f}{\partial y} = 0, \qquad \text{on } y = \pm b.$$

It follows from the discussion in Sec. 38 that one can build up the desired solution by forming an infinite series of particular solutions
$$f(x, y) = Ax + \sum_{n=1}^{\infty} A_n \sinh \frac{n\pi x}{b} \cos \frac{n\pi y}{b}.$$

The boundary condition on $y = \pm b$ is satisfied by each term of the series, while the satisfaction of the boundary condition on $x = \pm a$ is readily effected by noting the expansion
$$y^2 = \frac{b^2}{3} + \frac{4b^2}{\pi^2} \sum_{n=1}^{\infty} \frac{(-1)^n}{n^2} \cos \frac{n\pi y}{b}, \qquad -b \leq y \leq b.$$

The condition on the boundary $x = \pm a$ now takes the form

$$A + \sum_{n=1}^{\infty} \frac{n\pi}{b} A_n \cosh \frac{n\pi a}{b} \cos \frac{n\pi y}{b}$$

$$= -(1+\sigma)a^2 + \sigma \left[\frac{b^2}{3} + \frac{4b^2}{\pi^2} \sum_{n=1}^{\infty} \frac{(-1)^n}{n^2} \cos \frac{n\pi y}{b} \right],$$

and equating the coefficients of $\cos n\pi y/b$ leads to the result

$$f(x, y) = \left[-(1+\sigma)a^2 + \frac{1}{3}\sigma b^2 \right] x + \frac{4\sigma b^3}{\pi^3} \sum_{n=1}^{\infty} \frac{(-1)^n}{n^3} \frac{\sinh \frac{n\pi x}{b}}{\cosh \frac{n\pi a}{b}} \cos \frac{n\pi y}{b}.$$

The flexure function $\Phi(x, y)$ can now be found from the relation

$$\Phi(x, y) = f(x, y) + \tfrac{1}{6}(2 + \sigma)(x^3 - 3xy^2).$$

We shall dispense with the calculation of stresses, since such calculations contain no features of interest not already brought out in the preceding discussion. An elaborate discussion of the bending problem for a rectangular beam is given by Timoshenko,[1] who, however, approaches the problem in an entirely different way, by using an analogy between a certain stress function and the deflection of a stretched membrane under nonuniform pressure. We shall discuss this analogy in Sec. 60.

58. Conformal Mapping and the General Problem of Flexure; the Cardioid Section. The examples considered in the preceding sections have been illustrations of the specialized problem of flexure discussed in Sec. 53. The load $(W_x, 0, 0)$ was considered to be directed along a principal axis (which was also one of the two axes of symmetry) of the cross section of the beam. The analysis was also simplified by taking the centroid of the section as the point of application of the load. We consider now, as an illustration of the general problem of flexure of Sec. 52, the problem of bending of a beam whose cross section is bounded by a cardioid and thus has only one axis of symmetry. The load $(W_x, W_y, 0)$ will be considered to act at some point (x_0, y_0, l) of the end section. Upon this problem we shall bring to bear the

---

[1] S. Timoshenko, Theory of Elasticity, Sec. 92.

powerful weapon of analytic function theory, which was used earlier in the case of torsion of a beam.[1]

In Sec. 52, it was seen that the general case of flexure by a load $(W_x, W_y, 0)$ acting at any point (x_0, y_0, l) can be resolved into (1) a simpler flexure problem with α, the mean local twist, set equal to zero, and with the load applied at the center of flexure (x_{cf}, y_{cf}, l) and (2) a torsion problem with a twist α due to a couple of moment $W_y(x_0 - x_{cf}) - W_x(y_0 - y_{cf})$. The question of torsion of a cylinder was reduced in Sec. 35 to the boundary-value problem of finding the analytic function $\varphi(x, y) + i\psi(x, y)$ with

$$\psi = \tfrac{1}{2}(x^2 + y^2) \quad \text{on the boundary } C.$$

The simpler flexure problem (1) was seen in Sec. 52 to be equivalent to four boundary-value problems; that is, to the search for four analytic functions $\varphi_{11} + i\psi_{11}$, $\varphi_{22} + i\psi_{22}$, $\varphi_{12} + i\psi_{12}$, and $\varphi_{21} + i\psi_{21}$, with

$$\left. \begin{array}{ll} \dfrac{d\varphi_{11}}{d\nu} = x^2 \cos(x, \nu), & \dfrac{d\varphi_{22}}{d\nu} = y^2 \cos(y, \nu), \\ \psi_{21} = \tfrac{1}{3}x^3 + \text{const.}, & \psi_{12} = \tfrac{1}{3}y^3 + \text{const.} \end{array} \right\} \text{ on } C.$$

The boundary conditions on the normal derivatives of φ_{11} and φ_{22} can be replaced by conditions on the boundary values of the conjugate functions ψ_{11} and ψ_{22} by noting that

$$\frac{d\varphi_{11}}{d\nu} = \frac{\partial \varphi_{11}}{\partial x}\frac{dx}{d\nu} + \frac{\partial \varphi_{11}}{\partial y}\frac{dy}{d\nu} = \frac{\partial \psi_{11}}{\partial y}\frac{dy}{ds} + \frac{\partial \psi_{11}}{\partial x}\frac{dx}{ds} = \frac{d\psi_{11}}{ds}.$$

The condition on ψ_{11} now becomes

$$\frac{d\psi_{11}}{ds} = x^2 \cos(x, \nu) = x^2 \cos(y, s) \quad \text{on } C,$$

---

[1] For solutions of this problem by other methods, see W. M. Shephord, *Proceedings of the Royal Society* (London), ser. A, vol. 154 (1936), p. 500; A. C. Stevenson, "Flexure with Shear and Associated Torsion in Prisms of Uni-axial and Asymmetric Cross-sections," *Philosophical Transactions of the Royal Society* (London), ser. A, vol. 237 (1939), pp. 161–229; R. M. Morris, "Some General Solutions of St. Venant's Flexure and Torsion Problem I," *Proceedings of the London Mathematical Society*, ser. 2, vol. 46 (1940), pp. 81–98.

or
$$\psi_{11} = \int x^2 \frac{dy}{ds} ds = \int x^2 dy \quad \text{on } C.$$

Similarly, we have
$$\psi_{22} = -\int y^2 dx \quad \text{on } C.$$

The general problem of flexure is thus made to depend on the solution of five boundary-value problems of Dirichlet for the conjugate torsion function ψ, and the four flexure functions ψ_{11}, ψ_{22}, ψ_{21}, and ψ_{12}.

We have already seen, in Sec. 44, how to solve the boundary-value problem of the torsion of a beam of cross section R bounded by a curve C,

$$\nabla^2 \psi = 0 \quad \text{in } R, \qquad \psi = \tfrac{1}{2}(x^2 + y^2) \quad \text{on } C,$$

by mapping the region R on the interior of the unit circle $[\zeta] \leq 1$

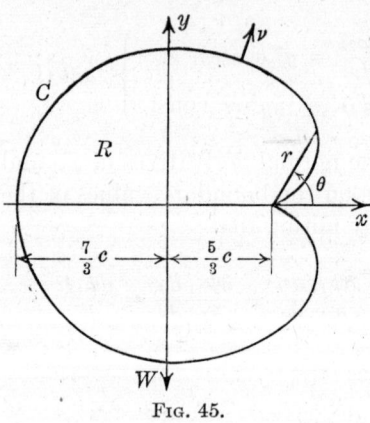

Fig. 45.

and applying the formula of Schwarz [Eq. (42.4)]. The same procedure can be used, of course, to write down the solution, in the form of an integral, for any problem of Dirichlet for any region that can be mapped conformally on the interior of a unit circle. In particular, the Schwarz integral affords solutions for the boundary-value problems of flexure. In this section, we shall see that in the case of the cardioid section, the complex integrals involved can be evaluated in a straightforward manner by making use of the theory of residues.

Consider a beam whose cross section is shown in Fig. 45, where the origin of the cartesian axes is at the centroid of the section. The polar and rectangular coordinates are related by

$$x = \frac{5c}{3} + r \cos t, \qquad y = r \sin t,$$

§58   EXTENSION, TORSION, AND FLEXURE OF BEAMS   245

and since the polar equation of the boundary C of the section is

$$r = 2c(1 - \cos t),$$

we can write

(58.1) $\begin{cases} x = \dfrac{5c}{3} + 2c \cos t(1 - \cos t) = \dfrac{5c}{3} - c + 2c \cos t - c \cos 2t, \\ y = 2c \sin t(1 - \cos t) = 2c \sin t - c \sin 2t. \end{cases}$

Then

$$x + iy = \frac{5c}{3} - c(1 - 2e^{it} + e^{i2t}) = \frac{5c}{3} - c(1 - e^{it})^2,$$

and it is seen that the analytic function

(58.2) $\qquad z = \omega(\zeta) \equiv \dfrac{5c}{3} - c(1 - \zeta)^2,$

with $\quad z = x + iy = \dfrac{5c}{3} + re^{it}, \qquad \zeta = \xi + i\eta = \rho e^{i\theta},$

maps the interior of the cardioid on the unit circle $|\zeta| \leq 1$, with $\theta = t$. The inverse transformation is

(58.3) $\qquad \zeta = 1 + i\left(\dfrac{r}{c}\right)^{1/2} e^{it/2}, \qquad 0 \leq t \leq 2\pi.$

The mapping function can be written as

$$\omega(\sigma) = \frac{5c}{3} - c(1 - \sigma)^2 = \frac{c}{3}(2 + 6\sigma - 3\sigma^2),$$

and we have

$$\omega(\sigma)\bar{\omega}\left(\frac{1}{\sigma}\right) = \frac{c^2}{9\sigma^2} f_1(\sigma),$$

where

$$f_1(\sigma) = -6 - 6\sigma + 49\sigma^2 - 6\sigma^3 - 6\sigma^4.$$

The complex torsion function is found from (44.5) to be given by

$$\varphi + i\psi = \frac{c^2}{18\pi} \int_\gamma \frac{f_1(\sigma)}{\sigma^2(\sigma - \zeta)} d\sigma = \frac{ic^2}{9}(R_1 + R_2),$$

where R_1 and R_2 are the residues of the integrand at $\sigma = 0$ and $\sigma = \zeta$, respectively, and where γ denotes the contour $|\zeta| = 1$. We have

$$R_1 = \frac{d}{d\sigma}\left[\frac{f_1(\sigma)}{\sigma - \zeta}\right]_{\sigma=0} = \frac{6}{\zeta} + \frac{6}{\zeta^2},$$

$$R_2 = \frac{f_1(\zeta)}{\zeta^2} = -\frac{6}{\zeta^2} - \frac{6}{\zeta} + 49 - 6\zeta - 6\zeta^2,$$

and

$$\varphi + i\psi = \frac{ic^2}{9}(49 - 6\zeta - 6\zeta^2)$$

$$= \frac{ic^2}{9}\left[37 - 18i\left(\frac{r}{c}\right)^{3/2} e^{it/2} + 6\frac{r}{c}e^{it}\right],$$

or

$$\varphi = 2r^{1/2}c^{3/2}\cos\frac{t}{2} - \frac{2}{3}rc\sin t,$$

$$\psi = 2r^{1/2}c^{3/2}\sin\frac{t}{2} + \frac{2}{3}rc\cos t.$$

Equation (44.8), for the moment of inertia I_0, takes the form

$$I_0 = \frac{1}{4i}\frac{2c^4}{27}\int_\gamma \frac{f_2(\sigma)}{\sigma^4}\,d\sigma,$$

where

$$f_2(\sigma) = 18 - 36\sigma - 177\sigma^2 + 495\sigma^3 - 220\sigma^4 - 128\sigma^5 + 36\sigma^6 + 12\sigma^7,$$

and hence

$$I_0 = \frac{\pi c^4}{27}R_3,$$

where

$$R_3 = \frac{1}{3!}\frac{d^3}{d\sigma^3}[f_2(\sigma)]_{\sigma=0} = 495$$

is the residue of the integrand at $\sigma = 0$. Thus,

(58.4) $$I_0 = \frac{55\pi c^4}{3}$$

is the polar moment of inertia of the cross section. Similarly, Eq. (44.7) yields

$$D_0 = -\frac{1}{4}\frac{i4c^4}{9}\int_\gamma \frac{f_3(\sigma)}{\sigma^5}\,d\sigma,$$

with

$$f_3(\sigma) = 2 + 3\sigma + \sigma^2 - 3\sigma^3 - 6\sigma^4 - 3\sigma^5 + \sigma^6 + 3\sigma^7 + 2\sigma^8.$$

§58   EXTENSION, TORSION, AND FLEXURE OF BEAMS   247

Then

$$D_0 = \frac{2\pi c^4}{9} R_4 = \frac{2\pi c^4}{9} \frac{1}{4!} \frac{d^4}{d\sigma^4} [f_3(\sigma)]_{\sigma=0} = -\frac{4\pi c^4}{3}.$$

The torsional rigidity D is

$$D = \mu(I_0 + D_0) = 17\mu\pi c^4.$$

The shearing stresses may be found either from Eq. (44.10) or from the relation

$$\tau_{zx} = \mu\tau \frac{\partial \Psi}{\partial y}, \qquad \tau_{zy} = -\mu\tau \frac{\partial \Psi}{\partial x},$$
$$\Psi = \psi - \tfrac{1}{2}(x^2 + y^2).$$

We turn now to the determination of the harmonic flexure function ψ_{21}, which takes the values

$$\psi_{21} = \tfrac{1}{3}x^3 + \text{const.} \qquad \text{on } C.$$

From

$$\cos t = \frac{1}{2}\left(\sigma + \frac{1}{\sigma}\right), \qquad \sigma = e^{i\theta} = e^{it}$$

and Eq. (58.1), we get

$$x = \frac{c}{3}[5 + 6(1 - \cos\theta)\cos\theta]$$
$$= \frac{c(-3 + 6\sigma + 4\sigma^2 + 6\sigma^3 - 3\sigma^4)}{6\sigma^2}$$

and

$$x^3 = \frac{c^3 f_4(\sigma)}{216\sigma^6},$$

with

$$f_4(\sigma) = -27 + 162\sigma - 216\sigma^2 - 54\sigma^3 - 441\sigma^4 + 828\sigma^5$$
$$+ 496\sigma^6 + 828\sigma^7 - 441\sigma^8 - 54\sigma^9 - 216\sigma^{10}$$
$$+ 162\sigma^{11} - 27\sigma^{12}.$$

The boundary condition becomes

$$\psi_{21} = \frac{c^3 f_4(\sigma)}{648\sigma^6} \qquad \text{on } \gamma,$$

and since ψ_{21} is the real part of $\frac{1}{i}(\varphi_{21} + i\psi_{21})$, we have by the Schwarz integral (42.4)

$$\frac{1}{i}(\varphi_{21} + i\psi_{21}) = \frac{1}{\pi i}\frac{c^3}{648}\int_\gamma \frac{f_4(\sigma)}{\sigma^6(\sigma - \zeta)}\,d\sigma,$$

or

$$\varphi_{21} + i\psi_{21} = \frac{ic^3}{324}(R_5 + R_6),$$

where R_5 and R_6 are the residues of the integrand at $\sigma = 0$ and $\sigma = \zeta$, respectively. We have

$$R_5 = \frac{1}{5!}\frac{d^5}{d\sigma^5}\left[\frac{f_4(\sigma)}{\sigma - \zeta}\right]_{\sigma=0}$$
$$= -\frac{828}{\zeta} + \frac{441}{\zeta^2} + \frac{54}{\zeta^3} + \frac{216}{\zeta^4} - \frac{162}{\zeta^5} + \frac{27}{\zeta^6},$$

and

$$R_6 = \frac{f_4(\zeta)}{\zeta^6}.$$

Then

$$\varphi_{21} + i\psi_{21} = \frac{ic^3}{324}(248 + 828\zeta - 441\zeta^2 - 54\zeta^3 - 216\zeta^4 + 162\zeta^5 - 27\zeta^6).$$

With the help of the inverse transformation (58.3), we find

$$\psi_{21} = \frac{1}{324}\left(500c^3 + 432r^{1/2}c^{5/2}\sin\frac{t}{2} + 684\,rc^2\cos t \right.$$
$$\left. + 162r^{5/2}c^{3/2}\sin\frac{3t}{2} + 189r^2c\cos 2t + 27r^3\cos 3t\right),$$

$$\varphi_{21} = \frac{1}{36}\left(48r^{1/2}c^{5/2}\cos\frac{t}{2} - 76rc^2\sin t + 18r^{5/2}c^{3/2}\cos\frac{3t}{2}\right.$$
$$\left. - 21r^2c\sin 2t - 3r^3\sin 3t\right).$$

Similarly, the function ψ_{12} can be determined by noting that (58.1) and the relations

$$\cos t = \frac{1}{2}\left(\sigma + \frac{1}{\sigma}\right), \qquad \sin t = \frac{1}{2i}\left(\sigma - \frac{1}{\sigma}\right),$$

§58 EXTENSION, TORSION, AND FLEXURE OF BEAMS

yield

$$y = \frac{c(1 - 2\sigma + 2\sigma^3 - \sigma^4)}{2i\sigma^2},$$

$$y^3 = \frac{ic^3}{8\sigma^6}(1 - 6\sigma + 12\sigma^2 - 2\sigma^3 - 27\sigma^4 + 36\sigma^5 - 36\sigma^7$$
$$+ 27\sigma^8 + 2\sigma^9 - 12\sigma^{10} + 6\sigma^{11} - \sigma^{12})$$
$$= \frac{ic^3}{8\sigma^6} f_5(\sigma).$$

The condition on the function ψ_{12} is

$$\psi_{12} = \frac{1}{3} \frac{ic^3}{8} \frac{f_5(\sigma)}{\sigma^6} \qquad \text{on } \gamma.$$

The Schwarz integral (42.4) affords the complex flexure function

$$\varphi_{12} + i\psi_{12} = \frac{ic^3}{24\pi} \int_\gamma \frac{f_5(\sigma)}{\sigma^6(\sigma - \zeta)} d\sigma = -\frac{c^3}{12}(R_7 + R_8),$$

where

$$R_7 = \frac{1}{5!} \frac{d^5}{d\sigma^5}\left[\frac{f_5(\sigma)}{\sigma - \zeta}\right]_{\sigma=0}, \qquad R_8 = \frac{f_5(\zeta)}{\zeta^6}.$$

Hence

$$\varphi_{12} + i\psi_{12} = \frac{c^3}{12}(36\zeta - 27\zeta^2 - 2\zeta^3 + 12\zeta^4 - 6\zeta^5 + \zeta^6),$$

from which it follows that

(58.6) $\begin{cases} \psi_{12} = \dfrac{1}{12}\left(6rc^2 \sin t - 6r^{3/2}c^{3/2}\cos\dfrac{3t}{2} - 3r^2 c \sin 2t - r^3 \sin 3t\right), \\ \varphi_{12} = \dfrac{1}{12}\left(14c^3 + 6rc^2 \cos t + 6r^{3/2}c^{3/2}\sin\dfrac{3t}{2} - 3r^2c\cos 2t - r^3 \cos 3t\right). \end{cases}$

The boundary values of the function ψ_{11} are themselves given in terms of a line integral, which must first be evaluated. From (58.1) we get

$$x^2 = \frac{c^2}{36\sigma^4}(9 - 36\sigma + 12\sigma^2 + 12\sigma^3 + 106\sigma^4 + 12\sigma^5$$
$$+ 12\sigma^6 - 36\sigma^7 + 9\sigma^8),$$

$$dy = \frac{ic}{\sigma^3}(1 - \sigma - \sigma^3 + \sigma^4)\, d\sigma,$$

and

$$\int x^2\, dy = \psi_{11}\Big|_\gamma = \frac{ic^3 f_6(\sigma)}{72\ \sigma^6},$$

with

$$f_6(\sigma) = -3 + 18\sigma - 24\sigma^2 + 6\sigma^3 - 139\sigma^4 + 284\sigma^5 \\ - 284\sigma^7 + 139\sigma^8 - 6\sigma^9 + 24\sigma^{10} - 18\sigma^{11} + 3\sigma^{12}.$$

From (42.4) we get

$$\varphi_{11} + i\psi_{11} = \frac{ic^3}{72\pi} \int_\gamma \frac{f_6(\sigma)}{\sigma^6(\sigma - \zeta)} d\sigma = \frac{-c^3}{36} (R_9 + R_{10}),$$

where

$$R_9 = \frac{1}{5!} \frac{d^5}{d\sigma^5} \left[\frac{f_6(\sigma)}{\sigma - \zeta} \right]_{\sigma=0}, \quad R_{10} = \frac{f_6(\zeta)}{\zeta^6},$$

and hence

$$\varphi_{11} + i\psi_{11} = \frac{c^3}{36} (284\zeta - 139\zeta^2 + 6\zeta^3 - 24\zeta^4 \\ + 18\zeta^5 - 3\zeta^6),$$

and

(58.7) $\quad \varphi_{11} = \dfrac{1}{36} \left(142c^3 + 130rc^2 \cos t + 30r^{3/2}c^{3/2} \sin \dfrac{3t}{2} \right.$

$\left. + 21r^2 c \cos 2t + 3r^3 \cos 3t \right).$

The last flexure function ψ_{22} is found in a similar way. The boundary values are calculated by observing that

$$y^2 = 4c^2(1 - \mu)^2(1 - \mu^2), \quad \mu = \cos t,$$
$$dx = 2c(1 - 2\mu) d\mu,$$

and

$$\int y^2 dx = \frac{4c^3}{3} (6\mu - 12\mu^2 + 8\mu^3 + 3\mu^4 - 6\mu^5 + 2\mu^6).$$

But $\mu = \tfrac{1}{2}(\sigma + 1/\sigma)$, and hence

$$-\int y^2 dx = \psi_{22}\bigg|_\gamma = -\frac{c^3}{24} \frac{f_7(\sigma)}{6},$$

where

$$f_7(\sigma) = 1 - 6\sigma + 12\sigma^2 + 2\sigma^3 - 57\sigma^4 + 132\sigma^5 - 136\sigma^6 \\ + 132\sigma^7 - 57\sigma^8 + 2\sigma^9 + 12\sigma^{10} - 6\sigma^{11} + \sigma^{12}.$$

§58 EXTENSION, TORSION, AND FLEXURE OF BEAMS 251

The Schwarz formula (42.4) yields

$$\varphi_{22} + i\psi_{22} = -\frac{c^3}{24\pi} \int_\gamma \frac{f_7(\sigma)}{\sigma^6(\sigma - \zeta)} d\sigma = -\frac{ic^3}{12}(R_{11} + R_{12}),$$

with

$$R_{11} = \frac{1}{5!}\frac{d^5}{d\sigma^5}\left[\frac{f_7(\sigma)}{\sigma - \zeta}\right]_{\sigma=0}, \qquad R_{12} = \frac{f_7(\zeta)}{\zeta^6},$$

or

$$\varphi_{22} + i\psi_{22} = \frac{ic^3}{12}(136 - 132\zeta + 57\zeta^2 - 2\zeta^3 - 12\zeta^4 + 6\zeta^5 - \zeta^6),$$

and

(58.8) $$\varphi_{22} = \frac{1}{12}\left(48r^{1/2}c^{5/2}\cos\frac{t}{2} + 24rc^2 \sin t - 10r^{3/2}c^{3/2}\cos\frac{3t}{2} - 3r^2 c \sin 2t - r^3 \sin 3t\right).$$

Before the stresses can be found, the constants K_x and K_y must first be evaluated. From (52.21) we get, since $I_{xy} = 0$,

$$K_x = \frac{W_x}{EI_y}, \qquad K_y = \frac{W_y}{EI_x}.$$

Now

$$I_x = \int\!\!\int_R y^2\, dx\, dy = -\tfrac{1}{3}\int_C y^3\, dx,$$

and from (58.1) we have

$$y = 2c \sin t\, (1 - \cos t), \qquad dx = -2c \sin t\, (1 - 2\cos t)\, dt.$$

Integration of $y^3\, dx$ from $t = 0$ to $t = 2\pi$ yields

$$I_x = \frac{21\pi c^4}{2}.$$

The moment of inertia I_y is found with the aid of Eq. (58.4) to be

$$I_y = I_0 - I_x = \frac{55\pi c^4}{3} - \frac{21\pi c^4}{2} = \frac{47\pi c^4}{6}.$$

The stresses can now be found from Eqs. (52.29).

The coordinates of the center of flexure will now be determined. The first term in the expression for x_{cf} in (52.28) becomes, with the help of Green's Theorem and Eqs. (52.22),

$$\int\int_R x \frac{\partial \varphi_2}{\partial y} dx\, dy = \int_C -x\varphi_2\, dx$$
$$= \int_C -x[(1+\sigma)\varphi_{22} + \sigma\varphi_{21}]\, dx.$$

This integral can be evaluated by noting that Eqs. (58.1) yield

$$x = \frac{c}{3}(5 + 6\cos t - 6\cos^2 t), \qquad dx = -2c(1 - 2\cos t)\sin t\, dt,$$

while the polar equation of the boundary C is $r = 2c(1 - \cos t)$, or

$$r^{1/2} = 2c^{1/2}\sin\frac{t}{2}, \qquad 0 \leq t \leq 2\pi.$$

Substitution of these expressions in Eqs. (58.5) and (58.8) yields the boundary values of the functions φ_{21} and φ_{22} in terms of the variable t. The integration indicated above is now carried out from $t = 0$ to $t = 2\pi$, with the result

$$\int\int_R x \frac{\partial \varphi_2}{\partial y} dx\, dy = (1+\sigma)\pi c^5 + \frac{\sigma 2\pi c^5}{9}$$
$$= \frac{(9 + 11\sigma)\pi c^5}{9}.$$

Similarly,

$$\int\int_R y \frac{\partial \varphi_2}{\partial x} dx\, dy = \int_C y[(1+\sigma)\varphi_{22} + \sigma\varphi_{21}]\, dy$$
$$= (1+\sigma)2\pi c^5 + \frac{\sigma\pi c^5}{3} = \frac{(6 + 7\sigma)\pi c^5}{3},$$
$$\int\int_R xy^2\, dx\, dy = -\tfrac{1}{3}\int_C xy^3\, dx = \pi c^5,$$
$$\int\int_R x^3\, dx\, dy = \tfrac{1}{4}\int_C x^4\, dy = \frac{-5\pi c^5}{9},$$

and finally, from Eq. (52.24),

$$x_{cf} = -\frac{2(3 + 4\sigma)}{63(1 + \sigma)}c, \qquad y_{cf} = 0.$$

Before concluding this section, it should be remarked that some of the foregoing results were obtained by W. M. Shepherd in 1936, and somewhat earlier by N. M. Mushtari. However, Mushtari's work was published in two journals that are not readily obtainable,[1] and not until after the appearance of Shepherd's paper did Mushtari publish a summary of his earlier papers. Mushtari considers the problems of torsion and flexure of beams whose boundaries of cross sections have the forms

$$r = a + b(1 + \cos \theta)$$

and

$$r^2 = a^2 + b^2 \cos 2\theta.$$

His method of solution consists essentially of assuming the complex torsion and flexure functions to have certain forms that involve integral and fractional powers of the complex variable z.

59. Bending of Circular Pipe. As an illustration of a simple application of the theory of analytic functions in determining the flexure function Φ (Sec. 53), consider a beam whose cross section is bounded by two concentric circles; that is, a pipe with inner radius r_i and outer radius r_0.

The complex flexure function $F(\mathfrak{z}) = \Phi + i\Psi$, being analytic and single-valued in the circular ring $r_i \leq r \leq r_0$, admits of an expansion in a Laurent series, so that

$$\Phi + i\Psi = \sum_{n=-\infty}^{\infty} (a_n + ib_n)\mathfrak{z}^n.$$

Setting $\mathfrak{z} = re^{i\theta}$, we obtain

$$\Phi + i\Psi = \sum_{-\infty}^{\infty} r^n(a_n + ib_n)e^{in\theta}$$

$$= \sum_{-\infty}^{\infty} r^n(a_n + ib_n)(\cos n\theta + i \sin n\theta),$$

or, separating real and imaginary parts,

(59.1) $$\begin{cases} \Phi = \sum_{-\infty}^{\infty} r^n(a_n \cos n\theta - b_n \sin n\theta), \\ \Psi = \sum_{-\infty}^{\infty} r^n(a_n \sin n\theta + b_n \cos n\theta). \end{cases}$$

---

[1] N. M. MUSHTARI, *Transactions of the Kazan Aviational Institute*, no. 1 (1933), pp. 17–32, and *Trans. KIIKS*, no. 1 (1935), pp. 53–67. These references are given by Mushtari in a paper published in *Applied Mathematics and Mechanics*, new ser., vol. 1 (1938), pp. 427–440. See also D. Z. AVAZASHVILI, "On the Application of Functions of a Complex Variable to the Torsion and Flexure Problems," *Applied Mathematics and Mechanics*, vol. 4, no. 1 (1940), pp. 129–134 (in Russian).

The constants a_n, b_n may be determined either from the boundary condition on the normal derivative of the function Φ [Eq. (53.3)] or from the boundary values of the function Ψ [Eq. (53.4)]. That is, we may solve either a problem of Neumann for the flexure function Φ or a Dirichlet problem for the conjugate function Ψ. The latter course will be followed, since the boundary condition on Ψ has already been given for a circular boundary [see Eq. (55.1)].

Rewriting Eq. (55.1) for the radii r_i and r_0 and using Eq. (59.1), we get

$$\sum_{-\infty}^{\infty} r_i^n (a_n \sin n\theta + b_n \cos n\theta) = -(\tfrac{3}{4} + \tfrac{1}{2}\sigma) r_i^3 \sin \theta + \tfrac{1}{4} r_i^3 \sin 3\theta,$$

$$\sum_{-\infty}^{\infty} r_0^n (a_n \sin n\theta + b_n \cos n\theta) = -(\tfrac{3}{4} + \tfrac{1}{2}\sigma) r_0^3 \sin \theta + \tfrac{1}{4} r_0^3 \sin 3\theta.$$

Comparing the coefficients of $\sin n\theta$ and $\cos n\theta$ gives a system of equations for the determination of the constants a_n and b_n. We have

$$-r_i^{-1} a_{-1} + r_i a_1 = -(\tfrac{3}{4} + \tfrac{1}{2}\sigma) r_i^3, \qquad -r_i^{-3} a_{-3} + r_i^3 a_3 = \tfrac{1}{4} r_i^3,$$
$$-r_0^{-1} a_{-1} + r_0 a_1 = -(\tfrac{3}{4} + \tfrac{1}{2}\sigma) r_0^3, \qquad -r_0^{-3} a_{-3} + r_0^3 a_3 = \tfrac{1}{4} r_0^3,$$
$$b_n = 0 \text{ if } n = \pm 1, \pm 2, \pm 3, \cdots,$$
$$a_n = 0 \text{ if } n = \pm 2, \pm 4, \pm 5, \pm 6, \cdots.$$

Solving these equations, we get

$$a_1 = -(\tfrac{3}{4} + \tfrac{1}{2}\sigma)(r_i^2 + r_0^2), \qquad a_{-1} = -(\tfrac{3}{4} + \tfrac{1}{2}\sigma) r_i^2 r_0^2,$$
$$a_3 = \tfrac{1}{4}, \qquad\qquad\qquad\qquad\quad a_{-3} = 0,$$

while the coefficients a_0 and b_0 are undetermined, since the boundary condition on Ψ involves an arbitrary constant.

Substituting these values in (59.1), we find that

$$\Phi = -\left(\frac{3}{4} + \frac{1}{2}\sigma\right)\left[(r_i^2 + r_0^2)r + \frac{r_i^2 r_0^2}{r}\right] \cos \theta + \frac{1}{4} r^3 \cos 3\theta + \text{const.},$$

$$\Psi = -\left(\frac{3}{4} + \frac{1}{2}\sigma\right)\left[(r_i^2 + r_0^2)r - \frac{r_i^2 r_0^2}{r}\right] \sin \theta + \frac{1}{4} r^3 \sin 3\theta + \text{const.}$$

The expressions for the stresses can be easily calculated with the aid of Eqs. (53.2).

If r_i is set equal to zero, we get the flexure functions for the solid circular beam discussed in Sec. 55.

PROBLEM

Calculate the stresses in a circular pipe of thickness t, fixed at one end and subjected to bending by an end load W, and show that the following approximate formulas are valid:

$$\tau_{zz} = \frac{-W(l-z)x}{\pi r_0^3 t},$$

$$(\tau_{zx})_{\max} = \frac{W}{\pi r_0 t},$$

$$(\tau_{zy})_{\max} = \frac{W}{2\pi r_0 t}.$$

60. Stress Functions and Analogies; Beams of Equilateral Triangular Section. We recall that in Sec. 52 the equilibrium equations (52.6) led to the definition of the stress function $F(x, y)$, in terms of which the stresses τ_{zx} and τ_{zy} were expressed in (52.7):

$$\tau_{zx} = \frac{\partial F}{\partial y} - \frac{1}{2} E K_x x^2, \qquad \tau_{zy} = -\frac{\partial F}{\partial x} - \frac{1}{2} E K_y y^2.$$

The function $F(x, y)$ was seen to be determined by the differential equation (52.8)

$$\nabla^2 F(x, y) = -2\mu\sigma K_y x + 2\mu\sigma K_x y - 2\mu\alpha,$$

and by the boundary condition (52.14)

$$\tau_{zx} \cos(x, \nu) + \tau_{zy} \cos(y, \nu) = 0.$$

In the course of the solution of the general flexure problem in Sec. 52, it was found convenient to phrase it not as a boundary-value problem for the determination of the function $F(x, y)$ but rather in terms of the torsion function φ and the flexure functions φ_1, φ_2 or φ_{11}, φ_{12}, φ_{21}, φ_{22}. In this section, the flexure problem will be stated in terms of a new stress function $T(x, y)$, which will be seen to be of value in certain problems.

We introduce the stress function $T(x, y)$ by defining

(60.1) $\qquad T(x, y) \equiv F(x, y) - \int R(x)\, dx - \int S(y)\, dy.$

The functions $R(x)$ and $S(y)$ may be so chosen as to yield either a simple boundary condition or a simple differential equation for

$T(x, y)$. The stresses can be written in terms of $T(x, y)$, with the aid of (52.7), as

$$(60.2) \quad \begin{cases} \tau_{zx} = \dfrac{\partial T}{\partial y} + S(y) - \dfrac{1}{2} EK_x x^2, \\ \tau_{zy} = -\dfrac{\partial T}{\partial x} - R(x) - \dfrac{1}{2} EK_y y^2, \end{cases}$$

while (52.8) yields the following differential equation for $T(x, y)$:

$$(60.3) \quad \nabla^2 T(x, y) = -2\mu\sigma K_y x - \frac{dR(x)}{dx} + 2\mu\sigma K_x y - \frac{dS(y)}{dy} - 2\mu\alpha.$$

The insertion of Eqs. (60.2) in (52.14) yields the boundary condition

$$(60.4) \quad \frac{\partial T}{\partial y}\frac{dy}{ds} + \frac{\partial T}{\partial x}\frac{dx}{ds} \equiv \frac{dT}{ds} = \left[\frac{1}{2} EK_x x^2 - S(y)\right]\frac{dy}{ds} \\ - \left[\frac{1}{2} EK_y y^2 + R(x)\right]\frac{dx}{ds},$$

where we make use of the relations

$$\cos(x, \nu) = \frac{dy}{ds}, \qquad \cos(y, \nu) = -\frac{dx}{ds}.$$

The functions $R(x)$, $S(y)$ may be prescribed arbitrarily. We choose them now to be any functions satisfying the relations

$$(60.5) \quad R(x) = -\tfrac{1}{2} EK_y y^2, \qquad S(y) = \tfrac{1}{2} EK_x x^2, \qquad \text{on } C;$$

then the condition on $T(x, y)$ becomes

$$\frac{dT}{ds} = 0 \qquad \text{on } C.$$

Thus, the function $T(x, y)$ is constant along the contour C, and since the choice of this constant cannot affect the stresses, we shall take it equal to zero. With this choice of the functions $R(x)$ and $S(y)$, the stress function $T(x, y)$ is determined by the differential equation (60.3) and by the condition

$$(60.6) \qquad\qquad T = 0 \qquad\qquad \text{on } C.$$

It is to be noted that the function $R(x)$ [or $S(y)$] may take any value along a portion of the boundary where $\dfrac{dx}{ds}$ $\left(\text{or } \dfrac{dy}{ds}\right)$ vanishes.

§60 EXTENSION, TORSION, AND FLEXURE OF BEAMS

It should be recalled that the constant of integration α was seen in Sec. 52 to be the mean value of the local twist $\dfrac{\partial \omega}{\partial z}$ over the section (or the value of the local twist at the centroid). As noted in that section, the constant α is to be chosen equal to zero if the load is applied at the center of flexure of the end section.

The stress function $T(x, y)$ can be given an interesting physical interpretation. Comparison of the differential equations (60.3) and (46.1) shows that $T(x, y)$ may be thought of as the deflection of an elastic membrane stretched, with tension t, over an opening of contour C in a rigid plane plate and distorted by a nonuniform load $p(x, y)$, where

$$\frac{-p(x, y)}{t} = -2\mu\sigma K_y x - \frac{dR(x)}{dx} + 2\mu\sigma K_x y - \frac{dS(y)}{dy} - 2\mu\alpha.$$

When the general flexure problem considered above is specialized to the case of bending by a load $(W_x, 0, 0)$ along a principal axis (Sec. 53), we have

$$K_x = \frac{W_x}{EI_y} = \frac{W_x}{2\mu(1 + \sigma)I_y}, \qquad K_y = 0.$$

The stress function $T(x, y)$ is determined by the conditions

(60.7)
$$\nabla^2 T(x, y) = \frac{\sigma}{1 + \sigma} \frac{W_x}{I_y} y - \frac{dS(y)}{dy} - 2\mu\alpha,$$
$$T = 0 \qquad \text{on } C,$$

where $S(y)$ is any function such that

(60.8)
$$S(y) = \frac{W_x}{2I_y} x^2 \qquad \text{on } C,$$

except that $S(y)$ may take any boundary value along a portion of the contour where $\dfrac{dy}{ds}$ is zero. The stresses are given by

(60.9)
$$\begin{cases} \tau_{zz} = \dfrac{\partial T}{\partial y} + S(y) - \dfrac{W_x}{2I_y} x^2, \\ \tau_{zy} = -\dfrac{\partial T}{\partial x}. \end{cases}$$

The position of the center of flexure can be found in terms of the function $T(x, y)$ by applying the definition given in Eq.

(52.26) and using Eqs. (60.2). We have

$$x_{cf}W_y - y_{cf}W_x = \iint_R \left\{ x\left[-\frac{\partial T}{\partial x} - R(x) - \frac{1}{2}EK_y y^2\right] \right.$$
$$\left. - y\left[\frac{\partial T}{\partial y} + S(y) - \frac{1}{2}EK_x x^2\right] \right\} dx\, dy$$
$$= \iint_R \left\{ 2T(x,y) + \frac{\partial}{\partial x}\left[-xT(x,y) - xyS(y) + \frac{1}{6}EK_x x^3 y\right] \right.$$
$$\left. - \frac{\partial}{\partial y}\left[yT(x,y) + xyR(x) + \frac{1}{6}EK_y xy^3\right] \right\} dx\, dy$$
$$= 2\iint_R T(x,y)\, dx\, dy$$
$$+ \int_C \left[yT(x,y) + xyR(x) + \frac{1}{6}EK_y xy^3\right] dx$$
$$+ \int_C \left[-xT(x,y) - xyS(y) + \frac{1}{6}EK_x x^3 y\right] dy,$$

where Green's Theorem was used in the last step. A reference to the boundary conditions (60.3) shows that this can be written as

$$(60.10) \quad x_{cf}W_y - y_{cf}W_x = 2\iint_R T(x,y)\, dx\, dy$$
$$- \tfrac{1}{3}E\left[K_y \int_C xy^3\, dx + K_x \int_C x^3 y\, dy\right],$$

wherein we are to set in the function $T(x,y)$ the constant α equal to zero. The coordinates x_{cf}, y_{cf} of the center of flexure are then found by comparing the coefficients of W_x and W_y. For the special case of bending by a load W_x along a principal axis, Eq. (60.10) becomes

$$(60.11) \quad y_{cf} = -\frac{2}{W_x}\iint_R T(x,y)\, dx\, dy + \frac{1}{3I_y}\int_C x^3 y\, dy.$$

As an illustration of the use of the stress function $T(x,y)$ in the solution of the flexure problem, we consider the bending of a beam of elliptical cross section with a contour given by the equation $x^2/a^2 + y^2/b^2 = 1$. Since on the boundary one has

$$x^2 = \frac{a^2}{b^2}(b^2 - y^2),$$

one can, evidently, choose

$$S(y) = \frac{W_x}{2I_y} \frac{a^2}{b^2} (b^2 - y^2).$$

From Eq. (60.7) it is seen that the function $T(x, y)$ is subject to the conditions

$$\nabla^2 T(x, y) = \frac{W_x}{I_y} \left(\frac{\sigma}{1 + \sigma} + \frac{a^2}{b^2} \right) y,$$

$$T(x, y) = 0 \quad \text{on} \quad \frac{x^2}{a^2} + \frac{y^2}{b^2} - 1 = 0.$$

The differential equation and boundary condition suggest that we seek a solution of the form

$$T(x, y) = ky \left(\frac{x^2}{a^2} + \frac{y^2}{b^2} - 1 \right),$$

and it is readily found that

$$T(x, y) = \frac{a^2[(1 + \sigma)a^2 + \sigma b^2]}{2(1 + \sigma)(3a^2 + b^2)} \frac{W_x}{I_y} y \left(\frac{x^2}{a^2} + \frac{y^2}{b^2} - 1 \right).$$

The stresses τ_{zx}, τ_{zy} can now be found from Eqs. (60.9). This method of solution should be compared with that applied in Sec. 56 to the same problem.

The stress function $T(x, y)$ will now be used to solve a special case of the flexure problem for a beam whose cross section is an equilateral triangle (Fig. 46).

The boundary of the triangular section can be written as

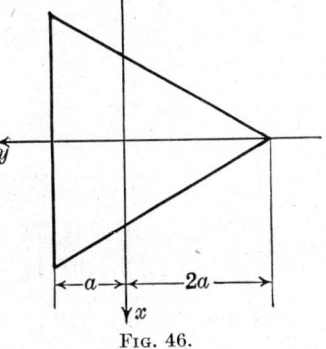

Fig. 46.

$$(y - a) \left(x + \frac{2a + y}{\sqrt{3}} \right) \left(x - \frac{2a + y}{\sqrt{3}} \right) = 0,$$

where the origin has been taken at the centroid of the section. Along the side $y = a$ we have $\frac{dy}{ds} = 0$, and hence no condition is imposed on the boundary values of the function $S(y)$ along this side, whereas we require that

$$S(y) = \frac{W_x}{2I_y} x^2 = \frac{W_x}{2I_y} \frac{(2a+y)^2}{3} \quad \text{on} \quad x = \pm \frac{2a+y}{\sqrt{3}}.$$

Therefore we take

$$S(y) \equiv \frac{W_x}{6I_y} (2a+y)^2,$$

and from (60.7) it follows that

$$\nabla^2 T(x,y) = \frac{W_x}{I_y} \left[\frac{2\left(\sigma - \frac{1}{2}\right)}{3(\sigma+1)} y - \frac{2}{3} a \right],$$

where we have set $\alpha = 0$ and are, accordingly, solving the problem of pure flexure by a load applied at the center of flexure.

The differential equation and boundary conditions on $T(x,y)$ can be readily satisfied when Poisson's ratio takes a particular value, namely, $\sigma = \frac{1}{2}$. In this case, we have

$$\nabla^2 T(x,y) = -\frac{2}{3} \frac{W_x}{I_y} a \quad \text{in } R$$

$$T(x,y) = 0 \quad \text{on} \quad \begin{cases} y = a \\ x^2 = \frac{1}{3}(2a+y)^2. \end{cases}$$

We try

$$T(x,y) = k[x^2 - \tfrac{1}{3}(2a+y)^2](y-a)$$

and find that the stress function is given by

$$(60.12) \quad T(x,y) = \frac{W_x}{6I_y}\left[x^2 - \frac{1}{3}(2a+y)^2\right](y-a).$$

Equation (60.11) now yields the position of the center of flexure. Straightforward calculations give

$$I_y = \frac{3\sqrt{3}\, a^4}{2}$$

$$\iint_R T(x,y)\, dx\, dy = \frac{3\sqrt{3}\, W_x a^5}{10 I_y} = \frac{a W_x}{5},$$

$$\int_C x^3 y\, dy = 2 \int_0^{a\sqrt{3}} x^3(\sqrt{3}\, x - 2a)\sqrt{3}\, dx = \frac{9\sqrt{3}\, a^5}{5},$$

and therefore $y_{cf} = 0$. Since the cross section is symmetrical about the y-axis, we see that the center of flexure is at the origin, and hence at the centroid of the section. Thus, the function $T(x, y)$, given above, furnishes the solution of the flexure problem for a beam of equilateral triangular section when the load W_x is applied at the centroid and when Poisson's ratio has the value of one-half.

The flexure function $T(x, y)$ was introduced, for the case of bending by a load along a principal axis, by Timoshenko[1] and was used by him to solve the flexure problem for a number of cross sections.

It will be recalled that when the functions $R(x)$ and $S(y)$ were introduced, it was remarked that they might be so chosen as to yield either a simple boundary condition or a simple differential equation for the function $T(x, y)$. The first course led to Timoshenko's stress function $T(x, y)$, discussed above, which can be interpreted physically as representing the deflection of an elastic membrane stretched over an opening of boundary C in a plane plate and subjected to a nonuniform load. We follow now the alternative course and choose $R(x)$ and $S(y)$ to make $T(x, y)$ a harmonic function.

Let us define

(60.13) $$\begin{cases} R(x) = -\mu\sigma K_y x^2 - \mu\alpha x, \\ S(y) = \mu\sigma K_x y^2 - \mu\alpha y. \end{cases}$$

We shall designate the function $T(x, y)$ defined by Eqs. (60.3) and (60.4) with this choice of $R(x)$ and $S(y)$ by $M(x, y)$. Then these equations become

(60.14) $$\nabla^2 M(x, y) = 0$$

and

$$\frac{dM}{ds} = [\mu(1 + \sigma)K_x x^2 - \mu\sigma K_x y^2 + \mu\alpha y]\frac{dy}{ds}$$
$$- [\mu(1 + \sigma)K_y y^2 - \mu\sigma K_y x^2 - \mu\alpha x]\frac{dx}{ds} \quad \text{on } C.$$

---

[1] S. TIMOSHENKO, *Proceedings of the London Mathematical Society*, ser. 2, vol. 20 (1922), p. 398.

An account of this work will be found in S. Timoshenko, Theory of Elasticity, Secs. 88–94.

This boundary condition can be integrated in part to give
(60.15) $\quad M = \frac{1}{2}\mu\alpha(x^2 + y^2)$
$$+ \mu K_x \left[-\frac{1}{3}\sigma y^3 - (1 + \sigma) \int_{(x_0,y_0)}^{(x,y)} x^2\, dy\right]$$
$$+ K_y \left[\frac{1}{3}\sigma x^3 - (1 + \sigma) \int_{(x_0,y_0)}^{(x,y)} y^2\, dx\right] + \text{const.} \quad \text{on } C.$$

The stresses τ_{zx}, τ_{zy} can be written from Eqs. (60.2) and (60.13) as
$$\tau_{zx} = \frac{\partial M}{\partial y} + \mu K_x[\sigma y^2 - (1 + \sigma)x^2] - \mu\alpha y,$$
$$\tau_{zy} = -\frac{\partial M}{\partial x} + \mu K_y[\sigma x^2 - (1 + \sigma)y^2] + \mu\alpha x.$$

In the case of bending by a load W_x along a principal axis, the formulas for the stresses become
$$\tau_{zx} = \frac{\partial M}{\partial y} + \frac{W_x}{2I_y}\left[\frac{\sigma}{1+\sigma} y^2 - x^2\right] - \mu\alpha y,$$
$$\tau_{zy} = -\frac{\partial M}{\partial x} + \mu\alpha x,$$

while $M(x, y)$ is subject to the condition
(60.16) $\quad M = \frac{1}{2}\mu\alpha(x^2 + y^2) - \frac{\sigma}{6(1+\sigma)}\frac{W_x}{I_y} y^3$
$$+ \frac{W_x}{2I_y} \int_{(x_0,y_0)}^{(x,y)} x^2\, dy \qquad \text{on } C.$$

One can interpret the determination of the harmonic function $M(x, y)$, subject to the condition (60.15) or (60.16) on C, in terms of a membrane analogy, as was done in connection with the torsion problem in Sec. 46. Thus, the solution of the flexure problem by means of the *membrane function $M(x, y)$* is mathematically identical with the determination of the deflection of an unloaded elastic membrane stretched across a closed space curve whose projection on the xy-plane is the contour C, and whose variable height above the plane is given by the boundary values of $M(x, y)$ [(60.15) or (60.16)]. This analogy, for the case in which the boundary values are given by Eq. (60.16), has been used by Griffith and Taylor,[1] among others, to obtain experi-

---
[1] A. A. GRIFFITH and G. I. TAYLOR, *Technical Report of The National Advisory Committee on Aeronautics* (Great Britain), vol. 3 (1917–1918), p. 950.
See S. Timoshenko, Theory of Elasticity, Sec. 95, for a description of the experimental procedure.

mental solutions of the flexure problem for beams whose cross sections are such that the problem does not yield readily to mathematical treatment. Neményi[1] has derived both $M(x, y)$ and another flexure function $F_I(x, y)$ as special cases of a more general flexure function and has discussed these two formulations of the flexure problem in the light of the membrane analogy and of numerical solutions, respectively.

There exists a very close connection between the membrane function $M(x, y)$ and the canonical flexure functions φ_{11}, φ_{12}, φ_{21}, φ_{22}, discussed in Sec. 52. A comparison of the boundary condition (60.15) with the boundary values taken by the torsion function ψ [Eq. (35.4)] and by the flexure functions ψ_1 and ψ_2 [Eqs. (52.17)] shows that

$$(60.17) \quad M(x, y) = \mu\alpha\psi(x, y) + \mu K_x \psi_1(x, y) + \mu K_y \psi_2(x, y) \quad \text{on } C.$$

Since a harmonic function is uniquely determined by its boundary values, it follows that Eq. (60.17) holds throughout the region R of the cross section. If we define the conjugate membrane function $L(x, y)$ so that $L + iM$ is analytic in R, then we can write

$$L + iM = \mu\alpha(\varphi + i\psi) + \mu K_x(\varphi_1 + i\psi_1) + \mu K_y(\varphi_2 + i\psi_2).$$

Eqs. (52.22) now furnish the following relation between the complex membrane function $L + iM$, on the one hand, and the torsion function $\varphi + i\psi$ and the canonical flexure functions $\varphi_{11} + i\psi_{11}, \cdots$ on the other hand:

$$\begin{aligned} L + iM = {}& \mu\alpha(\varphi + i\psi) + \mu K_x[(1 + \sigma)(\varphi_{11} + i\psi_{11}) \\ & - \sigma(\varphi_{12} + i\psi_{12})] + \mu K_y[(1 + \sigma)(\varphi_{22} + i\psi_{22}) \\ & + \sigma(\varphi_{21} + i\psi_{21})]. \end{aligned}$$

61. Technical Theory of the Bending of Beams. In the preceding sections of this chapter, we have considered the state of stress in a beam fixed at one end and loaded at the other end by a system of stresses equivalent to an arbitrary force and a couple; the stresses over the two ends were assumed to be distributed

---

[1] P. Neményi, "Über die Berechnung der Schubspannungen im gebogenen Balken," *Zeitschrift für angewandte Mathematik und Mechanik*, vol. 1 (1921), pp. 89–96.

in the particular manner prescribed by the solution of the problem, and the sides of the beam were assumed to be free of external force. The problem was seen to reduce to the task of solving a boundary-value problem of Dirichlet or Neumann.

In practice, however, a beam is often used to carry loads distributed over its sides; the beam may be supported at one or both ends, as well as at intermediate points, or it may be carried on a continuous elastic foundation, and the cross section is often of such form as to be analytically intractable. Exact solutions to such problems have been found in only a few cases. It is clear that the engineer, faced with the necessity of dealing with problems of beams, must make simplifying assumptions in order that solutions may be made readily available. Such assumptions and the resulting approximate theory of bending will now be considered briefly.

In Sec. 32 the bending of a beam of arbitrary cross section by end couples was studied. It was seen there that the curvature $1/R$ of the central line of the beam is related to the magnitude M of the applied couple by the Bernoulli-Euler law,

$$(61.1) \qquad \frac{1}{R} = \frac{M}{EI},$$

where I is the moment of inertia of the cross section about an axis through its centroid and perpendicular to the plane of the couple. The same relation was found in Sec. 54 to hold for a beam of arbitrary cross section fixed at one end and bent by an end load directed along a principal axis. The technical theory of the bending of beams assumes that the Bernoulli-Euler law is universally valid and may be applied to all problems involving initially straight girders.

It will be convenient to choose the x-axis to coincide with the line of centroids of the undeformed horizontal beam, and the positive y-axis will be taken downward; then the deflection of the central line of the beam after deformation is y. We shall consider only beams whose cross sections have an axis of symmetry. This will be taken as the y-axis, which is thus a principal axis of the section. The external loads will be assumed to act in the xy-plane. From symmetry it follows that this is also the plane of bending. Since the deflection derivatives contemplated in the linear theory of elasticity are small compared with unity (as are

§61 EXTENSION, TORSION, AND FLEXURE OF BEAMS

those that occur, for the most part, in practice), we can neglect the square of the slope dy/dx appearing in the formula

$$\frac{1}{R} = \frac{\pm \frac{d^2y}{dx^2}}{\left[1 + \left(\frac{dy}{dx}\right)^2\right]^{3/2}},$$

so that

(61.2) $$\frac{1}{R} = \pm \frac{d^2y}{dx^2}.$$

Consider now the portion of a beam to the right of section xx (Fig. 47). This section is subject both to the external forces (for example, weight of beams, external load, reactions due to

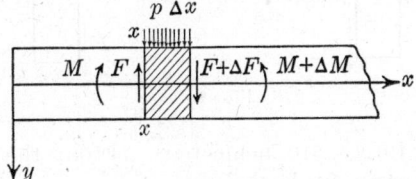

Fig. 47.

supports) and to the internal forces exerted by the material to the left and transmitted across the section xx. These internal reactions are equivalent to a force and to a couple. The couple will be assumed to have a bending moment M about the z-axis; furthermore, we shall be concerned only with the y-component F of the force. This bending moment (representing the action of the material to the left of section xx on that to the right) will be reckoned as positive[1] if the stresses producing it deform the central line of the beam so that it becomes concave upward (Fig. 48). Since for this form of the central line the center of curvature lies on the negative side of the y-axis (for positive M), we have, from (61.1) and (61.2),

(61.3) $$\frac{d^2y}{dx^2} = -\frac{M}{EI}.$$

The direction of positive shearing force F is as shown in Fig. 48.

[1] It should be noted that if the action of the material to the right of the section on that to the left is under discussion, then the directions of positive M and F are reversed. See the directions of $M + \Delta M$ and $F + \Delta F$ in Fig. 48.

The relations between load, shearing force, bending moment, and deflection may be found from the equations of static equilibrium as follows. Consider an element Δx of a beam (Fig. 47) carrying a distributed load $p(x)$, where $p(x)$ is the load per unit length of the beam. This element is subject to the load $p\,\Delta x$, the bending moments M, $M + \Delta M$, and the shearing forces F,

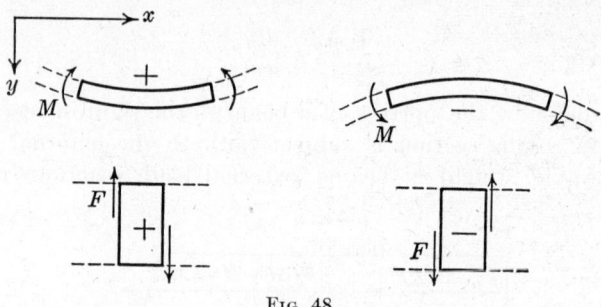

Fig. 48.

$F + \Delta F$ (body forces are neglected). From the condition of equilibrium of forces, we have

$$p\,\Delta x + \Delta F = 0,$$

or

(61.4) $$p(x) = -\frac{dF}{dx}.$$

From the vanishing of the resultant moment, we get

$$(F + \Delta F)\,\Delta x + M - (M + \Delta M) + p\,\Delta x \cdot \frac{\Delta x}{2} = 0,$$

or

$$F = \frac{dM}{dx},$$

and

(61.5) $$p(x) = -\frac{dF}{dx} = -\frac{d^2M}{dx^2}.$$

Eliminating the bending moment M from Eqs. (61.3) and (61.5), we find that

$$\frac{d^2}{dx^2}\left(EI\,\frac{d^2y}{dx^2}\right) = p(x).$$

If the cross section of the beam is uniform (as we shall henceforth assume), we have the equation

$$EI \frac{d^4y}{dx^4} = p(x)$$

for the determination of the deflection y of the central line.

The bending moment M, shearing force F, and load p are given in terms of the deflection y by the relations

(61.6) $\quad M = -EI \dfrac{d^2y}{dx^2}, \qquad F = -EI \dfrac{d^3y}{dx^3}, \qquad p = EI \dfrac{d^4y}{dx^4},$

while in terms of the bending moment, one has

(61.7) $\qquad F = \dfrac{dM}{dx}, \qquad p = -\dfrac{d^2M}{dx^2}.$

The constants of integration that appear in the solution of the above equations are determined by specifying the conditions of fixity at the ends of the beam. We have at the ends the following conditions:

1. Clamped (built-in) end: $y =$ known const., $\dfrac{dy}{dx} =$ known const.,
2. Simply supported (hinged) end: $y =$ known const.,

$$\frac{d^2y}{dx^2} = 0, \qquad (\text{from } M = 0),$$

3. Free end carrying no load:

$$\frac{d^2y}{dx^2} = \frac{d^3y}{dx^3} = 0, \qquad (\text{from } M = F = 0),$$

4. Free end carrying a load W: $\dfrac{d^2y}{dx^2} = 0, \; F = -EI \dfrac{d^3y}{dx^3} = W,$
5. Elastically clamped end (rigid support): $y =$ known const.,

$$\frac{d^2y}{dx^2} - k\frac{dy}{dx} = 0.$$

When the load p includes concentrated forces, then the condition of equilibrium of the element Δx (Fig. 49) requires that

$\Delta F = P$. In other words, the magnitude of the shearing force F $\left(\text{and hence of } \dfrac{dM}{dx}\right)$ changes by the amount P as a concentrated load P is passed. If graphs of shearing force and bending moment are drawn as functions of position x, then at a concentrated load the shearing-force diagram will have an ordinary discontinuity, while the bending-moment diagram will have a discontinuous slope.

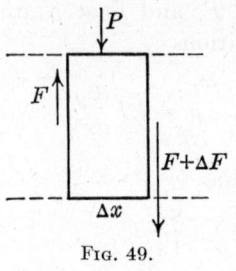

Fig. 49.

The normal stresses τ_{xx} that give rise to the bending moment can be found from the moment M by assuming a stress distribution over the cross section that varies linearly with y. It will be recalled that in the bending of a beam by end couples (Sec. 32), as well as in bending by an end load along a principal axis (Sec. 53), the normal stress was found to be given (in the notation of this section) by

$$(61.8) \qquad \tau_{xx} = \frac{M_z y}{I_z}.$$

This relation between bending stress and bending moment will be assumed to hold for all beam problems and will be used to determine the bending stress. The maximum stress is given by

$$(61.9) \qquad (\tau_{xx})_{\max} = \frac{(M_z)_{\max}}{I_z/y_{\max}} = \frac{(M_z)_{\max}}{Z},$$

where $y_{\max}$ is the distance from the centroid to the outermost fiber, and $Z = I_z/y_{\max}$ is called the *section modulus*. The values of both I_z and Z for commonly used sections are given in engineering handbooks.

The shearing force F can be found from the bending moment M [Eq. (61.7)], but the shearing stress cannot be determined without further information or assumption as to the distribution of stress across the section. The mean shearing stress F/A (where A is the cross-sectional area) can be readily found but may fall considerably short of the maximum shearing stress. In practice, the shearing stress is often neglected, as being small in comparison with the bending stress τ_{xx}. We shall return to these points when we compare the approximate solution of the problem

§61 EXTENSION, TORSION, AND FLEXURE OF BEAMS 269

of a cantilever carrying an end load with the exact solution found earlier.

As an illustration, we consider a beam carrying a uniform load p per foot run. From (61.6) and successive integrations, one finds

(61.10)
$$\begin{cases} EI\dfrac{d^4y}{dx^4} = p, \\ EI\dfrac{d^3y}{dx^3} = -F = px + a, \\ EI\dfrac{d^2y}{dx^2} = -M = \dfrac{1}{2}px^2 + ax + b, \\ EI\dfrac{dy}{dx} = \dfrac{1}{6}px^3 + \dfrac{1}{2}ax^2 + bx + c, \\ EIy = \tfrac{1}{24}px^4 + \tfrac{1}{6}ax^3 + \tfrac{1}{2}bx^2 + cx + d. \end{cases}$$

The constants of integration a, b, c, d are to be determined from the end conditions. We have for a

(61.11) clamped end (at $x = 0$), $c = d = 0$,

and for a

(61.12) simply supported end (at $x = 0$), $b = d = 0$.

The problem will now be specialized to that of a cantilever built in *at the origin* and carrying, in addition to the distributed load pl, a concentrated load W at the free end $x = l$. The *boundary* conditions at the free end are

$$\frac{d^2y}{dx^2} = 0, \qquad -EI\frac{d^3y}{dx^3} = W, \qquad (\text{at } x = l),$$

from which it follows that

$$a = -(pl + W), \qquad b = \tfrac{1}{2}l(pl + 2W),$$

and we have

$$EIy = \tfrac{1}{24}px^4 - \tfrac{1}{6}(pl + W)x^3 + \tfrac{1}{4}l(pl + 2W)x^2.$$

The deflection of a cantilever carrying only an end load W can be found by setting $p = 0$ in the last equation. This gives

(61.13)
$$\begin{cases} EIy = \tfrac{1}{6}Wx^2(3l - x), \\ EI\dfrac{dy}{dx} = \dfrac{1}{2}Wx(2l - x), \\ M = -EI\dfrac{d^2y}{dx^2} = -W(l - x), \\ F = -EI\dfrac{d^3y}{dx^3} = W. \end{cases}$$

These results may also be obtained in the following manner. From Eq. (61.7) it is seen that when the distributed load $p(x)$ vanishes, the bending moment is a linear function of x. In the present example, the moment vanishes at the free end, while at the root $x = 0$ the reaction of the support is seen (from considerations of static of equilibrium) to consist of a couple of moment $M = -Wl$ and a vertical force W. The bending moment is then given by

$$M = -W(l - x)$$

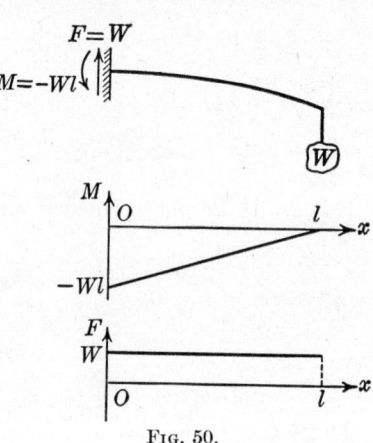

Fig. 50.

and is shown in Fig. 50. The shearing force F is given by the slope of the moment diagram (Fig. 50). The deflection of the beam may now be found by integrating the relation

$$EI\frac{d^2y}{dx^2} = -M = W(l - x)$$

and using the boundary conditions $y(0) = y'(0) = 0$.

The problem just considered is an example of a *statically determinate system;* that is, one in which the unknown reactions at the support (horizontal and vertical components of force, moment) may be found from the three equations of plane static equilibrium. Systems in which the number of reactive elements is greater than the number of equilibrium conditions are called *statically indeterminate;* in such systems, the elastic deflection of the bar must be taken into account. An example of a statically indeterminate system will be considered below.

§61 EXTENSION, TORSION, AND FLEXURE OF BEAMS

From (61.13) it is seen that the maximum deflection in a cantilever carrying an end load W occurs at the free end, where

$$EIy_{max} = \tfrac{1}{3}Wl^3, \qquad (at\ x = l),$$

and this agrees with the result found by the exact theory [Eqs. (54.2)]. The maximum bending moment occurs at the fixed end of the beam, where

$$M_{max} = -Wl, \qquad (at\ x = 0).$$

From (61.9) it follows that the maximum bending stress is

$$(\tau_{xx})_{max} = \frac{Wl}{Z} = \frac{Wly_{max}}{I_z}.$$

If the beam is of circular cross section, then $I_z = \pi a^4/4$, $Z = \pi a^3/4$, and

$$(\tau_{xx})_{max} = \frac{4Wl}{\pi a^3}, \qquad (at\ x = 0).$$

The mean shearing stress is given by

$$\frac{F}{area} = \frac{W}{\pi a^2}.$$

The exact solution[1] (Sec. 55) gives

$$(\tau_{xy})_{max} = \frac{3 + 2\sigma}{2(1 + \sigma)} \frac{W}{\pi a^2} \doteq \frac{4}{3} \frac{W}{\pi a^2},$$

$$(\tau_{xz})_{max} = (\tau_{xz})_{x=y=\sqrt{2}a/2} = \frac{1 + 2\sigma}{2(1 + \sigma)} \frac{W}{\pi a^2} \doteq \frac{2}{3} \frac{W}{\pi a^2}.$$

The neglect of shear stress in the engineering theory of thin beams is justified in this case, for we have

$$\frac{(\tau_{xy})_{max}}{(\tau_{xx})_{max}} = \frac{3 + 2\sigma}{8(1 + \sigma)} \frac{a}{l} \doteq \frac{1}{3} \frac{a}{l},$$

and the ratio of shearing stress to bending stress is seen to be of the order of the ratio of the lateral to the longitudinal dimensions of the beam.

---

[1] Note that the axis of the beam is directed along the x-axis in this section but along the z-axis in Sec. 55. The loading force is taken parallel to the y-axis in this section, while it was parallel to the x-axis in the earlier section.

The maximum shearing stress is found, in the technical theory of beams, by making some assumptions as to the distribution of shear across the section. In the case of a beam of rectangular cross section, for example, it is commonly assumed that the shearing stress is parallel to the shearing force F (that is, that $\tau_{xz} = 0$) and is uniformly distributed across the width of the beam, so that $\dfrac{\partial \tau_{xy}}{\partial z} = 0$.

Some information about the state of stress in a thin rectangular beam can be got by studying the results obtained in Sec. 56 for a beam whose cross section is bounded by the ellipse $y^2/a^2 + z^2/b^2 = 1$. When one axis of the ellipse is much smaller than the other, the beam resembles a bar of thin rectangular section. If $b \ll a$, then the shape of the beam is nearly that of a thin rectangular plank loaded in the direction of its longer side. In this case, we have, from Sec. 56,

$$(\tau_{xy})_{\max} \doteq \frac{4}{3} \frac{W}{A},$$

$$(\tau_{xz})_{\max} \doteq \frac{4}{3} \frac{W}{A} \frac{b}{2a} \ll (\tau_{xy})_{\max},$$

and the assumption of the technical theory regarding the vanishing of τ_{xz} is borne out. If, however, $a \ll b$, then the load acts along the shorter side of the section, and we have

$$(\tau_{xy})_{\max} \doteq \frac{2}{1+\sigma} \frac{W}{A} \doteq \frac{3}{2} \frac{W}{A},$$

$$(\tau_{xz})_{\max} \doteq \frac{2}{1+\sigma} \frac{W}{A} \frac{\sigma b}{a} \doteq \frac{3}{2} \frac{W}{A} \frac{b}{3a}.$$

Thus, in beams whose ratio of depth to width is small, the lateral shearing stress τ_{xz}, instead of vanishing, may exceed in magnitude the vertical shear τ_{xy}. A numerical comparison of values of the shearing stress as given by the exact and the technical theory for rectangular beams of various ratios of depth to width is given by Timoshenko.[1]

The calculations made above were for a cantilever carrying an end load W. The problem was seen to be statically determinate;

---

[1] S. Timoshenko, Theory of Elasticity, Sec. 92.

that is, the three equations of plane static equilibrium were sufficient to determine the three unknown reactive elements. As an example of a statically indeterminate system, we consider now the problem of a beam built in at the origin, simply supported at the end $x = l$, and bearing a uniform load p per unit length. In this case, the number of unknown reactive force components and moments is greater than three, and the elastic deformation of the bar must be considered. The boundary conditions at the supported end are

$$y(l) = y''(l) = 0,$$

and from Eqs. (61.10) and (61.11) we get

$$a = -\tfrac{5}{8}pl, \qquad b = \tfrac{1}{8}pl^2, \qquad c = d = 0,$$

and hence

$$\begin{aligned}EIy(x) &= \tfrac{1}{48}p(2x^4 - 5lx^3 + 3l^2x^2) \\ &= \tfrac{1}{48}px^2(3l - 2x)(l - x).\end{aligned}$$

The deflection at the center, which is about 4 per cent less than the maximum deflection, is given by

$$EIy\left(\frac{l}{2}\right) = \frac{pl^4}{192}.$$

62. Some Further Developments in Problems on Beams. In this section, we indicate briefly some directions in which the developments of this chapter have been extended.

It is obvious that Saint-Venant's theory of torsion and flexure provides exact solutions of the problems of twisting and bending only if the end stresses are distributed in a manner specified by the solutions themselves. If the external forces are distributed over the ends of the beams in some different fashion, then the actual state of stress will differ near the ends from that given by Saint-Venant's theory. These local irregularities may be expected, however, to be very small sufficiently far from the points of application of the forces. This problem has been reviewed by J. L. Synge.[1]

We have assumed throughout that the sides of the cylinder are free of external force. In practice, however, beams are com-

---

[1] J. L. SYNGE, "The Problem of Saint Venant for a Cylinder with Free Sides," *Quarterly of Applied Mathematics*, vol. 2 (1945), pp. 307–317.

monly used to carry distributed loads. Some special cases of exact solution of the problem of bending by loads distributed over the beam are contained in A. E. H. Love's A Treatise on the Mathematical Theory of Elasticity, in Chaps. XVI and XVII. As is to be expected, the problem of bending by distributed loads is a difficult one, and it turns out that the length of the central line of the beam, in general, is altered, and that the bending moment is no longer proportional to the curvature of the central line. In addition to the curvature $1/R = M/EI$ given by the Bernoulli-Euler formula, there is a corrective term that is the same for all cross sections of the beam. If the length of the beam is large in comparison with the greatest linear dimension of the cross section, this additional curvature is negligibly small. In some cases, however, such as those of flexure of plate girders in a suspension bridge, it becomes sufficiently important to be taken into account.

The case of torsion of a beam by forces applied to its sides has been treated by Riz and Zvolinskij.[1]

The general Saint-Venant problem of a bar having a small natural twist has been considered in a series of papers by Riz and, from a different standpoint, in an independent series by Lourie and Janelidze.[2]

[1] N. V. ZVOLINSKIJ and P. M. RIZ, "Torsion eines zylindrischen Stabes durch Kräfte die auf seiner Seitenfläche verteilt sind," *Bulletin de l'académie des sciences de l'U.R.S.S., classe des sciences mathématiques et naturelles*, no. 10 (1939), pp. 21–26. (In Russian with a German abstract.)

P. M. RIZ, "Concerning the Torsion of a Prismatic Bar by Axial Forces Distributed along its Side Surface," *Applied Mathematics and Mechanics*, new ser., vol. 4, no. 2 (1940), pp. 121–122. (In Russian with an English abstract.)

P. M. RIZ, "General Solution of Torsion Problem in Nonlinear Theory of Elasticity," *Applied Mathematics and Mechanics*, vol. 7 (1943), pp. 149–154. (In Russian with an English abstract.)

[2] P. RIZ, "On the Deformations of Naturally Twisted Bars," *Comptes rendus de l'académie des sciences de l'U.R.S.S.*, new ser., vol. 23 (1929), pp. 17–20, 441–444, 765–767.

A. I. LOURIE, "Flexion and Stability of Naturally Torsioned Rectilineal Rods," *Applied Mathematics and Mechanics*, new ser., vol. 2 (1938), pp. 55–68 (in Russian with English summary).

A. LOURIE and G. JANELIDZE, "On Saint Venant's Problem for Naturally Twisted Bars," *Comptes rendus de l'académie des sciences de l'U.R.S.S.*, new ser., vol. 24 (1939), pp. 24–27, 227–228; vol. 25 (1939), pp. 577–579; vol. 27 (1940), pp. 436–439.

§62   EXTENSION, TORSION, AND FLEXURE OF BEAMS

The rods considered in this chapter have been either cylinders, with cross section defined, say, by the equation

$$f(x, y) = 0$$

or they have been bodies of revolution. Consider, now, the surface defined by

$$f[x(1 - kz), y(1 - kz)] = 0,$$

where the parameter k is such that kz is small compared with unity. This equation defines the surface of a rod that is almost a cylinder. The torsion and bending of such rods have been discussed by Panov and Ruchadze.[1]

A problem similar to that just mentioned is obtained by dealing with rods whose surface is defined by

$$f(x, y + kz^2) = 0,$$

where k is again a small parameter. The rod has thus a slightly curved axis. The torsion and bending of such beams have been treated by Riz and Ruchadze.[2]

Comprehensive reviews of the Saint-Venant torsion problem, with extensive bibliographies, have been given by Pöschl and Higgins.[3]

---

[1] D. Y. PANOV, "Concerning the Torsion of Nearly Prismatic Rods," *Applied Mathematics and Mechanics*, new ser., vol. 2 (1938), pp. 159–180 (in Russian with English summary); *Comptes rendus de l'académie des sciences de l'U.R.S.S.*, new ser., vol. 20 (1938), pp. 251–253.

A. K. RUCHADZE, "Problem of Bending of Beams Near to Prismatical Ones," *Mitteilungen der Georgischen Abteilung der Akademie der Wissenschaften der U.S.S.R.*, vol. 1 (1940), pp. 577–582 (in Russian); *Applied Mathematics and Mechanics*, vol. 6 (1942), pp. 123–138 (in Russian with German summary).

[2] P. RIZ, "On Deformations of Bars with Slightly Curved Axis," *Comptes rendus de l'académie des sciences de l'U.R.S.S.*, new ser., vol. 24 (1939), pp. 110–113, 229–232.

A. K. RUCHADZE, "On the Problem of Deformation of a Beam with Slightly Curved Axis," *Mitteilungen der Georgischen Abteilung der Akademie der Wissenschaften der U.S.S.R.*, vol. 2 (1941), pp. 35–42 (in Russian).

[3] Th. PÖSCHL, "Bisherige Lösungen des Torsionsproblems," *Zeitschrift für angewandte Mathematik and Mechanik*, vol. 1 (1921), pp. 312–328.

T. J. HIGGINS, "A Comprehensive Review of Saint Venant's Torsion Problem," *American Journal of Physics*, vol. 10 (1942), pp. 248–259.

Compound beams composed of several elastic materials (reinforced concrete) have been discussed by Muschelišvili[1] and by Ruchadze and Vekoua.[2]

Lechnitzky[3] has recently considered the problems of tension, torsion, pure bending, and flexure for anisotropic beams. The most general anisotropy is assumed, the material being characterized by 21 elastic constants. Vantorin[4] has used the Ritz method to study the torsion of an anisotropic beam with a cross section somewhat similar to that of an airplane propeller

$$[y^2 = k^2 x(1 - x)^2].$$

The work of Palerino[5] and Platrier[6] should also be mentioned.

[1] N. Muschelišvili, "Sur le problème de torsion et de flexion des poutres élastiques composeés," *Bulletin de l'académie des sciences de l'U.R.S.S.*, ser. VII, no. 7 (1932), pp. 907–945 (in Russian).

[2] A. Ruchadze, "Sur la flexion d'un cylindre circulaire armé d'une barre longitudinale par une force transversale," *Bulletin de l'académie des sciences de l'U.R.S.S.*, ser. VII, no. 9 (1933), pp. 1297–1308 (in Russian).

A. Roukhadze and I. Vekoua, "Sur le problème de torsion et de flexion par une force transversale d'une poutre composée de deux matériaux élastiques limités par deux ellipses homofocales," *Applied Mathematics and Mechanics*, vol. 1 (1933), pp. 167–178 (in Russian with French summary).

A. Ruchadze, "Torsion und Verbiegung dürch Querkräfte eines elastischen Balkens, der aus zwei verschiedenen dürch Epitrochoiden abgegrenzten Materialen besteht," *Travaux de l'institut mathématique de Tbilissi*, vol. 1 (1937), pp. 125–139. (In Georgian with Russian summary.)

A. K. Ruchadze, "Zur Aufgabe der Biegung eines aus verschiedenen Werkstoffen zusammengesetzten Balkens," *Mitteilungen der Georgischen Abteilung der Akademie der Wissenschaften der U.S.S.R.*, vol. 1 (1940), pp. 107–114. (In Russian with German summary.)

[3] S. G. Lechnitzky, "Sur le problème d'équilibre élastique du cylindre homogène avec l'anisotropie arbitraire," *Applied Mathematics and Mechanics*, new ser., vol. 2 (1939), pp. 345–368 (in Russian with French summary).

S. G. Lechnitzky, "Equilibrium of an Anisotropic Cantilever Beam," *Applied Mathematics and Mechanics*, new ser., vol. 6 (1942), pp. 3–18 (in Russian with English summary).

[4] V. D. Vantorin, "Torsion of a Prism with a Cross-section Bounded by the Curve $y^2 = k^2 x(1 - x)^2$," *Applied Mathematics and Mechanics*, new ser., vol. 3 (1939), pp. 151–152 (in Russian with English summary).

[5] D. Palerino, "Sulle dilatazioni superficiali dei solidi elastici," *Atti della reale accademia nazionale dei lincei*, ser. 6, vol. 18 (1933), pp. 140–145.

[6] C. F. F. Platrier, "Généralisations du problème de Saint-Venant," *Proceedings of the Fifth International Congress for Applied Mechanics*, 1939, pp. 102–106.

CHAPTER V
VARIATIONAL METHODS

63. Introduction. The earlier chapters of this volume provide an illustration of one general mode of approach to the boundary-value problems of mathematical physics. It was based on the formulation of differential equations, with appropriate boundary conditions, and it has been used extensively in the preceding sections, where the differential equations of equilibrium and of compatibility have been set up and solved in certain special cases.

An entirely different approach to the problems of elasticity is provided by the methods of the calculus of variations. Variational procedures can serve two distinct purposes. One of these is concerned with the derivation of certain general theorems on potential energy of an elastic system and with conversion of the variational problem into the problem of solving an equivalent system of differential equations. The early sections of this chapter provide an illustration of this particular use of the calculus of variations. Another important use of variational methods is in obtaining approximate solutions of problems that are otherwise intractable. This is illustrated in the latter half of this chapter, where several procedures for deducing approximate solutions of the boundary-value problems of mathematical physics are outlined.

The torsion and flexure problems provide excellent illustrations of the use of the calculus of variations in the solution of the boundary-value problems of mathematical physics. In the preceding chapter, it was shown that the solution of the problem of a beam subjected to certain end stresses, statically equivalent to an arbitrary force and couple, can be made to rest upon the solution of one or more Dirichlet problems; that is, upon the search for a function harmonic within the region of the cross section of the cylinder, and assuming prescribed values on the boundary. As was pointed out in the discussion of physical analogues, many other physical problems can likewise be formulated in terms of the first boundary-value problem in Laplace's equation. The exact analytical solutions of such problems, however, have been found in relatively few cases, since the general methods developed in the preceding chapter presuppose the knowledge of the analytical function that maps the cross section of a beam conformally on the unit circle. In many problems of technical importance, the mapping function is either unknown or unmanageable, and one is obliged to turn to different methods of obtaining approximate solutions. A discussion of several of these alternative methods, followed by the illustration of their application to the torsion problem for a beam of rectangular cross section, is contained in the latter half of this chapter.

64. Minimum Potential Energy. In this section, a general theorem on the potential energy of the internal strain and the external load is established. This theorem is derived from a comparison of the potential energy associated with the equilibrium displacements u_i with the energy of other displacements satisfying the given boundary conditions but not the equations of equilibrium.

In Sec. 26, we defined the strain-energy density W and the strain energy $U = \int_\tau W \, d\tau$ of an elastic body occupying a region τ. The strain-energy density was given in terms of the stresses, strains, and elastic coefficients $c_{ij} = c_{ji}$ by the relations

[26.7]
$$W = \tfrac{1}{2} c_{ij} e_i e_j = \tfrac{1}{2} \tau_i e_i, \quad (i, j = 1, 2, \cdots, 6),$$
$$= \tfrac{1}{2} \tau_{ij} e_{ij}, \quad (i, j = 1, 2, 3),$$

where the stresses τ_i and strains e_i are defined by the relations

[21.1]
$$\begin{cases} \tau_1 = \tau_{11}, \ \tau_2 = \tau_{22}, \ \tau_3 = \tau_{33}, \ \tau_4 = \tau_{23}, \ \tau_5 = \tau_{31}, \ \tau_6 = \tau_{12}, \\ e_1 = e_{11}, \ e_2 = e_{22}, \ e_3 = e_{33}, \ e_4 = 2e_{23}, \ e_5 = 2e_{31}, \ e_6 = 2e_{12}. \end{cases}$$

The stress components can be obtained by differentiating the strain-energy density; that is,

[26.7]
$$\tau_i = \frac{\partial W}{\partial e_i} = c_{ij} e_j, \, (i, j = 1, 2, \cdots, 6).$$

Let us now compare the strain energy U, associated with displacements u_i that satisfy the equilibrium and boundary conditions, with the strain energy $U + \Delta U$, corresponding to any other set of displacements $u_i + u_i' \equiv u_i + \Delta u_i$.

§64 VARIATIONAL METHODS

From the equilibrium displacements u_i, one calculates the strains e_i and the strain-energy density $W(e) = \tfrac{1}{2}c_{ij}e_ie_j$. Similarly, from the displacements $u_i + u_i'$, one obtains

$$W + \Delta W = \tfrac{1}{2}c_{ij}(e_i + e_i')(e_j + e_j')$$
$$= W + \tfrac{1}{2}c_{ij}(e_ie_j' + e_i'e_j + e_i'e_j').$$

By interchanging the subscripts i, j in the term $\tfrac{1}{2}c_{ij}e_i'e_j$ and noting that the coefficients c_{ij} are symmetric, one finds that

$$\Delta W = c_{ij}e_ie_j' + W(e'),$$

where

$$W(e') = \tfrac{1}{2}c_{ij}e_i'e_j'$$

is the strain-energy density calculated from the strains e_i'. From Hooke's law, in the form (26.7), it is apparent that the first term in the expression for the strain-energy function ΔW can be written as

$$c_{ij}e_ie_j' = \tau_j e_j' = \tau_{ij}e_{ij}'$$
$$= \tfrac{1}{2}\tau_{ij}(u_{i,j}' + u_{j,i}').$$

An interchange of the subscripts i, j in $\tau_{ij}u_{j,i}'$ and the use of the symmetry of the stress components τ_{ij} gives

$$c_{ij}e_ie_j' = \tau_{ij}u_{i,j}' = (\tau_{ij}u_i')_{,j} - \tau_{ij,j}u_i'.$$

Since the displacements u_i lead to stresses that satisfy the equations of equilibrium $\tau_{ij,j} + F_i = 0$, the foregoing expression can be written as

$$c_{ij}e_ie_j' = (\tau_{ij}u_i')_{,j} + F_i u_i'.$$

The strain-energy difference ΔU can now be written as

$$\Delta U = \int_\tau \Delta W \, d\tau = \int_\tau c_{ij}e_ie_j' \, d\tau + \int_\tau W(e') \, d\tau$$
$$= \int_\tau (\tau_{ij}u_i')_{,j} \, d\tau + \int_\tau F_i u_i' \, d\tau + U(e'),$$

where

$$U(e') = \int_\tau W(e') \, d\tau = \int_\tau \tfrac{1}{2}c_{ij}e_i'e_j' \, d\tau.$$

Using the Divergence Theorem, one can write

$$\Delta U = \int_\Sigma \tau_{ij}u_i'\nu_j \, d\sigma + \int_\tau F_i u_i' \, d\tau + U(e'),$$

where Σ is the surface of the region τ. Inasmuch as the displacements u_i yield stresses that satisfy the conditions of equilibrium and the boundary conditions $\tau_{ij}\nu_j = \overset{\nu}{T}_i$, it follows that

$$(64.1) \quad \Delta U = \int_\Sigma \overset{\nu}{T}_i u'_i \, d\sigma + \int_\tau F_i u'_i \, d\tau + U(e').$$

This equation expresses the difference ΔU in strain energy between the energy associated with the displacements u_i, satisfying the equilibrium and boundary conditions, and the energy of the displacements $u_i + \Delta u_i \equiv u_i + u'_i$; the displacements u'_i are quite arbitrary at this point in the argument. For every choice of values for the functions $u'_i \equiv \Delta u_i$, there is obtained an expression relating the applied forces $\overset{\nu}{T}_i$, F_i, the displacements Δu_i, and the change ΔU in strain energy.

The hitherto arbitrary functions $u'_i \equiv \Delta u_i$ are specialized by the requirement that the displacements $u_i + u'_i$ satisfy the same boundary conditions that the equilibrium displacements u_i satisfy. It is supposed that the displacements are prescribed over the region Σ_u of the surface Σ, while the surface forces $\overset{\nu}{T}_i$ are prescribed over the remainder Σ_T of the surface Σ. Then the functions u'_i must vanish on Σ_u and Eq. (64.1) can be written as

$$(64.2) \quad \Delta U = \int_{\Sigma_T} \overset{\nu}{T}_i \, \Delta u_i \, d\sigma + \int_\tau F_i \, \Delta u_i \, d\tau + U(e').$$

Now in comparing the state characterized by displacements u_i, with associated strain energy U, and the state corresponding to displacements $u_i + \Delta u_i$, with associated strain energy $U + \Delta U$, it is observed that the body force F_i is the same in the two cases. Similarly, the surface force $\overset{\nu}{T}_i$ is the same on the region Σ_T, where the surface forces are prescribed. Hence one can write

$$\Delta U = \Delta \left(\int_\Sigma \overset{\nu}{T}_i u_i \, d\sigma + \int_\tau F_i u_i \, d\tau \right) + U(e'),$$

or

$$(64.3) \quad \Delta \left(U - \int_\Sigma \overset{\nu}{T}_i u_i \, d\sigma - \int_\tau F_i u_i \, d\tau \right) = U(e').$$

The potential energy V of the system is now defined as

$$(64.4) \quad V = U - \int_\Sigma \overset{\nu}{T}_i u_i \, d\sigma - \int_\tau F_i u_i \, d\tau,$$

where it is understood that the integral $\int_\Sigma \overset{\nu}{T}_i u_i \, d\sigma$ is to be extended only over that portion of the surface Σ on which the surface forces $\overset{\nu}{T}_i$ are prescribed. The potential energy V is thus the sum of the strain energy of the body and the potential energy of the external forces. Equation (64.3) can now be written as

(64.5) $$\Delta V = U(e').$$

The displacements $u_i + \Delta u_i$ are now further restricted to be *neighboring displacements* to the equilibrium displacements u_i. The term *neighboring displacements* is used in the sense that the functions Δu_i and the strains e'_i are required to be small compared with the equilibrium displacements and strains. Then $U(e')$, which can be written as

$$U(e') = \int_\tau W(e') \, d\tau = \int_\tau \frac{1}{2} c_{ij} e'_i e'_j \, d\tau$$
$$= \int_\tau \left(\frac{1}{2} c_{11} e'_{11} e'_{11} + \cdots \right) d\tau = \int_\tau \left[\frac{1}{2} c_{11} \left(\frac{\partial \Delta u_1}{\partial x_1}\right)^2 + \cdots \right] d\tau,$$

is of the second order in the small strains e'_i. Equation (26.9) shows, furthermore, that $U(e')$ takes only positive values provided the strains e'_i are not all zero. Hence

(64.6) $\quad \Delta V = 0 +$ positive, second-order terms in e'_i.

Equation (64.6) expresses the Theorem of Minimum Potential Energy.

THEOREM OF MINIMUM POTENTIAL ENERGY: *Of all displacements satisfying given boundary conditions, those that satisfy the equilibrium conditions make the potential energy V a minimum.*

According to the Theorem of Minimum Potential Energy, it is necessary to select from the set of all functions satisfying the boundary conditions those functions u_i that make the potential energy a minimum. The methods of the calculus of variations[1] may be applied to convert this variational problem into an equivalent problem of solving a differential equation with the prescribed boundary conditions. If the differential equation together with the appropriate boundary conditions can be solved, then the equilibrium displacements u_i can be determined exactly.

[1] See Sec. 65.

In many cases, however, the differential equation and boundary conditions are intractable. In this event, one may turn to a method of obtaining an approximate representation of the state of stress within the elastic body. Here the Theorem of Minimum Potential Energy plays an important role. Instead of looking for the minimal function u_i among the set of all functions that satisfy the given boundary conditions, one may narrow the choice to a smaller class of functions satisfying the boundary conditions, and select from it the functions that yield a least value for the potential energy V. In practice, it is assumed that the displacements can be written as functions that satisfy the boundary conditions and involve a number of parameters. These parameters are then determined so as to make the potential energy of the system a minimum. As a simple example of this approach, the problem of a cantilever beam under uniform load is discussed from the energy standpoint in Sec. 66.

As an illustration, the potential energy of a twisted shaft will now be calculated. It will be recalled (see Sec. 35) that the stresses are given in terms of the stress function Ψ by

$$\tau_{zx} = \mu\alpha \frac{\partial \Psi}{\partial y}, \qquad \tau_{zy} = -\mu\alpha \frac{\partial \Psi}{\partial x}, \qquad \tau_{xx} = \tau_{yy} = \tau_{zz} = \tau_{xy} = 0,$$

where μ is the shear modulus and α is the angle of twist per unit length. The strain-energy density is then calculated to be

$$W = \tfrac{1}{2}\tau_{ij}e_{ij} = \tau_{zx}e_{zx} + \tau_{zy}e_{zy}$$
$$= \frac{1}{2\mu}(\tau_{zx}^2 + \tau_{zy}^2) = \frac{1}{2}\mu\alpha^2 \left[\left(\frac{\partial \Psi}{\partial x}\right)^2 + \left(\frac{\partial \Psi}{\partial y}\right)^2\right],$$

and the strain energy takes the form

$$U = \int_\tau W\,d\tau = \frac{1}{2}\mu\alpha^2 l \int\int_R \left[\left(\frac{\partial \Psi}{\partial x}\right)^2 + \left(\frac{\partial \Psi}{\partial y}\right)^2\right] dx\,dy$$
$$= \frac{1}{2}\mu\alpha^2 l \int\int_R (\nabla\Psi)^2\,dx\,dy,$$

where R is the region of the cross section of the shaft. The potential energy of the applied torque is given by the integral

$$-\int_\Sigma \overset{\nu}{T}_i u_i\,d\sigma = -\int\int_R (\tau_{zx}u + \tau_{zy}v)\,dx\,dy$$
$$= -\alpha l \int\int_R (x\tau_{zy} - y\tau_{zx})\,dx\,dy,$$

where the expression (34.2) for the displacements has been used with $z = l$. This last integral represents the applied torque M_z, which can be written in terms of the unit twist α and the torsional rigidity D as (34.11)

$$M_z = D\alpha.$$

The torsional rigidity D and the stress function Ψ are connected, according to Eq. (35.10), by the relation

$$D = 2\mu \int\int_R \Psi \, dx \, dy.$$

Hence the potential energy of the applied torque can be written as

$$-\int_\Sigma \overset{\nu}{T}_i u_i \, d\sigma = -\alpha^2 l D$$
$$= -2\mu\alpha^2 l \int\int_R \Psi \, dx \, dy.$$

The total potential energy of the system is then

(64.7) $V = \tfrac{1}{2}\mu\alpha^2 l \int\int_R (\nabla\Psi)^2 \, dx \, dy - 2\mu\alpha^2 l \int\int_R \Psi \, dx \, dy$
$= \tfrac{1}{2}\mu\alpha^2 l \int\int_R [(\nabla\Psi)^2 - 4\Psi] \, dx \, dy,$

where

$$(\nabla\Psi)^2 \equiv \left(\frac{\partial\Psi}{\partial x}\right)^2 + \left(\frac{\partial\Psi}{\partial y}\right)^2.$$

The stress function Ψ is to be determined as that function which minimizes the expression (64.7) for the potential energy of the twisted shaft and satisfies the condition

$$\Psi = 0 \quad \text{on the boundary } C \text{ of } R.$$

It will be recalled that this boundary condition expresses the fact that no load is applied to the lateral faces of the cylinder. This minimal problem is amplified in the next section, where some elementary results from the calculus of variations are considered.

From the Theorem of Minimum Potential Energy, another general energy theorem can be derived immediately. It is, indeed, only a reformulation of the earlier result. Consider a body in equilibrium under the applied surface forces $\overset{\nu}{T}_i$ and body forces F_i, and with the equilibrium displacements u_i. Now let the body undergo a small additional deformation Δu_i. The applied loads will do work, and there will be a change in the internal strain energy of the body. According[1] to Eq. (64.2),

$$\Delta U = \int_{\Sigma_T} \overset{v}{T}_i \, \Delta u_i \, d\sigma + \int_{\tau} F_i \, \Delta u_i \, d\tau$$
$$+ \text{ second-order terms in the increments of strain.}$$

This equation expresses the Theorem of Virtual Work.

THEOREM OF VIRTUAL WORK: *If a body is in equilibrium under the action of prescribed body and surface forces, the work done by these forces in a small additional displacement, the virtual displacement Δu_i, is equal to the change in the internal strain energy, second-order terms in the increments of strain being neglected.*

The Theorem of Virtual Work and that of Minimum Potential Energy are seen to be merely different ways of stating the same principle.

The Theorem of Minimum Potential Energy was obtained by comparing the strain energy U of the equilibrium state, characterized by displacements u_i and strains e_{ij}, with the strain energy $U + \Delta U$ of a neighboring displacement state $u_i + \Delta u_i$ (or strain state $e_{ij} + \Delta e_{ij}$). A corresponding minimal principle may be derived by varying the stresses τ_{ij} rather than the strains e_{ij}. The calculation of the change ΔU in the strain energy is similar to that of the earlier case and will be outlined briefly.

Consider the equilibrium state with stresses τ_{ij}, strains e_{ij}, displacements u_i, and strain-energy density W. The latter can be written in terms of the stresses alone as

$$W = \frac{1}{2} \tau_{ij} e_{ij} = \frac{1}{2} \tau_{ij} \left[\frac{1+\sigma}{E} \tau_{ij} - \frac{\sigma}{E} \delta_{ij} \Theta \right]$$
$$= \frac{1+\sigma}{2E} \tau_{ij} \tau_{ij} - \frac{\sigma}{2E} \Theta^2,$$

where $\Theta \equiv \tau_{ii}$. Compare with this equilibrium state a neighboring state with stresses $\tau_{ij} + \tau'_{ij} \equiv \tau_{ij} + \Delta \tau_{ij}$ and hence with strain-energy density given by

$$W + \Delta W = \frac{1+\sigma}{2E} (\tau_{ij} + \tau'_{ij})(\tau_{ij} + \tau'_{ij}) - \frac{\sigma}{2E} (\Theta + \Theta')^2.$$

---

[1] It should be noted that the virtual displacements Δu_i must be compatible with the prescribed boundary conditions; that is, the displacements Δu_i vanish on the portion of the surface where the displacements are prescribed.

The increment in the strain-energy density is then

$$\Delta W = \frac{1+\sigma}{2E}(2\tau_{ij}\tau'_{ij} + \tau'_{ij}\tau'_{ij}) - \frac{\sigma}{2E}(2\Theta\Theta' + \Theta'^2)$$

$$= \frac{1+\sigma}{E}\tau_{ij}\tau'_{ij} - \frac{\sigma}{E}\Theta\Theta' + W(\tau'_{ij}),$$

where

$$W(\tau'_{ij}) = \frac{1+\sigma}{2E}\tau'_{ij}\tau'_{ij} - \frac{\sigma}{2E}\Theta'^2$$

is the strain-energy density calculated from the additional stresses $\tau'_{ij} \equiv \Delta\tau_{ij}$. Using Hooke's law, write

$$\begin{aligned}\Delta W &= e_{ij}\tau'_{ij} + W(\tau'_{ij}) \\ &= \tfrac{1}{2}(u_{i,j} + u_{j,i})\tau'_{ij} + W(\tau'_{ij}) \\ &= u_{i,j}\tau'_{ij} + W(\tau'_{ij}) \\ &= (u_i\tau'_{ij})_{,j} - u_i\tau'_{ij,j} + W(\tau'_{ij}).\end{aligned}$$

Consider now only such varied stress states as satisfy the equations of equilibrium. That is,

$$(\tau_{ij} + \tau'_{ij})_{,j} + F_i = 0,$$

from which it follows that

$$\tau'_{ij,j} = 0.$$

The change ΔW in the strain-energy density can now be written as

$$\Delta W = (u_i\tau'_{ij})_{,j} + W(\tau'_{ij}),$$

and the increment in the total strain energy is

$$(64.8) \quad \Delta U = \int_\tau \Delta W\, d\tau = \int_\tau (u_i\tau'_{ij})_{,j}\, d\tau + \int_\tau W(\tau'_{ij})\, d\tau$$

$$= \int_\Sigma u_i\tau'_{ij}\nu_j\, d\sigma + \int_\tau W(\tau'_{ij})\, d\tau.$$

It has been assumed that the varied stresses $\tau_{ij} + \tau'_{ij}$ satisfy the equations of equilibrium. Therefore, on that portion Σ_T of the surface Σ on which the surface forces $\overset{\nu}{T}_i$ are prescribed, one must have

$$(\tau_{ij} + \tau'_{ij})\nu_j = \overset{\nu}{T}_i.$$

as well as
$$\tau_{ij}\nu_j = \overset{\nu}{T}_i,$$
and hence
$$\tau'_{ij}\nu_j = 0 \qquad \text{on } \Sigma_T.$$

On the remainder Σ_u of the surface, where it is assumed that the displacements are prescribed, one has

$$\tau'_{ij}\nu_j = \overset{\nu}{T}'_i \equiv \Delta \overset{\nu}{T}_i \qquad \text{on } \Sigma_u,$$

where the functions $\Delta \overset{\nu}{T}_i$ are the additional surface forces on the surface Σ_u. Equation (64.8) can then be written as

$$(64.9) \qquad \Delta U = \int_{\Sigma_u} u_i \, \Delta \overset{\nu}{T}_i \, d\sigma + \int_\tau W(\tau'_{ij}) \, d\tau,$$

or, since the displacements u_i are not varied on Σ_u, as

$$(64.10) \qquad \Delta \left(U - \int_{\Sigma_u} u_i \overset{\nu}{T}_i \, d\sigma \right) = \int_\tau W(\tau'_{ij}) \, d\tau$$
$$= 0 + \text{second-order terms in the small stresses } \Delta\tau_{ij}.$$

Equation (64.10) expresses a minimal principle for stresses that can be stated in the following way:

Of all stress states satisfying the conditions of equilibrium in the interior and on that portion of the surface where the surface forces are prescribed, the actual state of stress is such as to minimize the expression[1]

$$V^* = U - \int_{\Sigma_u} u_i \overset{\nu}{T}_i \, d\sigma.$$

The quantity V^* is known as the *complementary energy* of the elastic system.

The Theorem of Virtual Work, which was seen to be a reformulation of the Theorem of Minimum Potential Energy, states that

$$(64.11) \qquad \Delta U = \int_{\Sigma_T} \overset{\nu}{T}_i \, \Delta u_i \, d\sigma + \int_\tau F_i \, \Delta u_i \, d\tau,$$

and Δu_i vanishes on Σ_u, where the displacements are prescribed.

[1] It will be recalled that the displacements u_i and not the surface forces $\overset{\nu}{T}_i$ are prescribed over the region Σ_u.

Equation (64.9) is now seen to yield a parallel result:

$$(64.12) \qquad \Delta U = \int_{\Sigma_u} u_i \, \Delta \overset{\nu}{T}_i \, d\sigma.$$

In this expression, $\Delta \overset{\nu}{T}_i \equiv \Delta \tau_{ij} \nu_j$ vanishes on Σ_T, where the surface forces are prescribed, and $\Delta \tau_{ij}$ satisfy the equations of equilibrium with zero body forces. In Eq. (64.11), the change in strain energy associated with the given variation is given by the integral of force multiplied by the increment of displacement; in (64.12), it is expressed as the integral of displacement multiplied by the increment of force.

PROBLEM

Consider the case in which there are no body forces and the displacements u_i are prescribed over the entire surface Σ. Show from Eq. (64.1) that the increment ΔU in strain energy is positive in this case and that the equilibrium displacements yield a minimum value for the strain energy U.

65. Variational Problem and Euler's Equation. This section is concerned with a brief treatment of the problem of determining the function $u(x, y)$ minimizing the integral

$$(65.1) \qquad I = \int \int_R F\left(x, y, u, \frac{\partial u}{\partial x}, \frac{\partial u}{\partial y}\right) dx \, dy,$$

where F is a known function of the arguments x, y, u, $\dfrac{\partial u}{\partial x}$, and $\dfrac{\partial u}{\partial y}$, possessing continuous first and second partial derivatives with respect to these arguments. The unknown function $u(x, y)$ is likewise assumed to be continuous of class $C^{(2)}$, with values prescribed on the boundary C of the simply connected region R.

Let us imagine the problem solved by the function $u(x, y)$, which thus minimizes the integral I, and let us compare the minimum value $I(u)$ with the value of the integral I corresponding to other functions. If the function $u(x, y)$ in Eq. (65.1) is replaced by $u(x, y) + \epsilon \eta(x, y)$ where ϵ is an arbitrary small parameter and $\eta(x, y)$ is any function that vanishes on the boundary C [and satisfies the same conditions of continuity as does $u(x, y)$], the integral I becomes a function of the parameter ϵ and has a minimum at $\epsilon = 0$. Hence

(65.2) $$\left.\frac{dI(\epsilon)}{d\epsilon}\right|_{\epsilon=0} = 0.$$

But
$$I(\epsilon) = \iint_R F(x, y, u + \epsilon\eta, u_x + \epsilon\eta_x, u_y + \epsilon\eta_y)\, dx\, dy,$$
and hence
$$\left.\frac{dI}{d\epsilon}\right|_{\epsilon=0} = \iint_R [F_u \eta(x, y) + F_{u_x}\eta_x + F_{u_y}\eta_y]\, dx\, dy = 0,$$

where the subscript u_x in F_{u_x}, for example, denotes the partial derivative of the function $F(x, y, u, u_x, u_y)$ with respect to u_x. This equation can be rewritten as

$$\iint_R \left[F_u - \frac{\partial F_{u_x}}{\partial x} - \frac{\partial F_{u_y}}{\partial y}\right] \eta\, dx\, dy$$
$$+ \iint_R \left[\frac{\partial}{\partial x}(F_{u_x}\eta) + \frac{\partial}{\partial y}(F_{u_y}\eta)\right] dx\, dy = 0,$$

and an application of Green's Theorem yields

$$\iint_R \left[F_u - \frac{\partial F_{u_x}}{\partial x} - \frac{\partial F_{u_y}}{\partial y}\right] \eta\, dx\, dy + \int_C \eta(-F_{u_y}\, dx + F_{u_x}\, dy) = 0.$$

Since η vanishes on the boundary C, the foregoing equation becomes

(65.3) $$\iint_R \left(F_u - \frac{\partial F_{u_x}}{\partial x} - \frac{\partial F_{u_y}}{\partial y}\right) \eta\, dx\, dy = 0,$$

and since the variation $\eta(x, y)$ is an arbitrary function,[1] one must have

(65.4) $$F_u - \frac{\partial F_{u_x}}{\partial x} - \frac{\partial F_{u_y}}{\partial y} = 0.$$

---

[1] This follows from the fundamental lemma of the calculus of variations, namely,

If $\iint_R \Phi(x, y, u)\eta(x, y)\, dx\, dy = 0$ for all functions $\eta(x, y)$ that vanish on the boundary C of the region R and that are continuous together with their first and second partial derivatives, and if Φ is continuous, then the function $\Phi(x, y, u)$ vanishes identically.

See. R. Courant and D. Hilbert, Methoden der Mathematischen Physik, vol. 1, Chap. IV, Sec. 3.

The differential equation (65.4) is known as the *Euler equation* for the variational problem defined by the integral (65.1), and it provides a necessary condition for an extremum (minimum, maximum, or stationary value) of the integral I. The minimal integral (65.1) and the associated Euler equation are written out below for some important special cases.

In Sec. 64, it was shown that the potential energy of a twisted shaft is given by the expression (64.7)

$$V = \frac{1}{2}\mu\alpha^2 l \iint_R [(\nabla\Psi)^2 - 4\Psi]\,dx\,dy$$
$$= \frac{1}{2}\mu\alpha^2 l \iint_R \left[\left(\frac{\partial\Psi}{\partial x}\right)^2 + \left(\frac{\partial\Psi}{\partial y}\right)^2 - 4\Psi\right] dx\,dy.$$

According to the Theorem of Minimum Potential Energy, the function corresponding to the state of equilibrium is obtained by minimizing the potential energy V, subject to the boundary condition

$$\Psi = 0 \qquad \text{on } C,$$

where C is the contour of the region R of the cross section of the shaft. The Euler equation (65.4) for this problem is seen to be

$$-4 - \frac{\partial}{\partial x}\left(2\frac{\partial\Psi}{\partial x}\right) - \frac{\partial}{\partial y}\left(2\frac{\partial\Psi}{\partial y}\right) = 0,$$

or

$$\nabla^2\Psi = -2,$$

and this is precisely the differential equation for the stress function found in Sec. 35 from the equations of equilibrium.

The Euler equation for the minimal problem characterized by the integral

$$I = \iint_R [(\nabla u)^2 + 2ku]\,dx\,dy = \text{minimum}$$

is found in the same way to be

$$\nabla^2 u = k.$$

If k is set equal to zero, one obtains the Laplace equation

$$\nabla^2 u = 0,$$

whose solution minimizes the integral

$$(65.5) \quad I = \iint_R (\nabla u)^2 \, dx \, dy = \iint_R \left[\left(\frac{\partial u}{\partial x}\right)^2 + \left(\frac{\partial u}{\partial y}\right)^2\right] dx \, dy.$$

The Euler equation for the minimal problem defined by

$$I = \int_{x_0}^{x_1} F\left(x, y, \frac{dy}{dx}\right) dx = \text{minimum}$$

is

$$(65.6) \quad F_y - \frac{d}{dx} F_{y'} = 0.$$

If two independent variables enter into the integrand, as in

$$I = \int_{x_0}^{x_1} F(x, y, z, y', z') \, dx = \text{minimum},$$

then the Euler equation must hold for each variable:

$$F_y - \frac{d}{dx} F_{y'} = 0,$$

$$F_z - \frac{d}{dx} F_{z'} = 0.$$

If the minimal problem involves higher derivatives, as in

$$I = \int_{x_0}^{x_1} F(x, y, y', y'', \cdots, y^{(n)}) \, dx = \text{minimum},$$

then the Euler equation is

$$(65.7) \quad F_y - \frac{d}{dx} F_{y'} + \frac{d^2}{dx^2} F_{y''} - \cdots + (-1)^n \frac{d^n}{dx^n} F_{y^{(n)}} = 0.$$

The Euler equation for the minimal problem

$$I = \iint_R F(x, y, u, u_x, u_y, u_{xx}, u_{xy}, u_{yy}) \, dx \, dy = \text{minimum}$$

is

$$(65.8) \quad F_u - \frac{\partial}{\partial x} F_{u_x} - \frac{\partial}{\partial y} F_{u_y} + \frac{\partial^2}{\partial x^2} F_{u_{xx}} + 2 \frac{\partial^2}{\partial x \, \partial y} F_{u_{xy}} + \frac{\partial^2}{\partial y^2} F_{u_{yy}} = 0.$$

It was shown above that the solution of Euler's equation (65.4), associated with the problem of minimizing the integral (65.1), will yield an extremal function $u(x, y)$. The problem of determining the minimizing function $u(x, y)$ may also be attacked directly, if it is possible to represent $u(x, y)$ by the expression

$$(65.9) \qquad u = \sum_{i=1}^{N} c_i f_i(x, y), \quad (i = 1, 2, \cdots, N),$$

where the functions $f_i(x, y)$ satisfy the same boundary and continuity conditions as $u(x, y)$ but are otherwise arbitrary. Substitution of (65.9) into (65.1) makes I a function of the coefficients $c_1, c_2, \cdots, c_N$, and the N equations

$$\frac{\partial I}{\partial c_i} = 0, \quad (i = 1, 2, \cdots, N),$$

may enable one to determine the constants c_i.

If the assumed representation of the function $u(x, y)$ is impossible, it may still be true that the procedure just indicated would yield an approximate solution of the problem. Both of these methods of determining the function $u(x, y)$ are illustrated in the following section, where a simple problem from the technical theory of beams is discussed.

PROBLEM

Show that a function that minimizes the integral

$$I = \iint (\nabla^2 u)^2 \, dx \, dy = \iint (u_{xx}^2 + 2u_{xx}u_{yy} + u_{yy}^2) \, dx \, dy$$

is a solution of the biharmonic equation

$$\nabla^4 u \equiv u_{xxxx} + 2u_{xxyy} + u_{yyyy} = 0.$$

66. An Application of the Theorem of Minimum Potential Energy. This section contains an illustration of the application of the Theorem of Minimum Potential Energy to the approximate determination of deflections and bending moments in a cantilever beam carrying a uniformly distributed load p per unit length. It will be recalled [cf. Eq. (61.10)] that the deflection of such a beam is given by

$$(66.1) \qquad EI y(x) = \tfrac{1}{24} p(x^4 - 4lx^3 + 6l^2 x^2).$$

The maximum deflection is

$$(66.2) \qquad y(l) = \frac{1}{8} \frac{pl^4}{EI},$$

and the maximum bending moment is

$$M(0) = -EIy''(0) = -\tfrac{1}{2}pl^2.$$

Equation (66.1) gives the "exact" solution of the problem—exact, that is, within the limits of the technical theory of beams. The function $y(x)$ satisfies the differential equation of equilibrium (61.6)

$$EI \frac{d^4y}{dx^4} = p$$

and the boundary conditions

$$y(0) = y'(0) = 0, \qquad y''(l) = y'''(l) = 0.$$

Consider now the formulation of the problem of a beam bent by a distributed load $p(x)$. According to the Theorem of Minimum Potential Energy, we are to seek the function $y(x)$ that minimizes the potential energy V, where

$$V = U - \int_\Sigma \overset{\nu}{T}_i u_i \, d\sigma - \int_\tau F_i u_i \, d\tau,$$

or, in the present problem,[1]

(66.3) $$V = U - \int_0^l py \, dx.$$

The strain energy U can be expressed in terms of the deflection $y(x)$. The bending stress is given by (61.8) as

$$\tau_{xx} = \frac{My}{I},$$

and the bending strain is

$$e_{xx} = \frac{\tau_{xx}}{E} = \frac{My}{EI}.$$

The strain-energy density is then given by

$$W = \frac{1}{2} \tau_{xx} e_{xx} = \frac{M^2 y^2}{2EI^2},$$

if one neglects the contribution of the shear stresses[2] to the

---

[1] The weight of the beam is neglected in this discussion. It is assumed that the ends of the beam are clamped, hinged, or free, and hence that the support forces do not contribute to the potential energy V.

[2] See the discussion of the magnitude of shear stress in Sec. 61.

strain energy. The strain energy per unit length is found by integrating the energy density W over the cross section R to give

$$\int_R W \, d\sigma = \frac{M^2}{2EI^2} \int_R y^2 \, d\sigma = \frac{M^2}{2EI}$$
$$= \frac{(-EIy'')^2}{2EI} = \frac{1}{2} EI(y'')^2.$$

The total strain energy is then

(66.4) $$U = \int_0^l \tfrac{1}{2} EI(y'')^2 \, dx,$$

and the potential energy can be written as

$$V = \int_0^l V_0 \, dx,$$

where, from (66.3) and (66.4),

(66.5) $$V_0 = \tfrac{1}{2} EI(y'')^2 - py.$$

According to the Theorem of Minimum Potential Energy, the solution of the problem is given by that function $y(x)$ which minimizes the potential energy V and satisfies the boundary conditions on the displacement $y(x)$.[1]

Either of two courses may now be followed. On the one hand, the methods of the calculus of variations may be used to find the differential equation satisfied by the deflection function $y(x)$. From Eq. (65.7), it is seen that the Euler equation for the variational problem

$$V = \int_0^l V_0(x, y, y', y'') \, dx = \text{minimum}$$

is

(66.6) $$\frac{d^2}{dx^2}\left(\frac{\partial V_0}{\partial y''}\right) - \frac{d}{dx}\left(\frac{\partial V_0}{\partial y'}\right) + \frac{\partial V_0}{\partial y} = 0.$$

For the problem of a beam bent by a distributed load $p(x)$ per unit length, the Euler equation is found from (66.5) and (66.6) to be

(66.7) $$\frac{d^2}{dx^2}(EIy'') - p = 0,$$

[1] See the discussion of geometric and dynamic boundary conditions in Sec. 68.

which is precisely the equation of equilibrium found in Sec. 61 by a different method. Energy methods may often be used to advantage in this way in setting up the equations of equilibrium.[1]

Equation (66.7) together with the appropriate boundary conditions gives necessary conditions for the determination of that function which yields a minimum value for the potential energy V.

Instead of the course outlined above, a more direct attack may be chosen. Write

$$(66.8) \quad y = c_1 f_1(x) + c_2 f_2(x) + \cdots + c_N f_N(x),$$

where the functions $f_j(x)$ are arbitrary, subject to the restriction that $\sum_{j=1}^{N} c_j f_j(x)$ shall satisfy the boundary conditions and be reasonably representative of the true deflection. The coefficients c_j are then chosen so as to make the potential energy a minimum. That is, the N conditions

$$\frac{\partial V}{\partial c_j} = 0, \quad (j = 1, 2, \cdots, N),$$

yield N equations for the coefficients $c_1, c_2, \cdots, c_N$. Thus, instead of inspecting all functions to find the one that minimizes the potential energy V, we look only among the smaller class of functions that can be represented according to (66.8) as a linear combination of previously chosen functions $f_j(x)$.

In the problem of a cantilever with no end load, the boundary conditions are

$$y(0) = y'(0) = 0, \quad y''(l) = y'''(l) = 0.$$

A suitable approximate deflection function $y(x)$ can be found by writing

$$EI y'''(x) = c \cos \frac{\pi x}{2l},$$

which satisfies the end condition $y'''(l) = 0$. By successive integrations and application of the boundary conditions, one obtains

---

[1] See, for example, Chap. 2, Sec. VI, Anwendung der Minimalprinzipe zur Aufstellung der Differentialgleichungen in besonderen Fällen, in E. Trefftz, Handbuch der Physik, vol. 6.

$$(66.9) \quad \begin{cases} EIy''(x) = -\dfrac{2cl}{\pi}\left(1 - \sin\dfrac{\pi x}{2l}\right), \\ EIy'(x) = -\dfrac{2cl}{\pi}\left[x - \dfrac{2l}{\pi}\left(1 - \cos\dfrac{\pi x}{2l}\right)\right], \\ EIy(x) = -\dfrac{2cl}{\pi}\left[\dfrac{1}{2}x^2 - \dfrac{2l}{\pi}\left(x - \dfrac{2l}{\pi}\sin\dfrac{\pi x}{2l}\right)\right]. \end{cases}$$

The potential energy in the case of a uniformly distributed load p per unit length, and corresponding to this choice of deflection $y(x)$, is given by

$$V = \int_0^l \left[\dfrac{1}{2}EI(y'')^2 - py\right] dx$$
$$= c^2 \dfrac{l^3(3\pi - 8)}{EI\pi^3} + c\dfrac{pl^4(\pi^3 - 6\pi^2 + 48)}{3EI\pi^4}.$$

The minimizing condition

$$\dfrac{\partial V}{\partial c} = 0$$

yields

$$c = -pl\dfrac{\pi^3 - 6\pi^2 + 48}{6\pi(3\pi - 8)}.$$

The maximum deflection is found from (66.9) to be

$$y(l) = -\dfrac{2cl}{\pi EI}\left[\dfrac{1}{2}l^2 - \dfrac{2l}{\pi}\left(l - \dfrac{2l}{\pi}\right)\right]$$
$$= \dfrac{pl^4}{EI}\dfrac{(\pi^3 - 6\pi^2 + 48)(\pi^2 - 4\pi + 8)}{6\pi^4(3\pi - 8)}$$
$$= 0.12603\dfrac{pl^4}{EI},$$

which is 0.8 per cent greater than the exact value

$$y(l) = \dfrac{1}{8}\dfrac{pl^4}{EI}.$$

The maximum bending moment is

$$M(0) = -EIy''(0) = \dfrac{2cl}{\pi}$$
$$= \dfrac{-pl^2(\pi^3 - 6\pi^2 + 48)}{3\pi^2(3\pi - 8)}$$
$$= -0.469pl^2,$$

which is 6 per cent less than the exact value

$$M(0) = -\tfrac{1}{2}pl^2.$$

The approximate solution just found does not satisfy the differential equation of equilibrium. In fact,

$$EI\frac{d^4y}{dx^4} = -\frac{\pi c}{2l}\sin\frac{\pi x}{2l}$$

$$= 1.157p\sin\frac{\pi x}{2l}.$$

D. Williams[1] has compared the bending moment curves obtained from various approximate deflection functions $y(x)$ and has illustrated the effect of failure to satisfy all of the boundary conditions[2] as well as the effect of satisfying false boundary conditions.

PROBLEMS

1. Solve the problem of a cantilever beam under uniform load p per unit length by means of the Theorem of Minimum Potential Energy. For the approximate displacement take $EIy(x) = cx^2(3l - x)$, a function that satisfies all the boundary conditions except the vanishing of the shear force at the free end $x = l$. Find the value of the constant c, the maximum displacement, and the maximum bending moment, and compare with the exact results.

2. Illustrate the theorem of Eq. (26.12) by showing that the strain energy stored in a cantilever beam of length l bent by a uniformly distributed load pl is $\dfrac{p^2l^5}{40EI}$, which is one-half the numerical value of the potential energy of the external forces.

3. Show that the potential energy V associated with the exact solution of a minimal problem is always less than that corresponding to the approximate solution of the same minimal problem. Verify this conclusion numerically in the case of a cantilever bent by a uniformly distributed load pl, and compare the exact potential energy $V = -\dfrac{p^2l^5}{40EI}$ with that corresponding to the approximate solution found in Sec. 66.

---

[1] "The Use of the Principle of Minimum Potential Energy in Problems of Static Equilibrium," Aeronautical Research Committee (Great Britain), *Reports and Memoranda* 1827 (1938).

[2] See problems at the end of this section.

67. Theorems of Work and Reciprocity.

We derive now a very general reciprocal expression relating the equilibrium states of a body under different applied loads.

Consider two equilibrium states of an elastic body: *one with displacements u_i due to the body forces F_i and surface forces $\overset{\nu}{T}_i$, and the other with displacements u'_i due to body forces F'_i and surface forces $\overset{\nu}{T}'_i$.* Let us calculate the work that would be done by the unprimed forces, F_i, $\overset{\nu}{T}_i$, if they acted through the primed displacements u'_i. This work can be written, with the help of the equations of equilibrium, as

$$\int_\Sigma \overset{\nu}{T}_i u'_i \, d\sigma + \int_\tau F_i u'_i \, d\tau = \int_\Sigma \tau_{ij} \nu_j u'_i \, d\sigma - \int_\tau \tau_{ij,j} u'_i \, d\tau$$
$$= \int_\tau (\tau_{ij} u'_i)_{,j} \, d\tau - \int_\tau \tau_{ij,j} u'_i \, d\tau,$$

where the Divergence Theorem has been used. Carrying out the indicated differentiation leads to

$$(67.1) \quad \int_\Sigma \overset{\nu}{T}_i u'_i \, d\sigma + \int_\tau F_i u'_i \, d\tau = \int_\tau \tau_{ij} u'_{i,j} \, d\tau$$
$$= \int_\tau (\lambda \delta_{ij} \vartheta + 2\mu e_{ij}) u'_{i,j} \, d\tau \quad \text{(by Hooke's law)}$$
$$= \int_\tau [\lambda \vartheta u'_{i,i} + \mu(u_{i,j} + u_{j,i}) u'_{i,j}] \, d\tau$$
$$= \int_\tau (\lambda \vartheta \vartheta' + \mu u_{i,j} u'_{i,j} + \mu u_{j,i} u'_{i,j}) \, d\tau.$$

With the exception of the last term, the integrand is obviously symmetric in the primed and unprimed variables. But the last term can be written, by interchange of i, j, as

$$\mu u_{j,i} u'_{i,j} = \mu u_{i,j} u'_{j,i} = \mu u'_{j,i} u_{i,j},$$

and we see that the integral, and hence the original expression, is symmetric with respect to the primed and unprimed states. That is, we can write

$$(67.2) \quad \int_\Sigma \overset{\nu}{T}_i u'_i \, d\sigma + \int_\tau F_i u'_i \, d\tau = \int_\Sigma \overset{\nu}{T}'_i u_i \, d\sigma + \int_\tau F'_i u_i \, d\tau.$$

This theorem can be expressed in words.

RECIPROCAL THEOREM OF BETTI AND RAYLEIGH: *If an elastic body is subjected to two systems of body and surface forces, then* the work that would be done by the first system $\overset{\nu}{T_i}$, F_i in acting through the displacements u'_i due to the second system of forces is equal to the work that would be done by the second system $\overset{\nu}{T'_i}$, F'_i in acting through the displacements u_i due to the first system of forces.

The Reciprocal Theorem can be written in terms of the stresses and strains by modifying Eq. (67.1). We observe that the integrand on the right-hand side of (67.1) can be written as

$$\tau_{ij} u'_{i,j} = \tau_{ij}[\tfrac{1}{2}(u'_{i,j} + u'_{j,i}) + \tfrac{1}{2}(u'_{i,j} - u'_{j,i})]$$
$$= \tau_{ij} e'_{ij} + \tau_{ij} \omega'_{ij}.$$

But the last term vanishes[1] on account of the skew symmetry of the rotation components, $\omega'_{ij} = -\omega'_{ji}$. Hence we have merely

$$\tau_{ij} u'_{i,j} = \tau_{ij} e'_{ij},$$

and Eq. (67.1) takes the form

(67.3) $$\int_\Sigma \overset{\nu}{T_i} u'_i \, d\sigma + \int_\tau F_i u'_i \, d\tau = \int_\tau \tau_{ij} e'_{ij} \, d\tau$$

with

$$\int_\tau \tau_{ij} e'_{ij} \, d\tau = \int_\tau \tau'_{ij} e_{ij} \, d\tau.$$

Equation (67.3) is an alternative form of the Reciprocal Theorem, which is thus seen to be a generalization of Eq. (26.12).

The Reciprocal Theorem can also be deduced by means of the following argument. First subject an elastic body to the force system I: $\overset{\nu}{T_i}$, F_i. The resulting displacements are denoted by u_i; the work done, by $U_{I,I}$. On the elastic body thus strained, superpose the force system II: $\overset{\nu}{T'_i}$, F'_i. The additional displacements are the same as though force system I were absent. The additional work done consists of two parts, the work $U_{II,II}$ done by force system II acting through displacements II, and the work

---

[1] This can be seen by interchanging i, j. Thus,

$$\tau_{ij} \omega'_{ij} = \tau_{ji} \omega'_{ji} = -\tau_{ij} \omega'_{ij}$$

or

$$2\tau_{ij} \omega'_{ij} = 0.$$

$U_{\mathrm{I,II}}$ done by force system I acting through displacements II. The total work is thus given by

$$U_{\mathrm{I,I}} + U_{\mathrm{II,II}} + U_{\mathrm{I,II}}.$$

If the force systems were applied in the reverse order, II and I, then the total work would be

$$U_{\mathrm{II,II}} + U_{\mathrm{I,I}} + U_{\mathrm{II,I}}.$$

But the final state of the body is independent of the order in which the loads are applied.[1] Hence we must have

$$U_{\mathrm{I,I}} + U_{\mathrm{II,II}} + U_{\mathrm{I,II}} = U_{\mathrm{II,II}} + U_{\mathrm{I,I}} + U_{\mathrm{II,I}}$$

or

$$U_{\mathrm{I,II}} = U_{\mathrm{II,I}};$$

that is,

$$\int_\Sigma \overset{\nu}{T}_i u'_i \, d\sigma + \int_\tau F_i u'_i \, d\tau = \int_\Sigma \overset{\nu}{T}'_i u_i \, d\sigma + \int_\tau F'_i u_i \, d\tau.$$

The value of the Reciprocal Theorem lies in the fact that for every choice of values for the variables of the primed state $\overset{\nu}{T}'_i$, F'_i, u'_i, one obtains a theorem relating the applied forces $\overset{\nu}{T}_i$, F_i, and displacements u_i in an elastic body. This is exemplified in the problems at the end of this section and in the theorems we now proceed to derive as special cases of the Reciprocal Theorem.

We write out now an important specialization of the Reciprocal Theorem to the case of a body deformed by concentrated forces. For the sake of concreteness, we shall speak of a beam bent by point loads. Assume that the body forces vanish, and write the Reciprocal Theorem in the form

$$\int_\Sigma \overset{\nu}{T}_i u'_i \, d\sigma = \int_\Sigma \overset{\nu}{T}'_i u_i \, d\sigma.$$

Consider two equilibrium states of a beam, one with load $p(x)$ and deflection $y(x)$, and another with load $p'(x)$ and correspond-

---

[1] It should be noted that the Reciprocal Theorem depends only on the linearity of the equations of equilibrium and hence on the principle of superposition. We assume that the forces F_i, $\overset{\nu}{T}_i$ do not depend on the displacements u'_i and that the displacements u_i do not affect the forces F'_i, $\overset{\nu}{T}'_i$.

ing deflection $y'(x)$. By the Reciprocal Theorem, we have

(67.4) $$\int_0^l py'\,dx = \int_0^l p'y\,dx.$$

Let the beam be loaded only by concentrated forces $P_1, P_2, \cdots$ applied at the points $x_1, x_2, \cdots$, and denote by α_{ij} *the transverse displacement at x_i due to a unit transverse force applied at x_j.* We choose for the load system p the concentrated force P_1; then the corresponding displacements y at x_1 and x_2 are

$$y_1 = \alpha_{11}P_1 \text{ at } x_1, \qquad y_2 = \alpha_{21}P_1 \text{ at } x_2.$$

Similarly, for the load p' in (67.4), we take the force P_2; the associated displacements y' at the points x_1, x_2 are

$$y_1' = \alpha_{12}P_2 \text{ at } x_1, \qquad y_2' = \alpha_{22}P_2 \text{ at } x_2.$$

According to the Reciprocal Theorem, we have

$$P_1 y_1' = P_2 y_2,$$

or

$$P_1 \alpha_{12} P_2 = P_2 \alpha_{21} P_1;$$

that is,

(67.5) $$\alpha_{12} = \alpha_{21}.$$

The quantity α_{ij} is called the *influence coefficient* (designated so by Maxwell) for transverse deflection at x_i due to a force applied at x_j. The symmetry of the influence coefficients

$$\alpha_{ij} = \alpha_{ji}$$

is seen to be a special case of the Reciprocal Theorem.[1]

Consider now the effect of varying a force T_i applied at a point P on the surface of an elastic body. To this end, denote by Σ' a portion of the surface Σ that includes the point P as an inner point. The remainder of the surface Σ will be denoted by $\Sigma - \Sigma'$. In the Reciprocal Theorem, we choose

$$F_i' = F_i,$$

$$\overset{\nu}{T_i'} = \begin{cases} \overset{\nu}{T_i} & \text{on } \Sigma - \Sigma', \\ \overset{\nu}{T_i} + \Delta \overset{\nu}{T_i} & \text{on } \Sigma', \end{cases}$$

---

[1] For a discussion of the Maxwell influence coefficients and of determinants with these coefficients as elements, see C. B. Biezeno and R. Grammel, Technische Dynamik, Chap. II, Secs. 9–11.

§67 VARIATIONAL METHODS

and get[1]

(67.6) $$\int_\tau F_i u'_i \, d\tau + \int_\Sigma \overset{\nu}{T}_i u'_i \, d\sigma = \int_\tau F_i u_i \, d\tau \\ + \int_\Sigma \overset{\nu}{T}_i u_i \, d\sigma + \int_{\Sigma'} (\Delta \overset{\nu}{T}_i) u_i \, d\sigma.$$

Now by (26.12), the strain energy U associated with the original load system $\overset{\nu}{T}_i, F_i$ is

$$2U = \int_\tau F_i u_i \, d\tau + \int_\Sigma \overset{\nu}{T}_i u_i \, d\sigma,$$

while that corresponding to the varied state $\overset{\nu}{T}'_i, F'_i$ is

$$2U' = \int_\tau F'_i u'_i \, d\tau + \int_\Sigma \overset{\nu}{T}'_i u'_i \, d\sigma \\ = \int_\tau F_i u'_i \, d\tau + \int_\Sigma \overset{\nu}{T}_i u'_i \, d\sigma + \int_{\Sigma'} (\Delta \overset{\nu}{T}_i) u'_i \, d\sigma.$$

Equation (67.6) can thus be written as

$$2U' - \int_{\Sigma'} (\Delta \overset{\nu}{T}_i) u'_i \, d\sigma = 2U + \int_{\Sigma'} (\Delta \overset{\nu}{T}_i) u_i \, d\sigma,$$

or

$$2(U' - U) \equiv 2\Delta U = \int_{\Sigma'} (\Delta \overset{\nu}{T}_i)(u_i + u'_i) \, d\sigma.$$

When the region Σ' is small, we have, approximately,

$$2\Delta U \doteq (\Delta \overset{\nu}{T}_i)(u_i + u'_i) \Sigma'.$$

We denote the increment of force acting on the area Σ' by

$$\Delta T_i = (\Delta \overset{\nu}{T}_i) \Sigma'.$$

Then

$$\frac{\Delta U}{\Delta T_i} \doteq \frac{1}{2}(u_i + u'_i),$$

and letting ΔT_i approach zero, we get

(67.7) $$\frac{dU}{dT_i} = u_i.$$

Equation (67.7) expresses the *Theorem of Castigliano*.[2]

[1] It is assumed that the body is rigidly supported and hence that the supporting forces do no work.

[2] For further discussion of this important theorem, see C. B. Biezeno and R. Grammel, Technische Dynamik, Chap. II, Sec. 8.

The relations between the various energy theorems of elasticity have been discussed by D. Williams and Th. Pöschl.[1]

PROBLEMS

1. Consider a beam loaded by concentrated forces P_1 at x_1 and P_2 at x_2, and let y_j be the deflection at x_j. Calculate the additional deflections dy_j and the change dU in the strain energy corresponding to a change dP_1 in the force P_1. Show that

$$\alpha_{12} = \frac{\partial^2 U}{\partial P_1 \partial P_2},$$

and hence that

$$\alpha_{12} = \alpha_{21}.$$

2. Show that the influence coefficients for a cantilever beam are

$$\alpha_{12} = \alpha_{21} = \begin{cases} \dfrac{1}{6EI} x_1^2(3x_2 - x_1), & x_1 < x_2, \\ \dfrac{1}{6EI} x_2^2(3x_1 - x_2), & x_2 < x_1, \end{cases}$$

where α_{ij} is the deflection at x_i due to a unit load at x_j.

Show that the deflection of a cantilever beam bent by an end load P is given by

$$y(x) = P\alpha_{xl} = \frac{P}{6EI} x^2(3l - x),$$

and from Eq. (66.4) verify that the strain energy stored in the beam is

$$U = \frac{l^3}{6EI} P^2 = \frac{3EI}{2l^3} \delta^2,$$

where δ is the end deflection $y(l)$. Show that

$$\frac{dU(P)}{dP} = \delta, \qquad \frac{dU(\delta)}{d\delta} = P.$$

3. In the Reciprocal Theorem, take $F_i' = 0$, $\tau_{ij}' = \delta_{ij}$. Show that

$$\overset{\nu}{T}_i' = \tau_{ij}'\nu_j = \nu_i, \quad \Theta' = \tau_{ii}' = 3, \quad e_{ij}' = \frac{1-2\sigma}{E}\delta_{ij}, \quad \text{and} \quad u_i' = \frac{1-2\sigma}{E} x_i.$$

---

[1] D. Williams, "The Relations between the Energy Theorems Applicable in Structural Theory," *Philosophical Magazine*, ser. 7, vol. 26 (1938), pp. 617–635.

Th. Pöschl, "Über die Minimalprinzipe der Elastizitätstheorie," *Bauingenieur* (1936), pp. 160–164.

See also A. J. S. Pippard, Strain Energy Methods of Stress Analysis.

§67 VARIATIONAL METHODS 303

Insert these expressions in (67.2), and derive the following expression for the change in volume ΔV_0 in an elastic body under the action of surface forces $\overset{\nu}{T}_i$ and body forces F_i:

$$\Delta V_0 \equiv \int_\tau \vartheta \, d\tau = \frac{1-2\sigma}{E}\left(\int_\Sigma \overset{\nu}{T}_i x_i \, d\sigma + \int_\tau F_i x_i \, d\tau\right).$$

4. Fill in the details of the following direct calculation of the change in volume ΔV_0 of an elastic body τ under the action of surface forces $\overset{\nu}{T}_i$ and body forces F_i:

$$\Delta V_0 = \int_\tau \vartheta \, d\tau = \frac{1-2\sigma}{E}\int_\tau \tau_{ii} \, d\tau$$

$$= \frac{1-2\sigma}{E}\int_\tau \tau_{ij} x_{i,j} \, d\tau$$

$$= \frac{1-2\sigma}{E}\int_\tau (\tau_{ij} x_{i,j} + \tau_{ij,j} x_i + F_i x_i) \, d\tau$$

$$= \frac{1-2\sigma}{E}\int_\tau [(\tau_{ij} x_i)_{,j} + F_i x_i] \, d\tau$$

$$= \frac{1-2\sigma}{E}\left[\int_\Sigma \overset{\nu}{T}_i x_i \, d\sigma + \int_\tau F_i x_i \, d\tau\right].$$

5. Show that the average value of a strain component e_{11}, say, throughout an elastic body subjected to surface forces $\overset{\nu}{T}_i$ and body forces F_i, can be found from the Reciprocal Theorem in the form (67.3) by putting $\tau'_{11} = 1$, other $\tau_{ij} = 0$. Derive the formula

$$\int_\tau e_{11} \, d\tau = \frac{1}{E}\int_\Sigma (\overset{\nu}{T}_1 x_1 - \sigma \overset{\nu}{T}_2 x_2 - \sigma \overset{\nu}{T}_3 x_3) \, d\sigma$$

$$+ \frac{1}{E}\int_\tau (F_1 x_1 - \sigma F_2 x_2 - \sigma F_3 x_3) \, d\tau.$$

6. Show that the average deflection of a cantilever beam due to a concentrated load P applied at a point x_0 is equal to the deflection at x_0 produced by the load P distributed uniformly over the length of the beam. Neglect the weight of the beam.

7. In the Reciprocal Theorem, take for the primed system of forces and displacements those of the problem of a beam under tension by end forces. That is, derive the expression

$$\iint_{z=0} (\sigma x \tau_{zx} + \sigma y \tau_{zy}) \, dx \, dy + \iint_{z=l} (-\sigma x \tau_{zx} - \sigma y \tau_{zy} + l\tau_{zz}) \, dx \, dy$$
$$= -E \iint_{z=0} w \, dx \, dy + E \iint_{z=l} w \, dx \, dy,$$

which is valid for the stress system in any beam free of body forces and loaded at the ends. Verify the Reciprocal Theorem by taking the longitudinal displacement w and the stresses τ_{ij} to be those of the problem of bending by end couples.

68. The Rayleigh-Ritz Method. It was demonstrated in Secs. 65 and 66 that functions minimizing certain integrals can be obtained by solving the differential equations of Euler, and it was also indicated how these minimizing functions might be deduced by approaching the variational problem directly. One of the direct methods of attack on the variational problems, stemming from ideas developed by Lord Rayleigh and W. Ritz,[1] is outlined below.

It will be recalled that the equilibrium displacements u_i were characterized in Sec. 64 as those functions that minimize the potential energy (64.4)

$$V = U - \int_{\Sigma_T} \overset{\nu}{T}_i u_i \, d\sigma - \int_{\tau} F_i u_i \, d\tau$$

of the elastic system, and that are contained in a set of all functions satisfying the boundary conditions on that portion of the surface Σ where the displacements are prescribed. The region Σ_T in Eq. (64.4) denotes the portion of the surface Σ on which the external forces are specified, and hence need not be taken into account in determining the set of admissible functions.

According to the Theorem of Minimum Potential Energy, the exact solution of a problem is to be found by examining all functions satisfying the boundary conditions on the portion of the surface where the displacements are specified, and selecting only those that minimize the potential energy V. Since this procedure is, in general, very difficult, one might hope to obtain an approximate solution by selecting from the set of all admissible functions a certain subset. For example, one may assume

---

[1] See W. Ritz, *Journal für reine und angewandte Mathematik*, vol. 135 (1909), pp. 1–61.

that the equilibrium displacements u, v, w can be represented with sufficient accuracy by the approximate displacements

$$u_N = \sum_{i=1}^{N} a_i P_i(x, y, z), \quad v_N = \sum_{i=1}^{N} b_i Q_i(x, y, z), \quad w_N = \sum_{i=1}^{N} c_i R_i(x, y, z).$$

The functions P_i, Q_i, R_i are assumed to satisfy the same boundary conditions as do the equilibrium displacements u, v, w and the same continuity conditions but are otherwise unrestricted.[1] If the approximate displacements are inserted in the expression (64.4) for the potential energy V, the latter becomes a function V_N of the parameters a_i, b_i, c_i. Since the strain-energy density is a quadratic function of the strains e_{ij}, which in turn are linear functions of the derivatives of displacements, it is clear that V_N is a quadratic function of the parameters a_i, b_i, c_i. The minimizing conditions

$$(68.1) \quad \frac{\partial V_N}{\partial a_i} = 0, \quad \frac{\partial V_N}{\partial b_i} = 0, \quad \frac{\partial V_N}{\partial c_i} = 0, \quad (i = 1, 2, \cdots, N)$$

are therefore linear equations for the determination of the unknown constants a_i, b_i, c_i. The approximate displacements u_N, v_N, w_N can thus be determined. The direct approach to variational problems just outlined is known as the Rayleigh-Ritz method.

The question of the convergence of the function u_N, v_N, w_N and their derivatives to the equilibrium displacements and displacement derivatives has been treated by E. Trefftz[2] and others.

As an illustration of the application of the Rayleigh-Ritz method, consider the problem of torsion of a cylindrical shaft. It will be recalled (Secs. 35, 64) that the stresses in a twisted shaft are given in terms of the stress function Ψ by

$$\tau_{zx} = \mu\alpha \frac{\partial \Psi}{\partial y}, \quad \tau_{zy} = -\mu\alpha \frac{\partial \Psi}{\partial x}, \quad \tau_{xx} = \tau_{yy} = \tau_{zz} = \tau_{xy} = 0,$$

where μ is the shear modulus, and α is the angle of twist per unit length. The strain energy U can be written as

---

[1] Alternatively, the required boundary conditions may be imposed on the functions $a_1 P_1$, $b_1 Q_1$, $c_1 R_1$, if the remaining functions P_i, Q_i, R_i are so chosen that they vanish on the region Σ_u where the displacements are prescribed.

[2] "Konvergenz und Fehlerabschätzung beim Ritzschen Verfahren," *Mathematische Annalen*, vol. 100 (1928), p. 503.

$$U = \int_\tau (\tau_{zx} e_{zx} + \tau_{zy} e_{zy}) \, d\tau$$
$$= \tfrac{1}{2}\mu\alpha^2 l \iint_R (\nabla\Psi)^2 \, dx \, dy,$$

where R is the region of the cross section of the shaft, and where

$$(\nabla\Psi)^2 \equiv \left(\frac{\partial\Psi}{\partial x}\right)^2 + \left(\frac{\partial\Psi}{\partial y}\right)^2.$$

The potential energy of the applied torque is

$$-\int_\Sigma \overset{\nu}{T}_i u_i \, d\sigma = -\iint_R (\tau_{zx} u + \tau_{zy} v) \, dx \, dy$$
$$= -2\mu\alpha^2 l \iint_R \Psi \, dx \, dy,$$

and the total potential energy of the system takes the form

(68.2) $$V = \tfrac{1}{2}\mu\alpha^2 l \iint_R [(\nabla\Psi)^2 - 4\Psi] \, dx \, dy.$$

The stress function Ψ satisfies the differential equation

$$\nabla^2 \Psi = -2 \qquad \text{in } R,$$

together with the condition

$$\Psi = 0 \quad \text{on the boundary } C \text{ of } R,$$

which expresses the fact that no load is applied to the lateral faces of the cylinder.

In applying the Rayleigh-Ritz method to obtain an approximate solution of the torsion problem, set up an approximate torsion function

(68.3) $$\Psi_N = \sum_{i=1}^{N} c_i f_i(x, y),$$

with the stipulation that

$$f_i(x, y) = 0 \qquad \text{on } C.$$

For convenience, drop the constant $\tfrac{1}{2}\mu\alpha^2 l$ in Eq. (68.2) and minimize instead the "energy integral"

(68.4) $$I \equiv \iint_R [(\nabla\Psi)^2 - 4\Psi] \, dx \, dy,$$

§68 VARIATIONAL METHODS

which is proportional to the potential energy of the system. When (68.3) is inserted in (68.4), there is obtained

(68.5) $\quad I_N(c_1, \cdots, c_N) = \iint_R [(\nabla \Psi_N)^2 - 4\Psi_N]\, dx\, dy.$

The minimizing conditions (68.1) can be written as

$$\frac{\partial I_N}{\partial c_i} = 2 \iint_R \nabla \Psi_N \cdot \nabla \frac{\partial \Psi_N}{\partial c_i}\, dx\, dy - 4 \iint_R \frac{\partial \Psi_N}{\partial c_i}\, dx\, dy = 0,$$

and from (68.3) it follows that

$$\iint_R \nabla \Psi_N \cdot \nabla f_i\, dx\, dy = 2 \iint_R f_i\, dx\, dy.$$

This equation can be written, with the help of Green's Theorem,[1] in the form

(68.6) $\quad \int_C f_i \frac{\partial \Psi_N}{\partial n}\, ds - \iint_R f_i \nabla^2 \Psi_N\, dx\, dy = 2 \iint_R f_i\, dx\, dy.$

Since the functions $f_i(x, y)$ are assumed to vanish on the boundary C, it follows that

(68.7) $\quad \iint_R f_i \nabla^2 \Psi_N\, dx\, dy = -2 \iint_R f_i\, dx\, dy,$

or, taking into account (68.3), one obtains

(68.8) $\quad \sum_{j=1}^N c_j \iint_R f_i \nabla^2 f_j\, dx\, dy = -2 \iint_R f_i\, dx\, dy,$

$$(i = 1, 2, \cdots, N).$$

These N equations determine the N unknown coefficients c_i and hence the approximate stress function Ψ_N and the approximate energy integral I_N. In practice, the integral (68.5) is evaluated in terms of the constants c_i, which are then determined from the N relations

$$\frac{\partial I_N}{\partial c_i} = 0.$$

---

[1] Recall that

$$\iint_R \nabla u \cdot \nabla v\, dx\, dy = \int_C u \frac{\partial v}{\partial n}\, ds - \iint_R u\, \nabla^2 v\, dx\, dy$$

It is not difficult to show that the value I_N just obtained always exceeds the true value I. Define the error function $\delta_N(x, y)$ by the equation

$$\delta_N(x, y) = \Psi(x, y) - \Psi_N(x, y),$$

whereupon Eq. (68.5) takes the form

$$I_N = \iint_R \{[\nabla(\Psi - \delta_N)]^2 - 4(\Psi - \delta_N)\} \, dx \, dy,$$

or

$$I_N = \iint_R [(\nabla \Psi)^2 - 4\Psi] \, dx \, dy - 2 \iint_R \nabla \Psi \cdot \nabla \delta_N \, dx \, dy$$
$$+ 4 \iint_R \delta_N \, dx \, dy + \iint_R (\nabla \delta_N)^2 \, dx \, dy.$$

With the help of Green's Theorem and Eq. (68.4), this can be written as

$$I_N = I - 2 \left(\int_C \delta_N \frac{\partial \Psi}{\partial n} \, ds - \iint_R \delta_N \nabla^2 \Psi \, dx \, dy \right)$$
$$+ 4 \iint_R \delta_N \, dx \, dy + \iint_R (\nabla \delta_N)^2 \, dx \, dy.$$

But the error function δ_N vanishes on the boundary C, and the true stress function Ψ satisfies the differential equation $\nabla^2 \Psi = -2$; hence the value of the approximate energy integral is

$$I_N = I + \iint_R (\nabla \delta_N)^2 \, dx \, dy,$$

and thus,

(68.9) $$I_N \geq I.$$

The Ritz method will be illustrated by its application to the problem of torsion of a beam of rectangular cross section formed by the lines $x = \pm A$ and $y = \pm B$. We take only a single term in the approximating function [Eq. (68.3)], and write

$$\Psi_1 = c(x^2 - A^2)(y^2 - B^2).$$

It is readily found that

$$(\nabla \Psi_1)^2 - 4\Psi = 4c^2[x^2(y^2 - B^2)^2 + (x^2 - A^2)^2 y^2]$$
$$- 4c(x^2 - A^2)(y^2 - B^2),$$

and

(68.10) $$\tfrac{1}{16} I_1 = \tfrac{8}{45} c^2 A^3 B^3 (A^2 + B^2) - \tfrac{4}{9} c A^3 B^3.$$

From the minimizing condition $\dfrac{dI_1}{dc} = 0$, it follows that

$$c = \frac{5}{4}\frac{1}{A^2 + B^2}.$$

The torsional rigidity D is given by

$$\frac{1}{\mu}D = 2\iint_R \Psi\, dx\, dy,$$

and the Rayleigh-Ritz method gives the approximate result

$$\frac{1}{\mu}D_1 = 2\iint_R \Psi_1\, dx\, dy = \frac{5}{18}\frac{(b/a)^2}{1 + (b/a)^2}a^3 b.$$

The maximum shearing stress $\tau_{\max}$ occurs at the midpoints of the long side; that is, at $x = \pm A$, $y = 0$. We find[1]

$$-\frac{1}{\mu\alpha}(\tau_{zy})_1 = \frac{\partial \Psi_1}{\partial x} = 2cx(y^2 - B^2),$$

and

$$\frac{1}{2\mu\alpha}(\tau_{\max})_1 = \frac{5}{4}\frac{(b/a)^2}{1 + (b/a)^2}A.$$

The approximate and exact[2] values of the torsional rigidity and maximum shear stress are compared in the following table:

$\dfrac{b}{a}$	Approx. $\dfrac{1}{\mu}\dfrac{D_1}{a^3 b}$	Exact $\dfrac{1}{\mu}\dfrac{D}{a^3 b}$	Error	Approx. $\dfrac{1}{2\mu\alpha}\dfrac{(\tau_{\max})_1}{A}$	Exact $\dfrac{1}{2\mu\alpha}\dfrac{\tau_{\max}}{A}$	Error
1	0.1389	0.1406	-1.2%	0.625	0.675	-7.4%
10	0.275	0.312	-11.9%	1.238	1.000	$+23.8\%$

[1] We denote by $(\tau_{\max})_N$ the value of the approximate shearing stress τ_N at the point where the exact shearing stress τ takes its maximum value. Hence $(\tau_{\max})_N$ does not necessarily equal max τ_N.

[2] See S. Timoshenko's Theory of Elasticity, Sec. 78, where the results are given in terms of parameters k and k_1. In our notation these are defined by

$$\frac{1}{\mu}D = k_1 a^3 b, \qquad \frac{1}{2\mu\alpha}\tau_{\max} = kA.$$

If greater accuracy is desired, further approximations may be taken of the form

$$\Psi_N = (x^2 - A^2)(y^2 - B^2)(c_1 + c_2 x^2 + c_3 y^2 + c_4 x^2 y^2 + \cdots).$$

For the torsion of a beam of a square section, one may take, for example,

$$\Psi_2 = (x^2 - A^2)(y^2 - A^2)[c_1 + c_2(x^2 + y^2)].$$

Then

$$\begin{aligned} I_2 &= \iint_R [(\nabla \Psi_2)^2 - 4\Psi_2]\, dx\, dy \\ &= {}^{64}\!/_{4725} A^6 (420 A^2 c_1^2 + 288 A^4 c_1 c_2 \\ &\quad + 176 A^6 c_2^2 - 525 c_1 - 210 A^2 c_2). \end{aligned}$$

The conditions $\dfrac{\partial I_2}{\partial c_1} = 0$, $\dfrac{\partial I_2}{\partial c_2} = 0$ give

$$280 A^2 c_1 + 96 A^4 c_2 = 175,$$
$$144 A^2 c_1 + 176 A^4 c_2 = 105,$$

from which one finds

$$A^2 c_1 = {}^{1295}\!/_{2216}, \qquad A^4 c_2 = {}^{525}\!/_{4432}.$$

The second approximation to the torsional rigidity is given by

$$\frac{1}{\mu} D_2 = 2 \iint_R \Psi_2\, dx\, dy = \frac{350}{2493} a^4 = 0.1404 a^4,$$

while the exact value is $D/\mu = 0.1406 a^4$. The second approximation gives, for the maximum shearing stress, the value

$$\frac{1}{2\mu\alpha}(\tau_{\max})_2 = -\frac{1}{2}\left(\frac{\partial \Psi_2}{\partial x}\right)_{\substack{x=A\\y=0}} = \frac{3115}{4432} A = 0.7028 A,$$

which is 4.1 per cent greater than the true value $0.675 A$.

Instead of using polynomials for the approximate stress function, one may write, for example,

$$\Psi_1 = c \cos \frac{\pi x}{a} \cos \frac{\pi y}{b}.$$

This choice of the approximate stress function Ψ_1 leads to

$$(\nabla \Psi_1)^2 - 4\Psi_1 = \pi^2 c^2 \left(\frac{1}{a^2} \sin^2 \frac{\pi x}{a} \cos^2 \frac{\pi y}{b} + \frac{1}{b^2} \cos^2 \frac{\pi x}{a} \sin^2 \frac{\pi y}{b} \right)$$
$$- 4c \cos \frac{\pi x}{a} \cos \frac{\pi y}{b}$$

and
$$\frac{1}{4} I_1 = \pi^2 c^2 \frac{ab}{16} \left(\frac{1}{a^2} + \frac{1}{b^2}\right) - 4c \frac{ab}{\pi^2}.$$

The minimizing condition $\frac{dI_1}{dc} = 0$ yields

$$c = \frac{32}{\pi^4} \frac{a^2 b^2}{a^2 + b^2},$$

from which it follows that

$$\frac{1}{\mu} D_1 = \frac{256}{\pi^6} \frac{(b/a)^2}{1 + (b/a)^2} a^3 b,$$

and

$$\frac{1}{2\mu\alpha} (\tau_{max})_1 = \frac{32}{\pi^3} \frac{(b/a)^2}{1 + (b/a)^2} A.$$

In this case, the approximate and exact values of the torsional rigidity and maximum shear stress are as follows:

$\frac{b}{a}$	$\frac{1}{\mu} \frac{D}{a^3 b}$			$\frac{1}{2\mu\alpha} \frac{\tau_{max}}{A}$		
	Approx.	Exact	Error	Approx.	Exact	Error
1	0.1331	0.1406	-5.3%	0.516	0.675	-23.6%
10	0.264	0.312	-15.4%	1.022	1.000	$+2.2\%$

The Ritz method provides an upper bound I_N [see (64.6)] to the value of the integral I. The matter of establishing the convergence of I_N to I presents some difficulty. This convergence, however, does not guarantee the approach of Ψ_N to Ψ or, even less, the approach of the derivatives of Ψ_N to those of Ψ. Moreover, the Ritz method itself furnishes no information as to the degree of approximation at any stage of the process. These questions of convergence, the improvement of convergence, and practical considerations on the choice of the functions $f_i(x, y)$ are discussed by R. Courant,[1] who also applies the Ritz method

---

[1] R. Courant, "Variational Methods for the Solution of Problems of Equilibrium and Vibrations," *Bulletin of the American Mathematical Society*, vol. 49 (1943), pp. 1–23.

to the torsion of a shaft of multiply connected cross section. For the particular case of the torsion problem, W. J. Duncan[1] has given an estimate of the error in Ψ_N.

PROBLEMS

1. Consider a beam bent by a load $p(x)$, and with potential energy

$$V = \int_0^l [\tfrac{1}{2}EI(y'')^2 - py]\,dx.$$

Introduce the approximate deflection

$$y_N = \sum_{i=1}^{N} c_i f_i(x),$$

where the functions $f_i(x)$ satisfy all the boundary conditions. Show that if the ends of the beam are clamped, hinged, or free, then the minimizing condition can be written in the form

$$\frac{\partial V_N}{\partial c_i} = \int_0^l [(EIy_N'')'' - p]f_i\,dx = 0.$$

2. In the problem of a cantilever bent by an end load P, take for the assumed deflection curve the function

$$EIy(x) = ax^2 + bx^3,$$

which satisfies the geometrical boundary conditions

$$y(0) = y'(0) = 0.$$

Determine the constants a, b from the Principle of Minimum Potential Energy. Note that the dynamical boundary conditions are automatically satisfied in the process of minimizing the potential energy, since the assumed function $y(x)$ is capable of representing the true deflection.

In the same problem, take for the assumed deflection the function

$$EIy(x) = P(-\tfrac{1}{6}x^3 + \tfrac{1}{2}lx^2 + Ax + B),$$

which satisfies the dynamic boundary conditions, shear force $F = P$, bending moment $M = 0$. Can the constants A and B be determined by minimizing the potential energy?

[1] W. J. Duncan, "On the Torsion of Cylinders of Symmetrical Section," *Proceedings of the Royal Society* (London), ser. A, vol. 136 (1932), pp. 95–113.

69. Galerkin's Method.
In Sec. 65, the problem of minimizing the integral[1] (65.1)

$$I = \int\int_R F\left(x, y, u, \frac{\partial u}{\partial x}, \frac{\partial y}{\partial y}\right) dx\, dy$$

was seen to lead to the equation

(69.1) $$\int\int_R L(u)\eta(x, y)\, dx\, dy = 0,$$

where

(69.2) $$L(u) \equiv F_u - \frac{\partial F_{u_x}}{\partial x} - \frac{\partial F_{u_y}}{\partial y} = 0$$

is the Euler equation associated with the minimal problem and $\eta(x, y)$ is an arbitrary function that vanishes on the boundary C of the region R. Equation (69.1) can be interpreted as stating that the Euler expression $L(u)$ is orthogonal to every variation $\epsilon\eta(x, y)$ that vanishes on the boundary C. Otherwise expressed, the variation in I vanishes for an arbitrary variation $\epsilon\eta(x, y)$ in the function $u(x, y)$.

It is reasonable to assume that an approximate solution u_N can be obtained by choosing

(69.3) $$u_N(x, y) = \sum_{i=1}^{N} c_i f_i(x, y),$$

where the functions $f_i(x, y)$ satisfy the boundary conditions imposed on $u(x, y)$. Equation (69.1) will, in general, no longer be valid for an arbitrary variation $\epsilon\eta(x, y)$, since this would imply that $u_N(x, y)$ satisfies the Euler equation (69.2) and hence is the exact solution of the boundary-value problem. Although the variation in I does not now vanish for an arbitrary variation $\epsilon\eta(x, y)$, yet the constants c_i in (69.3) can be so chosen that the variation in I is zero for the N variations $\epsilon f_i(x, y)$. If, now, N becomes infinite and the functions $f_i(x, y)$ form a complete[2]

---

[1] For the sake of concreteness, this discussion is phrased in terms of the problem of Eq. (65.1). The extension to other cases is immediate.

[2] A set of functions $f_i(x)$, for example, forms a complete system if any piecewise continuous function $u(x)$ can be approximated by a sum $\sum_{1}^{N} c_i f_i(x)$ in such a way that the mean square error $\int [f - \sum_{1}^{N} c_i f_i(x)]^2 dx$ can be made arbitrarily small by suitable choice of N. See R. Courant and D. Hilbert, Methoden der Mathematischen Physik, vol. 1, Chap. II, Sec. 1.

system of functions, then the set of all relations

(69.4) $$\iint_R L(u_N)f_i \, dx \, dy = 0, \quad (i = 1, 2, \cdots, N),$$

becomes equivalent to the arbitrary relation (69.1).

The Euler expression $L(u)$ vanishes when the function $u(x, y)$ is the exact solution of the problem. The magnitude of the expression $L(u_N)$ therefore can be taken as a measure of the error associated with the approximation u_N; we shall refer to $L(u_N)$ as the error function and write

$$e_N(x, y) = L(u_N).$$

Equation (69.4) then states that the error function is orthogonal to each of the N functions $f_i(x, y)$. These N conditions determine the N coefficients c_i and hence the approximate solution u_N. It should be noted that the method outlined above proceeds directly from the differential equation $L(u) = 0$ to Eqs. (69.4), for the determination of the coefficients c_i. It is thus unnecessary to form the energy integral I, the approximate integral I_N, and the N derivatives $\partial I_N/\partial c_i$. This method was proposed by Galerkin[1] in 1915.

Galerkin's method will now be illustrated by application to the torsion problem. The method is, indeed, implicit in the development of Sec. 68; it is necessary merely to rewrite Eq. (68.7), getting

(69.5) $$\iint_R (\nabla^2 \Psi_N + 2)f_i \, dx \, dy = 0, \quad (i = 1, 2, \cdots, N).$$

The error function in this case is

$$e_N(x, y) = \nabla^2 \Psi_N + 2.$$

Galerkin's method can be given a physical interpretation by thinking of the error function $e_N(x, y) = L(u_N)$ as a generalized force, and the functions $\epsilon f_i(x, y)$ as virtual displacements. Equation (69.4) then expresses the vanishing of the associated virtual work. In the membrane analogy for torsion, Ψ is interpreted as a quantity proportional to the membrane deflection, $\nabla^2 \Psi_N + 2$ is proportional to the unbalanced pressure applied

---

[1] B. G. GALERKIN, "Series Solutions of Some Problems of Elastic Equilibrium of Rods and Plates," *Vestnik Inzhenerov*, vol. 1 (1915), pp. 879–908 (in Russian).

to constrain the membrane to take the shape specified by Ψ_N, and Eq. (69.5) expresses the vanishing of the virtual work done by the unbalanced load $\nabla^2 \Psi_N + 2$ in the displacement $\epsilon f_i(x, y)$.

In the application of Galerkin's method to the torsion of a shaft of rectangular section, take, as in Sec. 64,

$$\Psi_1 = c(x^2 - A^2)(y^2 - B^2).$$

Equation (69.5) becomes

$$\iint_R \{2c[(x^2 - A^2) + (y^2 - B^2)] + 2\}(x^2 - A^2)(y^2 - B^2)\,dx\,dy = 0,$$

from which it follows that

$$c = \frac{5}{4}\frac{1}{A^2 + B^2},$$

as was found by the Rayleigh-Ritz method.

Galerkin's method has been applied by Duncan[1] to a number of problems including that of torsion.

A new method of obtaining approximate solutions of the boundary-value problems, constituting a generalization of the Ritz method, was proposed in 1933 by L. V. Kantorovitch.[2] The essence of Kantorovitch's generalization consists in the following. In obtaining an approximate solution of the equation $L(u) = 0$ in the form [see Eq. (68.3)]

$$u_N(x, y) = \sum_{i=1}^{N} c_i f_i(x, y),$$

the constants c_i are replaced by unknown functions of one variable [say, $c_i(x)$], and an application of the minimum principle (see Sec. 68) leads to a system of ordinary differential equations for the functions $c_i(x)$.

This method is intimately related to Galerkin's method,[3] and

---

[1] W. J. Duncan, "Application of the Galerkin Method to the Torsion and Flexure of Cylinders and Prisms," *Philosophical Magazine*, ser. 7, vol. 25 (1938), pp. 636–649.

[2] L. V. Kantorovitch, "One Direct Method of Approximate Solution of the Problem of Minimum of a Double Integral," *Bulletin of the Academy of Sciences, U.S.S.R.*, no. 5 (1933). (In Russian.)

[3] L. V. Kantorovitch, "Application of Galerkin's Method to the So-called Procedure of Reduction to Ordinary Differential Equations," *Applied Mathematics and Mechanics*, vol. 6 (1942), pp. 31–40.

has been used effectively by T. Tchepova and N. C. Arutinyan[1] to obtain approximate solutions of the torsion problem for beams of polygonal cross section.

70. The Error Function. The error function e_N was defined in the preceding section by the relation

$$e_N = L(u_N),$$

where

$$L(u) = 0$$

is the differential equation for the solution u of the problem, and where

$$u_N = \sum_{i=1}^{N} c_i f_i$$

is the approximate solution involving the unknown constants $c_1, c_2, \cdots, c_N$. In the torsion problem, for example,

$$L(\Psi) \equiv \nabla^2 \Psi + 2,$$

and

$$e_N = L(\Psi_N) = \nabla^2 \Psi_N + 2.$$

In Galerkin's method, the N coefficients c_i are determined from the N conditions that the error function e_N be orthogonal to each of the functions f_i, in terms of which the approximate solution is expanded. That is, it is required that

(70.1) $$\int e_N f_i \, d\sigma = 0, \quad (i = 1, 2, \cdots, N).$$

As N becomes infinite, the set of all conditions (70.1) becomes equivalent to the condition that the error function $e_N = L(u_N)$ be orthogonal to an arbitrary variation and hence that[2] $L(u) = 0$. The error function thus goes to zero as the approximate solution approaches the exact one.

Instead of requiring that the error function be orthogonal to each of the function f_i, we may impose on e_N some other set of N conditions that ensure that e_N approaches zero as N becomes

---

[1] T. Tchepova, "Approximate Solution of the Torsion Problem of Certain Prismatic Rods," *Applied Mathematics and Mechanics*, vol. 1 (1937), pp. 254–261. N. C. Arutinyan, "Approximate Solution of the Problem of Torsion of Bars Having a Polygonal Cross Section," *Applied Mathematics and Mechanics*, vol. 6 (1942), pp. 19–40.

[2] Compare Eqs. (65.3) and (65.4).

infinite. Several such methods of obtaining approximate solutions will now be considered.

The most obvious way of requiring that the error function e_N be small is to demand that it vanish at N points of the region. In this procedure, called the *collocation method*, the error is collocated or assigned at N stations, and the N equations for the coefficients c_i are obtained directly without carrying out any integrations.

The process of collocation will be illustrated by its application to the torsion problem for a beam of square cross section. We take as a first approximation

(70.2) $$\Psi_1 = c_1(x^2 - A^2)(y^2 - A^2)$$

and find that

(70.3) $\quad e_1(x, y) = \nabla^2 \Psi_1 + 2 = 2 + 2c_1[(x^2 - A^2) + (y^2 - A^2)].$

The condition $e_1(0, 0) = 0$ yields

$$c_1 = \frac{1}{2A^2}.$$

As a second approximation, take

$$\Psi_2 = (x^2 - A^2)(y^2 - A^2)[c_1 + c_2(x^2 + y^2)],$$

and require that

$$e_2(0, 0) = e_2\left(\frac{A}{2}, \frac{A}{2}\right) = 0.$$

Then

$$c_1 = \frac{25}{42A^2}, \quad c_2 = \frac{2}{21A^4}.$$

The approximate values of the torsional rigidity and shearing stress calculated from these approximations are compared with the exact values:

$$\frac{1}{\mu}\frac{D_1}{a^4} = 0.1111, \qquad \frac{1}{2\mu\alpha}\frac{(\tau_{\max})_1}{A} = 0.500,$$

$$\frac{1}{\mu}\frac{D_2}{a^4} = 0.1365, \qquad \frac{1}{2\mu\alpha}\frac{(\tau_{\max})_2}{A} = 0.690,$$

$$\cdots\cdots\cdots\cdots\cdots\cdots\cdots\cdots\cdots\cdots\cdots\cdots$$

$$\frac{1}{\mu}\frac{D}{a^4} = 0.1406, \qquad \frac{1}{2\mu\alpha}\frac{\tau_{\max}}{A} = 0.675.$$

The collocation method minimizes the error function e_N by requiring that it vanish at N points of the region. In the *method of least squares*, the constants c_i are found by requiring that the mean square error be as small as possible. That is, the condition

$$\int e_N^2 \, d\sigma = \text{minimum}$$

or

(70.4)
$$\int e_N \frac{\partial e_N}{\partial c_i} \, d\sigma = 0, \quad (i = 1, 2, \cdots, N)$$

affords N equations for the determinations of the N constants c_i.

It will be instructive to apply the method of least squares to the torsion problem for a beam of square cross section, and choose the same approximating function [Eq. (70.2)] as was used in illustrating the collocation method. Equation (70.4) becomes, in this case,

$$\int_0^A \int_0^A \{2 + 2c_1[(x^2 - A^2) + (y^2 - A^2)]\} 2[(x^2 - A^2) + (y^2 - A^2)] \, dx \, dy = 0,$$

from which it is found that

$$c_1 = \frac{15}{22} \frac{1}{A^2}.$$

The exact values of the torsional rigidity and maximum shearing stress are compared in the following table with the values found as a first approximation by the method of least squares:

$$\frac{1}{\mu} \frac{D_1}{a^4} = 0.1515, \qquad \frac{1}{2\mu\alpha} \frac{(\tau_{\max})_1}{A} = 0.682,$$

$$\frac{1}{\mu} \frac{D}{a^4} = 0.1406, \qquad \frac{1}{2\mu\alpha} \frac{\tau_{\max}}{A} = 0.675.$$

Another process for finding the approximate solution of a boundary-value problem is suggested by the analogy between the stress function for torsion $\Psi(x, y)$ and the deflection $z(x, y)$ of a membrane under pressure p and tension T and stretched over a plane closed curve C. We recall from Sec. 46 that the stress function is determined as the solution of the boundary-value problem (46.3, 46.4)

$$\nabla^2 \Psi = -2, \qquad \text{in } R,$$
$$\Psi = 0, \qquad \text{on } C,$$

while the membrane deflection $z(x, y)$ is found from the relations (46.1)

$$\nabla^2 z = -\frac{p}{T} \qquad \text{in } R,$$
$$z = 0 \qquad \text{on } C.$$

We proceed to find an approximate stress function Ψ_N by writing

(70.5) $$\Psi_N = \sum_{i=1}^{N} c_i f_i(x, y),$$

with $\qquad f_i = 0 \qquad \text{on } C.$

The approximate stress function Ψ_N will not, in general, satisfy the differential equation (46.3), and $\nabla^2 \Psi_N$ will equal not -2 but some function[1] $p_N(x, y)$,

The function $p_N(x, y)$ can be interpreted, in terms of the membrane analogy, as the nonuniform pressure necessary to constrain the membrane to take the form defined by the function $\Psi_N(x, y)$.

If the collocation method were applied to this problem, one would demand that the "approximate loading function" $p_N(x, y)$ equal the "true load" of -2 at N points of the region R. Instead of such local conditions, one may impose a different set of conditions. The region R may be divided into N regions R_i, over each of which it is required that the total approximate load equal the total true load. That is, the N coefficients c_i of Eq. (70.4) are to be determined from the N conditions

$$\iint_{R_i} \nabla^2 \Psi_N \, dx \, dy = \iint_{R_i} \nabla^2 \Psi \, dx \, dy, \quad (i = 1, 2, \cdots, N),$$

or

(70.6) $$\iint_{R_i} \nabla^2 \Psi_N \, dx \, dy = -2 \cdot (\text{area of region } R_i).$$

---

[1] Since the error function e_N is defined, in this case, by

$$\nabla^2 \Psi_N + 2 = e_N,$$

we see that the approximate loading function p_N is given in terms of the error function by the relation

$$p_N = e_N - 2.$$
$$\nabla^2 \Psi_N = p_N(x, y).$$

Thus, instead of starting with a prescribed "pressure" -2 and solving for the deflection (or stress function), we have inverted the problem. We start with an assumed stress function Ψ_N, involving N arbitrary coefficients, and calculate the corresponding approximate loading function $p_N(x, y)$. The constants c_i are then so determined that the approximate and true loading functions are equal in the mean over each of N subregions R_i of the section R.

In applying the method outlined above to the torsion of a rectangular beam, choose, as a first approximation,

$$\Psi_1 = c_1(x^2 - A^2)(y^2 - B^2)$$

and take the region R_1 to be the entire region R of the rectangle. Equation (70.6) becomes

$$2c_1 \int_0^B \int_0^A (x^2 - A^2 + y^2 - B^2)\, dx\, dy = -2AB,$$

from which it follows that

$$c_1 = \frac{3}{2(A^2 + B^2)}$$

and $\quad \dfrac{1}{\mu} D_1 = \dfrac{1}{3} \dfrac{(b/a)^2}{1 + (b/a)^2} a^3 b = 0.1667 a^4,\quad$ (square section),

$$\frac{1}{2\mu\alpha}(\tau_{\max})_1 = \frac{3}{2}\frac{(b/a)^2}{1+(b/a)^2} A = 0.75 A, \quad \text{(square section)},$$

as against the exact values of $0.1406 a^4$ and $0.675 A$, respectively (for a square section).

As a further approximation, take (for a square section)

$$\Psi_3 = (x^2 - A^2)(y^2 - A^2)[c_1 + c_2(x^2 + y^2) + c_3 x^2 y^2].$$

From considerations of symmetry, it is clear that for the three regions R_{ij} one may take

R_{11}: $\qquad 0 \le x \le \dfrac{A}{2},\qquad 0 \le y \le \dfrac{A}{2},$

R_{12}: $\qquad \dfrac{A}{2} \le x \le A,\qquad 0 \le y \le \dfrac{A}{2},$

R_{22}: $\qquad \dfrac{A}{2} \le x \le A,\qquad \dfrac{A}{2} \le y \le A.$

The three conditions

$$\iint_{R_{ij}} \nabla^2 \Psi_3 \, dx \, dy = -2 \iint_{R_{ij}} dx \, dy = -\frac{A^2}{2}$$

yield the equations

$$440A^2c_1 - 186A^4c_2 - 17A^6c_3 = 240,$$
$$320A^2c_1 + 564A^4c_2 + 19A^6c_3 = 240,$$
$$200A^2c_1 + 594A^4c_2 + 235A^6c_3 = 240,$$

which are solved by

$$A^2c_1 = {}^{149}\!/\!{}_{252}, \qquad A^4c_2 = {}^{20}\!/\!{}_{252}, \qquad A^6c_3 = {}^{80}\!/\!{}_{252}.$$

The torsional rigidity is given approximately by

$$\frac{1}{\mu} D_3 = 2 \iint_R \Psi_3 \, dx \, dy = 0.1413a^4$$

and is 0.5 per cent greater than the exact value, $0.1406a^4$. The maximum shearing stress is given approximately by

$$\frac{1}{2\mu\alpha} (\tau_{\max})_3 = 0.671A,$$

a value that is 0.6 per cent less than the exact result, $0.675A$.

The method of this section was introduced by Biezeno and Koch,[1] who have applied it to problems of thin plates and elastically supported beams. Biezeno[2] has observed that this procedure may be applied to the general problem of elastic equilibrium.

[1] C. B. BIEZENO and J. J. KOCH, "Over een nieuwe methode ter berekening van vlakke platen met toepassing op enkele voor de techniek belangrijke belastingsgevallen," *De Ingenieur*, vol. 38 (1923), pp. 25–36.

C. B. BIEZENO, "Graphical and Numerical Methods for Solving Stress Problems," *Proceedings of the First International Congress of Applied Mechanics, Delft, 1924*, pp. 3–17.

C. B. BIEZENO and R. GRAMMEL, Technische Dynamik, Chap. III, Sec. 9.

[2] C. B. BIEZENO, "Over een vereenvoudiging en over een uitbreiding van de methode van Ritz," *Christiaan Huygens Int. Math. Tids.*, vol. 3 (1923), pp. 69–75.

See also the preceding reference, Chap. III, Sec. 10.

Courant has pointed out[1] that both the method of Biezeno and Koch, and that of Galerkin, are special cases of a more general procedure for approximating the solution of a boundary-value problem. In order to see this, we write Eq. (70.6) in the form

$$\iint_{R_i} (\nabla^2 \Psi_N + 2)\, dx\, dy = 0.$$

But $\nabla^2 \Psi_N + 2$ is merely the error function $e_N = L(\Psi_N)$, and if the functions $\eta_i(x, y)$ are defined by

$$\eta_i(x, y) = \begin{cases} 1 \text{ in } R_i \\ 0 \text{ elsewhere in } R, \end{cases}$$

then one can write[2]

$$\iint_R L(\Psi_N)\eta_i\, dx\, dy = 0, \quad (i = 1, 2, \cdots, N).$$

In Galerkin's method, the constants c_i are determined from the N conditions

$$\iint_R L(\Psi_N)f_i\, dx\, dy = 0, \quad (i = 1, 2, \cdots, N),$$

where the functions f_i are those in terms of which the approximate solution Ψ_N is expanded,

$$\Psi_N = \sum_{i=1}^{N} c_i f_i.$$

Now the solution Ψ of the problem satisfies the relation [see Eq. (69.1)]

(70.7) $$\iint_R L(\Psi)\eta\, dx\, dy = 0$$

for any function $\eta(x, y)$ vanishing on C, and from this we get the differential equation of the problem

$$L(\Psi) = 0.$$

---

[1] In the discussion of Biezeno's paper, "Graphical and Numerical Methods for solving Stress Problems," *Proceedings of the First International Congress of Applied Mechanics, Delft,* 1924, pp. 3–17.

[2] We retain the notation Ψ used for the stress function for torsion. The discussion, however, applies to boundary-value problems in general.

That is, the error function vanishes for the exact solution Ψ. Courant observed that the vanishing of the integral (70.7) for an arbitrary variation $\epsilon\eta(x, y)$ can be replaced by the condition that this integral vanish for each function $\eta_i(x, y)$ of a complete set. Since the arbitrary variation $\epsilon\eta(x, y)$ may be expanded in terms of the functions $\eta_i(x, y)$, the two alternative conditions are equivalent. In Galerkin's method, the set of functions η_i is taken to be the collection of functions f_i, in terms of which Ψ_N is expanded. In the procedure of Biezeno and Koch, each function $\eta_i(x, y)$ vanishes over R except in the subregion R_i, where it takes the value unity.

PROBLEM

Consider the problem of a cantilever beam under uniform load, and take for the approximate deflection $y_1(x)$, the function given in (66.9). Define the error function $e_1(x)$ by

$$e_1(x) = EI \frac{d^4 y_1}{dx^4} - p.$$

Compare the exact and approximate values of the maximum deflection and moment when the constant c_1 in the function y_1 is found by collocating the error at the center $[e_1(l/2) = 0]$ and at the end $[e_1(l) = 0]$ by setting the mean error equal to zero $\left[\int_0^l e_1(x)\, dx = 0\right]$ or by minimizing the square of the error $\left[\int_0^l e_1^2\, dx = \text{minimum}\right]$.

71. Estimates of Error in the Minimal Integral and the Stress Function for Torsion.

The methods of Ritz, Galerkin, and Biezeno and Koch and the procedures of collocation and least squares enable one to obtain expressions that, it is hoped, approximate more or less closely to the solution of a boundary-value problem. The problem of convergence of the approximating function and its derivatives to the exact solution has not been discussed. Even if it were known that a method of approximate solution implied convergence, the success or failure of the method would still depend upon the rapidity of convergence. Indeed, if an estimate of error in the approximating function Ψ_N, its integral or its derivatives, is available, the corresponding method of solution may be practically useful (provided the error is within the tolerance allowed), whether or not the approximate solution converges. In this section and the next, methods are considered that make possible the estimation of the error in the approximate torsional rigidity D_N and in the approximate stress function Ψ_N.

The torsion problem has been formulated as one of minimizing the integral

$$I(\Psi) = \iint_R [(\nabla\Psi)^2 - 4\Psi]\, dx\, dy,$$

subject to the condition

$$\Psi = 0 \qquad \text{on } C.$$

The minimum value I of this integral is closely related to the torsional rigidity (35.10)

$$D = 2\mu \iint_R \Psi\, dx\, dy.$$

For, with the aid of Green's Theorem, Eq. (68.4) can be written as

$$I = \int_C \Psi \frac{\partial \Psi}{\partial n}\, ds - \iint_R \Psi \nabla^2 \Psi\, dx\, dy - 4 \iint_R \Psi\, dx\, dy,$$

or, since

$$\nabla^2 \Psi = -2 \qquad \text{in } R,$$

and

$$\Psi = 0 \qquad \text{on } C,$$

we have

(71.1) $$I = -2 \iint_R \Psi\, dx\, dy.$$

Hence the torsional rigidity D and the minimum value I of the energy integral are related by the expression

(71.2) $$D = -\mu I.$$

It was shown in Sec. 68 that the approximate value of the energy integral

$$I_N = \iint_R [(\nabla \Psi_N)^2 - 4\Psi_N]\, dx\, dy$$

always exceeds the true minimum $I_N \geq I$. Hence the corresponding approximate torsional rigidity $D_N = -\mu I_N$ will fall short of the exact value. The Ritz method, however, gives no estimate of the error in D_N. In this section, an ingenious scheme is considered whereby the true minimum is approached from below rather than from above; the error in D_N may therefore be estimated and made as small as desired.

We begin by considering the following formal inequality:

(71.3) $\quad (\nabla\Psi)^2 - 4\Psi + (\nabla w)^2 - 4x\dfrac{\partial w}{\partial y} + 4x^2 + 4\dfrac{\partial}{\partial x}(x\Psi)$

$$+ 2\left[\dfrac{\partial}{\partial x}\left(w\dfrac{\partial\Psi}{\partial y}\right) - \dfrac{\partial}{\partial y}\left(w\dfrac{\partial\Psi}{\partial x}\right)\right] \geq 0,$$

with the equality holding if, and only if, both

(71.4) $\quad \dfrac{\partial\Psi}{\partial x} = \dfrac{\partial w}{\partial y} - 2x \quad$ and $\quad \dfrac{\partial\Psi}{\partial y} = -\dfrac{\partial w}{\partial x}.$

This follows immediately upon rewriting the inequality (71.3) as

$$\left(\dfrac{\partial\Psi}{\partial x} - \dfrac{\partial w}{\partial y} + 2x\right)^2 + \left(\dfrac{\partial\Psi}{\partial y} + \dfrac{\partial w}{\partial x}\right)^2 \geqq 0.$$

The function Ψ is now taken to be the stress function of the torsion problem. We integrate the inequality (71.3) over the region R and observe that the last three terms of (71.3) drop out upon integration, since

$$\int\int_R 4\dfrac{\partial}{\partial x}(x\Psi)\,dx\,dy + 2\int\int_R\left[\dfrac{\partial}{\partial x}\left(w\dfrac{\partial\Psi}{\partial y}\right) - \dfrac{\partial}{\partial y}\left(w\dfrac{\partial\Psi}{\partial x}\right)\right]dx\,dy$$

$$= 4\int_C x\Psi\,dy + 2\int_C w\left(\dfrac{\partial\Psi}{\partial x}\,dx + \dfrac{\partial\Psi}{\partial y}\,dy\right)$$

$$= 4\int_C x\Psi\,dy + 2\int_C w\,d\Psi = 0.$$

(Ψ vanishes on the boundary C.) If the functional $J(w)$ is defined by

(71.5) $\quad J(w) = -\int\int_R\left[(\nabla w)^2 - 4x\dfrac{\partial w}{\partial y} + 4x^2\right]dx\,dy,$

then the result of the integration can be written as

$$I(\Psi) - J(w) \geqq 0,$$

or

(71.6) $\quad\quad\quad\quad\quad J(w) \leqq I(\Psi),$

and this inequality holds for any suitable $w(x, y)$. If we take for $w(x, y)$ a function for which (71.4) is satisfied, then the equality holds in (71.3), and

$$J(w) = I(\Psi).$$

Hence the maximum of the integral $J(w)$ is identical with the minimum of the integral $I(\Psi)$, and the true minimum I can readily be inclosed between an upper bound I_N and a lower bound J_N.

We apply the ideas outlined above to the torsion of a beam of square section, and note that in this case Eq. (71.5) becomes

$$J(w) = -\iint_R \left[(\nabla w)^2 - 4x \frac{\partial w}{\partial y} \right] dx\, dy - \frac{1}{3} a^4,$$

or

$$J(w) = L(w) - \tfrac{1}{3} a^4,$$

where

(71.7) $\qquad L(w) = -\iint_R \left[(\nabla w)^2 - 4x \frac{\partial w}{\partial y} \right] dx\, dy.$

Thus, it is necessary only to maximize the integral $L(w)$.

Since only bounds on the error in the integral J_N (or I_N) are involved, and not in the approximating function

$$w_N = \sum_{i=1}^{N} k_i g_i(x, y),$$

it is not necessary to inquire into the boundary conditions on the functions $g_i(x, y)$. The condition (71.6) will furnish an estimate of the error in J_N, whatever function w_N be substituted in (71.5). However, since the Euler equation [see (65.4)] for the variational problem (71.5) is

$$\nabla^2 w = 0,$$

one may choose for the function $g_i(x, y)$ polynomial solutions of Laplace's equation, namely, the real and imaginary parts of $(x + iy)^n$. Now if a term of the form $w = kx^r y^s$ be substituted in (71.7), it will be found that the condition $\dfrac{dL}{dk} = 0$ leads only to the vacuous result $k = 0$, unless both r and s are odd. Accordingly, we choose the functions $g_i(x, y)$ from among the polynomials

$$xy,\ x^3y - xy^3,\ 3x^5y - 10x^3y^3 + 3xy^5,\ \cdots.$$

For a first approximation, we take

$$w_1 = k_1 xy$$

and get

$$L(w_1) = -\tfrac{1}{6}a^4(k_1^2 - 2k_1).$$

The condition $\dfrac{dL}{dK_1} = 0$ yields

$$k_1 = 1, \qquad L_1 = \tfrac{1}{6}a^4, \qquad J_1 = -\tfrac{1}{6}a^4 = -0.1667 a^4.$$

As a second approximation, we choose

$$w_2 = k_1 xy + k_2(x^3y - xy^3)$$

and find

$$L(w_2) = -\frac{8A^4}{105}[35(k_1^2 - 2k_1) + 4A^2(9A^2 k_2^2 + 7k_2)].$$

From

$$\frac{\partial L}{\partial k_1} = 0, \qquad \frac{\partial L}{\partial k_2} = 0,$$

we have

$$k_1 = 1, \qquad k_2 = -\frac{7}{18A^2}, \qquad L_2 = \frac{26}{135}a^4,$$
$$J_2 = \tfrac{26}{135}a^4 - \tfrac{1}{3}a^4 = -\tfrac{19}{135}a^4 = -0.1407 a^4.$$

A second approximation to I from above was found[1] by the Ritz method to be $I_2 = -\dfrac{1}{\mu}D_2 = -0.1404 a^4$, while a second approximation from below is now found to be

$$J_2 = -0.1407 a^4.$$

Thus without knowledge of the exact value of D ($0.1406\mu a^4$), it is certain that it lies between $0.1404\mu a^4$ and $0.1407\mu a^4$ and that the error in either figure is at most $\dfrac{(1407 - 1404)}{1404}$ or 0.2 per cent.

[1] See problem discussed in Sec. 68.

The idea of setting up an auxiliary integral whose maximum is equal to the minimum of the original integral is due to Friedrichs[1] and was applied to the torsion problem by Basu,[2] who considered, as an example, a beam of square section.

W. J. Duncan[3] has given a method for estimating the error in Ψ_N, a method that may be applied to a boundary-value problem involving a Poisson equation,

$$\nabla^2 \Psi = \rho = \text{const.} \quad \text{in } R,$$
$$\Psi = 0 \quad \text{on } C.$$

Let Ψ_N be the approximate solution satisfying the boundary condition

$$\Psi_N = 0 \quad \text{on } C,$$

and define the error function $e(x, y)$, as in Sec. 65, by the relation

$$\nabla^2 \Psi_N = \rho + e(x, y).$$

The maximum and minimum values in R of $e(x, y)$ are denoted by e_M and e_m respectively. Then from

$$\nabla^2 \left[\left(1 + \frac{e_M}{\rho}\right) \Psi - \Psi_N \right] = e_M - e(x, y) \geq 0 \quad \text{in } R,$$

and

$$\left(1 + \frac{e_M}{\rho}\right) \Psi - \Psi_N = 0 \quad \text{on } C,$$

it follows[4] that

(71.8) $$\left(1 + \frac{e_M}{\rho}\right) \Psi - \Psi_N \leq 0 \quad \text{in } R.$$

---

[1] K. Friedrichs, "Ein Verfahren der Variationsrechnung des Minimum eines Integral als das Maximum eines änderen Ausdruckes darzustellen," *Nachrichten von der Gesellschaft der Wissenschaften zu Göttingen, Mathematisch-physikalische Klasse* (1929), pp. 13–20.

[2] N. M. Basu, "On an Application of the New Methods of the Calculus of Variations to Some Problems in the Theory of Elasticity," *Philosophical Magazine*, ser. 7, vol. 10 (1930), pp. 886–896; "On the Torsion Problem of the Theory of Elasticity," *Philosophical Magazine*, ser. 7, vol. 10 (1930), pp. 896–904.

[3] W. J. Duncan, "On the Torsion of Cylinders of Symmetrical Section," *Proceedings of the Royal Society*, (London), ser. A, vol. 136 (1932), pp. 95–113; "The Torsion and Flexure of Cylinders and Tubes," Aeronautical Research Committee (Great Britain), *Reports and Memoranda* 1444 (1932).

[4] We use here the theorem that

if $z(x, y) = 0$ on C, and if $\nabla^2 z \geq 0 (\leq 0)$ in R, then $z(x, y) \leq 0 (\geq 0)$ in R.

Similarly, from

$$\nabla^2\left[\Psi_N - \left(1 + \frac{e_m}{\rho}\right)\Psi\right] = e(x,y) - e_m \geq 0 \quad \text{in } R,$$

and

$$\Psi_N - \left(1 + \frac{e_m}{\rho}\right)\Psi = 0 \quad \text{on } C,$$

it follows that

(71.9) $$\Psi_N - \left(1 + \frac{e_m}{\rho}\right)\Psi \leq 0 \quad \text{in } R.$$

The inequalities (71.8) and (71.9) can be written as

(71.10) $$\left(1 + \frac{e_M}{\rho}\right)\Psi \leq \Psi_N \leq \left(1 + \frac{e_M}{\rho}\right)\Psi \quad \text{in } R.$$

When applied to the torsion problem (where $\rho = -2$), Eq. (71.10) furnishes the following estimate of the error in the approximate stress function Ψ_N:

$$\frac{1}{2}e_m \leq \frac{\Psi - \Psi_N}{\Psi} \leq \frac{1}{2}e_M.$$

72. Relaxation of the Boundary Conditions. We recall that the torsional rigidity D is given by $D = -\mu I$, where I is the minimum value of the energy integral (68.4). Since the Ritz (or Galerkin) method yields an approximation I_N that exceeds the true value [see (68.9)], the corresponding value of the torsional rigidity $D_N = -\mu I_N$ will fall below the exact value. In 1928 Trefftz[1] suggested a procedure (inspired by the work of Courant) that allows the value I to be approached from below and thus furnishes bounds on the error in the torsional rigidity.

The torsion problem has been considered thus far as the boundary-value problem of determining the function Ψ such that

$$\nabla^2\Psi = -2 \quad \text{in } R,$$
$$\Psi = 0 \quad \text{on } C.$$

---

[1] E. TREFFTZ, "Konvergenz und Fehlerabschätzung beim Ritzschen Verfahren," *Mathematische Annalen*, vol. 100 (1928), pp. 503–521; "Ein Gegenstück zum Ritzshen Verfahren," *Proceedings of the Second International Congress for Applied Mechanics, Zürich*, 1927, pp. 131–137.

If one sets

(72.1) $\psi(x, y) = \Psi(x, y) + \tfrac{1}{2}(x^2 + y^2)$,

then the determination of the conjugate torsion function ψ is a problem of Dirichlet, with

(72.2) $\begin{cases} \nabla^2 \psi = 0 & \text{in } R, \\ \psi = \tfrac{1}{2}(x^2 + y^2) & \text{on } C. \end{cases}$

In each of the methods discussed thus far, the stress function has been approximated by a finite sum of functions that satisfy the boundary conditions but not the differential equation. We follow now the alternative procedure of representing the approximate stress function by a sum of functions each of which satisfies the differential equation but not the boundary conditions. The function $\psi(x, y)$ will accordingly be approximated by a sum of harmonic functions

$$\psi_N = \sum_{i=1}^{N} c_i f_i(x, y)$$

with

$$\nabla^2 f_i(x, y) = 0 \qquad \text{in } R.$$

Instead of requiring that Eqs. (72.2) be rigorously satisfied (which is possible only if ψ_N is the exact solution), the boundary conditions are to be relaxed to some weaker conditions that can also be satisfied by the function $\psi_N(x, y)$. The precise boundary conditions to be imposed are suggested by the formulation of the problem as one in the calculus of variations.

A reference to Eq. (65.5) shows that the variational problem associated with the Laplace equation of (72.2) is that of finding a stationary value for the integral

$$K(\psi) = \iint_R (\nabla \psi)^2 \, dx \, dy,$$

subject to

$$\psi = \tfrac{1}{2}(x^2 + y^2) \qquad \text{on } C.$$

If we put

$$\delta = \psi - \psi_N$$

and

$$K(\delta) = \iint_R (\nabla \delta)^2 \, dx \, dy,$$

then we can write

$$K(\psi) = \iint_R [\nabla(\psi_N + \delta)]^2 \, dx \, dy$$
$$= K(\psi_N) + K(\delta) + 2 \iint_R \nabla\psi \cdot \nabla\delta \, dx \, dy$$
$$= K(\psi_N) + K(\delta) + 2 \int_C \delta \frac{\partial \psi_N}{\partial n} \, ds - 2 \iint_R \delta \nabla^2 \psi_N \, dx \, dy,$$

or

(72.3) $$K(\psi_N) = K(\psi) - K(\delta) - 2 \int_C (\psi - \psi_N) \frac{\partial \psi_N}{\partial n} \, ds.$$

Now notice that, since $K(\delta) \geq 0$, we can have

(72.4) $$K(\psi_N) = K(\psi) - K(\delta)$$

and hence

(72.5) $$K(\psi_N) \leq K(\psi),$$

provided that the last term in (72.3) is made to vanish. In other words, the value of the minimal integral can be approached from below, instead of from above as in the Ritz method. The vanishing of the last term in (72.3), written in the form

$$-2 \sum_{i=1}^{N} c_i \int_C (\psi - \psi_N) \frac{\partial f_i}{\partial n} \, ds,$$

can be ensured if we choose the N coefficients c_i to satisfy the N conditions

$$\int_C (\psi - \psi_N) \frac{\partial f_i}{\partial n} \, ds = 0$$

or

(72.6) $$\int_C \left[\frac{1}{2}(x^2 + y^2) - \psi_N \right] \frac{\partial f_i}{\partial n} \, ds = 0, \quad (i = 1, 2, \cdots, N).$$

Equations (72.6) constitute a weaker boundary condition than the true one

$$\tfrac{1}{2}(x^2 + y^2) - \psi = 0 \qquad \text{on } C,$$

for the conditions (72.6) are satisfied not only by the exact solution ψ but also by the approximate solution ψ_N.

The Trefftz procedure may be looked upon as one either for maximizing $K(\psi_N)$ or for minimizing $K(\delta)$, since, by (72.4), $K(\psi_N)$ and $-K(\delta)$ differ only by the value $K(\psi)$ (constant for a given problem). This property of Trefftz's method of minimizing $K(\delta)$ can be shown directly, and one could, in fact, take this as the starting point of the development rather than the idea of relaxing the boundary conditions. For one has

$$0 = \frac{\partial K(\delta)}{\partial c_i} = 2 \iint_R \nabla(\psi - \psi_N) \cdot \frac{\partial}{\partial c_i} \nabla(\psi - \psi_N) \, dx \, dy$$

$$= -2 \iint_R \nabla(\psi - \psi_N) \cdot \nabla f_i \, dx \, dy$$

$$= -2 \int_C (\psi - \psi_N) \frac{\partial f_i}{\partial n} \, ds + 2 \iint_R (\psi - \psi_N) \nabla^2 f_i \, dx \, dy$$

$$= -2 \int_C (\psi - \psi_N) \frac{\partial f_i}{\partial n} \, ds,$$

and these are just the conditions imposed on the coefficients c_i in (72.6).

The relation between $K(\psi)$ and the torsional rigidity can be found by using (72.1), (68.4), and Green's Theorem. One has

$$K(\psi) = \iint_R \left[\nabla \Psi + \frac{1}{2} \nabla(x^2 + y^2) \right]^2 dx \, dy$$

$$= \iint_R (\nabla \Psi)^2 \, dx \, dy + \iint_R \nabla \Psi \cdot \nabla(x^2 + y^2) \, dx \, dy$$

$$+ \iint_R (x^2 + y^2) \, dx \, dy$$

$$= \iint_R (\nabla \Psi)^2 \, dx \, dy + \int_C \Psi \frac{\partial(x^2 + y^2)}{\partial n} \, ds$$

$$- \iint_R \Psi \nabla^2(x^2 + y^2) \, dx \, dy + \iint_R (x^2 + y^2) \, dx \, dy$$

$$= \iint_R [(\nabla \Psi)^2 - 4\Psi] \, dx \, dy + \iint_R (x^2 + y^2) \, dx \, dy,$$

or

(72.7) $$K(\psi) - j_0 = I(\Psi),$$

where

$$j_0 = \iint_R (x^2 + y^2) \, dx \, dy$$

§72   VARIATIONAL METHODS

is the polar moment of inertia of the section. From (72.5) and (72.7), it is seen that as $K(\psi_N)$ approaches $K(\psi)$ from below, so $K(\psi_N) - j_0$ approaches $I(\Psi)$ from below. From (71.2) it follows that

$$\mu[j_0 - K(\psi_N)] \text{ approaches } D = -\mu I(\Psi) \text{ from above.}$$

We apply the method outlined above to the torsion of a beam of square section. For the functions $f_i(x, y)$, we choose the harmonic polynomials obtained by taking the real and imaginary parts of $(x + iy)^n$. The first two polynomials that satisfy the requirements of symmetry (even in x, symmetric in x, y) are

$$f_1 = 1, \quad f_2 = x^4 - 6x^2y^2 + y^4;$$

accordingly, we take

$$\psi_2 = c_1 + c_2(x^4 - 6x^2y^2 + y^4).$$

The constant c_1 is not determined by the condition (72.6). It can be fixed by requiring that the mean error on the boundary vanish; that is,

$$\int_C [\tfrac{1}{2}(x^2 + y^2) - \psi_2] \, ds = 0.$$

This furnishes the equation

$$15c_1 - 12A^4 c_2 = 10A^2.$$

From (72.6), we have

$$\int_0^A \left[\frac{1}{2}(A^2 + y^2) - c_1 - c_2(A^4 - 6A^2 y^2 + y^4) \right] \frac{\partial}{\partial x}(x^4 - 6x^2 y^2 + y^4)_{x=A} \, dy = 0,$$

or

$$c_2 = -\frac{7}{72A^2}.$$

The approximate stress function Ψ can now be written as

$$\Psi = \psi_2 - \tfrac{1}{2}(x^2 + y^2)$$
$$= \frac{53}{90} A^2 - \frac{7}{72A^2}(x^4 - 6x^2 y^2 + y^4) - \frac{1}{2}(x^2 + y^2),$$

and the corresponding value of the torsional rigidity is

$$\frac{1}{\mu} D_2 = 2 \int\int_R \Psi_2 \, dx \, dy = \frac{19}{135} a^4 = 0.1407 a^4,$$

which is almost identical with the exact value $0.1406a^4$. The maximum shearing stress is given approximately by

$$\frac{1}{2\mu\alpha} (\tau_{\max})_2 = -\frac{1}{2}\left(\frac{\partial \Psi}{\partial x}\right)_{\substack{x=A \\ y=0}} = \frac{25}{36} A = 0.694 A,$$

which is 2.8 per cent greater than the exact value, $0.675A$.

73. The Method of Finite Differences. The various methods for solving boundary-value problems that have been outlined in the preceding sections have a common origin in the variational formulation of the problems. One objection to any such variational procedure is that the approach of the integral $I(\Psi_n)$ to $I(\Psi)$ does not, in general, ensure the convergence of Ψ_n and its derivatives to the solution Ψ and the corresponding derivatives.[1] The foregoing methods are analytical and demand analytical expressions for the boundary and the boundary values. These methods are not applicable to problems involving highly irregular regions or problems in which the boundary values are given numerically. Moreover, it is often difficult to make a judicious choice of the coordinate functions $f_i(x, y)$ in $\Psi_n = \sum_{1}^{n} c_i f_i$, for even the first approximation should be fairly good, and the computation must not be too extensive.

In this section, an alternative procedure is discussed, which replaces derivatives by finite differences and the continuous region R by a set of net points. The differential equation is replaced by a difference equation that may be solved by an iterative procedure. If the mesh of the net is made sufficiently fine, then it can be shown that not only does the solution of the difference equation approach the solution of the original problem, but the difference quotients also converge to the corresponding derivatives of the solution.

[1] R. COURANT, "Variational Methods for the Solution of Problems of Equilibrium and Vibration," *Bulletin of the American Mathematical Society,* vol. 49 (1943), pp. 1–23.

§73 VARIATIONAL METHODS

The method of finite differences will be illustrated by applying it to the Dirichlet problem of determining the function $\psi(x, y)$ defined by

(73.1) $\qquad \begin{cases} \nabla^2 \psi = 0 & \text{in } R, \\ \psi = g(s) & \text{on } C, \end{cases}$

We lay down a square net over the region R, assign known values to the net points on the boundary, and approximate (guessed) values at interior points. The differential equation of the system (73.1) is replaced by a difference equation, which we proceed to derive.

In the neighborhood of any interior point of R (taken, for the moment, as the origin of coordinates), we can write

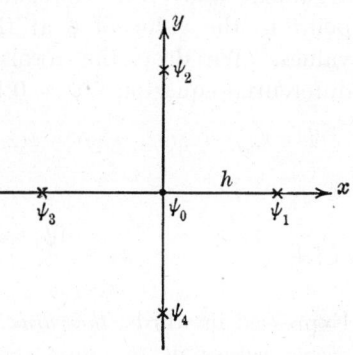

Fig. 51.

(73.2) $\psi(x, y) = \psi_0 + \alpha_{10}x + \alpha_{01}y + \alpha_{20}x^2 + \alpha_{02}y^2 + \alpha_{11}xy + \cdots$
$$= \sum_{i=0}^{\infty} \sum_{j=0}^{\infty} \alpha_{ij} x^i y^j.$$

The value of the function ψ at the origin is precisely

$$\psi(0, 0) = \psi_0 = \alpha_{00},$$

while at the neighboring net points to the right and left one has (Fig. 51)

$$\psi_1 = \psi(h, 0) = \sum_{i=0}^{\infty} \alpha_{i0} h^i = \psi_0 + \alpha_{10}h + \alpha_{20}h^2 + \cdots,$$
$$\psi_3 = \psi(-h, 0) = \psi_0 - \alpha_{10}h + \alpha_{20}h^2 - \cdots,$$

and

$$\psi_1 + \psi_3 = 2\psi_0 + 2\alpha_{20}h^2 + 2\alpha_{40}h^4 + \cdots.$$

Similarly,

$$\psi_2 + \psi_4 = 2\psi_0 + 2\alpha_{02}h^2 + 2\alpha_{04}h^4 + \cdots.$$

Since the value of the Laplacian at the origin is

$$(\nabla^2 \psi)_0 = 2\alpha_{20} + 2\alpha_{02},$$

one can write

(73.3) $$\frac{(\psi_1 + \psi_2 + \psi_3 + \psi_4 - 4\psi_0)}{h^2} = (\nabla^2\psi)_0 + \text{terms in } h^2.$$

As the choice of the origin of coordinates is not essential to the argument above, the foregoing expression relates $\nabla^2\psi$ at any point to the value of ψ at that point and to the neighboring values. We drop the terms in h^2 and replace the Laplace differential equation $\nabla^2\psi = 0$ by the Laplace difference equation

$$\psi(x+h, y) + \psi(x-h, y) + \psi(x, y+h) + \psi(x, y-h) - 4\psi(x, y) = 0,$$

or

(73.4) $$\psi_0 = \frac{(\psi_1 + \psi_2 + \psi_3 + \psi_4)}{4}.$$

Expressed in words, *the value of $\psi(x, y)$ at any point is the mean of its values at the four immediate neighboring points*. This difference equation is equivalent to a set of linear equations for the values of ψ at interior points in terms of the prescribed boundary values. The number of variables is usually so large, however, that direct solution is out of the question. Instead, we may resort to an iterative procedure.

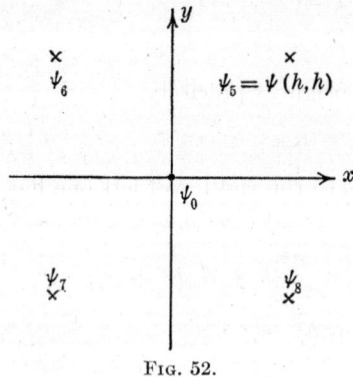

Fig. 52.

The simplest (but most laborious) way of solving the Laplace difference equation is to guess at the proper values for the interior points of the network; this guess is then corrected by traversing the net, replacing each interior value by the mean of its four immediate neighbors. Repeated traverses of the net will give interior values that converge to the values of the solution function $\psi(x, y)$. The convergence, however, is so slow as to require almost unlimited manpower in order to secure sufficiently accurate results. Fortunately, various procedures are available for improving the rapidity of convergence.

§73   VARIATIONAL METHODS   337

Instead of expressing the value of ψ as the mean of the four immediate neighbors, Eq. (73.4), the four diagonal neighbors may be used[1] (Fig. 52). We have from (73.2)

$$\psi_5 + \psi_6 + \psi_7 + \psi_8 = 4\psi_0 + 4(\alpha_{20} + \alpha_{02})h^2 + \text{terms in } h^4,$$

or, neglecting terms in h^2 compared with unity,

$$2(\nabla^2\psi)_0 = \frac{(\psi_5 + \psi_6 + \psi_7 + \psi_8 - 4\psi_0)}{h^2}.$$

The Laplace difference equation can also be written, then, as

(73.5) $$\psi_0 = \frac{(\psi_5 + \psi_6 + \psi_7 + \psi_8)}{4}.$$

If both the immediate and the diagonal neighbors are used, then the formula

(73.6) $$20\psi_0 = 4(\psi_1 + \psi_2 + \psi_3 + \psi_4) + (\psi_5 + \psi_6 + \psi_7 + \psi_8)$$

gives the value of ψ_0 for any seventh-order harmonic polynomial.[2]

The slowness of convergence of the original process [Eq. (73.4)] is explained by the fact that on any one traverse, an interior value is made to depend only on its immediate neighbors, and the effect of the prescribed boundary values moves inland very slowly as successive traverses are made. The boundary values may be made effective at a greater distance by the following procedure,

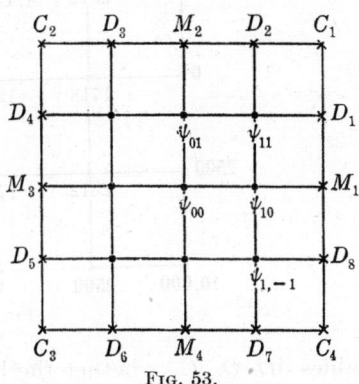

FIG. 53.

which uses nine interior points ψ_{ij} and 16 boundary points M_i, D_i, C_i (Fig. 53). The value at the center is first found from[3]

(73.7) $$\psi_{00} = \tfrac{1}{16}[D_1 + D_2 + D_3 + D_4 + D_5 + D_6 + D_7 + D_8 \\ + 2(M_1 + M_2 + M_3 + M_4)].$$

[1] This follows from invariance with respect to rotation.
[2] G. H. SHORTLEY and R. WELLER, *Journal of Applied Physics*, vol. 9 (1938), p. 345.
[3] G. H. SHORTLEY, R. WELLER, and B. FRIED, "Numerical Solution of Laplace's and Poisson's Equations," *Ohio State University Studies*, Engineering Ser. (1940), p. 11.

The corner values, such as ψ_{11}, are obtained from the diagonal neighbors, so that

$$\psi_{11} = \frac{(\psi_{00} + c_1 + M_1 + M_2)}{4},$$

while values such as ψ_{10} make use of the immediate neighbors

$$\psi_{10} = \frac{(\psi_{00} + M_1 + \psi_{11} + \psi_{1,-1})}{4}.$$

This same procedure may be applied, of course, to find the value of ψ on any block of nine points in terms of the surrounding

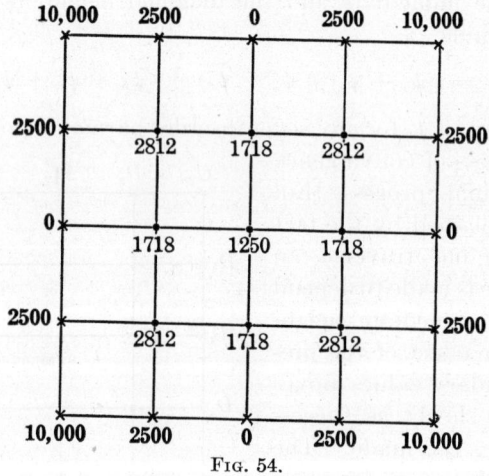

Fig. 54.

values M_i, D_i, C_i, whether the latter lie on the boundary of the region R or not.

The method of finite differences will now be applied to the torsion problem for a beam of square cross section with side length $2A$. The conjugate torsion function $\psi(x, y)$ is defined by

$$\nabla^2 \psi = 0 \qquad \text{in } R,$$
$$\psi = \tfrac{1}{2}(x^2 + y^2) \qquad \text{on } C.$$

We introduce the variables $X = x/A$, $Y = y/A$ and put

(73.8) $$\Omega = 10^4 \left(\frac{2}{A^2} \psi - 1 \right).$$

§73 VARIATIONAL METHODS

Then the function Ω is subject to the conditions

$$\nabla^2 \Omega = 0 \qquad \text{in } R,$$
$$\Omega = \begin{cases} 10^4\, Y^2 & \text{on } X = \pm 1, \\ 10^4\, X^2 & \text{on } Y = \pm 1. \end{cases}$$

A coarse net is now laid down over the square section (Fig. 54). Equation (73.7) gives $\Omega(0, 0) = 1250$, while $\Omega(\tfrac{1}{2}, \tfrac{1}{2}) = 2812$ is derived from its diagonal neighbors by (73.5) and $\Omega(\tfrac{1}{2}, 0) = 1718$ from its immediate neighbors by (73.4). From symmetry considerations, it is seen that only one octant of the section need be considered. The values of Ω in this octant are now improved by using Eq. (73.6) to give, in order, $\Omega(\tfrac{1}{2}, \tfrac{1}{2}) = 2250$, $\Omega(\tfrac{1}{2}, 0) = 1572$, and $\Omega(0, 0) = 1708$. Fig. 55 shows these values as well as those of the third approximation resulting from a further application of Eq. (73.6) to the net values in the same order as before.

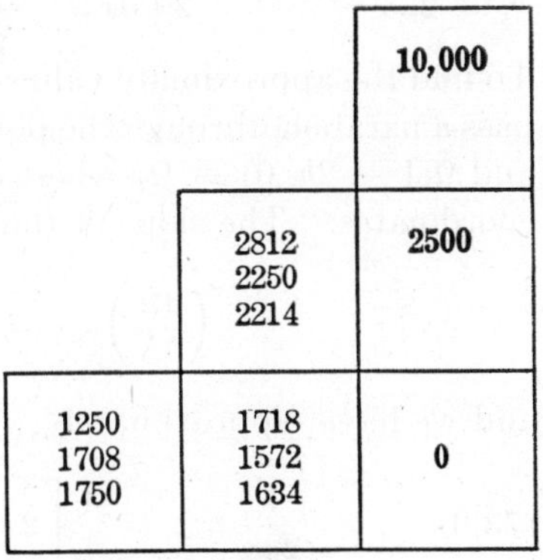

Fig. 55.

The values found above can be checked against those given by the exact solution of the problem in Sec. 38. From (38.10) and (73.8), it follows that

$$\Omega(X, Y) = 10^4 \left[1 + Y^2 - X^2 - 8 \sum_{n=0}^{\infty} \frac{(-1)^n}{N^3} \frac{\cosh NY}{\cosh N} \cos NX \right],$$

where $N = (2n + 1)\pi/2$. The exact values at the net points used above are found to be

$$\Omega(0, 0) = 1787.4, \qquad \Omega(0, \tfrac{1}{2}) = 1673.6, \qquad \Omega(\tfrac{1}{2}, \tfrac{1}{2}) = 2245.8,$$

and it is seen that the approximate values are in error at most by 2 per cent.

The torsional rigidity is given by

$$\frac{1}{\mu} D = 2 \int\!\!\int_R \Psi \, dx \, dy = 2 \int\!\!\int_R \left[\psi - \frac{1}{2}(x^2 + y^2) \right] dx \, dy$$
$$= A^2 \int\!\!\int_R (1 + 10^{-4}\Omega) \, dx \, dy - \frac{8}{3} A^4,$$

or
$$\frac{1}{\mu a^4} D = \frac{1}{12} + \frac{10^{-4}}{16} \int_{-1}^{1} \int_{-1}^{1} \Omega \, dX \, dY.$$

The maximum shear stress occurs at a midpoint of the boundary and is given by

$$\frac{1}{2\mu\alpha} \tau_{\max} = -\frac{1}{2}\left[\frac{\partial \psi}{\partial x} - x\right]_{\substack{x=A \\ y=0}} = \frac{A}{4}\left[2 - 10^{-4} \frac{\partial \Omega}{\partial X}\right]_{\substack{X=1 \\ Y=0}}.$$

To find the approximate value of the shear stress at this point, we pass a parabola through the points $\Omega(1, 0) = 0$, $\Omega(1 - h, 0) \equiv \Omega_1$, and $\Omega(1 - 2h, 0) \equiv \Omega_2$, where h is the mesh of the net in the X, Y coordinates. The slope of the parabola at $(1, 0)$ is given by

$$\left(\frac{\partial \Omega}{\partial X}\right)_{\substack{X=1 \\ Y=0}} = \frac{1}{2h}(\Omega_2 - 4\Omega_1),$$

and we have, approximately,

(73.9) $$\frac{1}{2\mu\alpha} \tau_{\max} = \frac{A}{4}\left[2 - \frac{10^{-4}}{2h}(\Omega_2 - 4\Omega_1)\right].$$

The approximate values of Ω given in Fig. 55 yield $\Omega_1 = 1634$, $\Omega_2 = 1750$, $h = \frac{1}{2}$, and

$$\frac{1}{2\mu\alpha} \tau_{\max} = 0.6196A,$$

which is 8.1 per cent below the exact value $0.675A$. The numerical values of Ω can be obtained by Simpson's rule to give

$$\int_{-1}^{1} \int_{-1}^{1} \Omega \, dX \, dY = \frac{80{,}246}{9},$$

and

$$\frac{1}{\mu a^4} D = 0.1391,$$

which is 1 per cent below the true value 0.1406.

If the exact solution of this problem were not known, it would have been necessary to proceed with the iterative procedure until the net values remained sensibly constant. Before continuing the process, however, we introduce a labor-saving modification.

§73 VARIATIONAL METHODS

Denote by $\Omega^{(0)}$ an approximate solution and by Ω the exact solution of the difference equation at a given net point. Then one can write

$$\Omega^{(0)} = \Omega + \epsilon^{(0)},$$

where $\epsilon^{(0)}$ is the error at the net point in the solution of the difference (not the differential) equation. A single traverse of the net yields an improved value $\Omega^{(1)}$ and error $\epsilon^{(1)}$:

$$\Omega^{(1)} = \Omega + \epsilon^{(1)}.$$

We denote by $\delta^{(1)}$ the change in Ω in one traverse; that is,

$$\delta^{(1)} = \Omega^{(1)} - \Omega^{(0)} = \epsilon^{(1)} - \epsilon^{(0)}.$$

Now replace the original boundary-value problem of determining the function Ω with given boundary values by the problem of determining the difference function δ, which vanishes at the boundary net points. In other words, the differences $\delta^{(1)}$ are improved in the same way [for example, by Eq. (73.4)] as were the original values Ω, and successive traverses yield the improved differences $\delta^{(2)}, \delta^{(3)}, \cdots$. The final values of Ω are obtained by adding to $\Omega^{(1)}$ the improved differences:

$$\Omega^{(n)} = \Omega^{(1)} + \delta^{(2)} + \delta^{(3)} + \cdots + \delta^{(n)}.$$

For

$$\delta^{(2)} = \Omega^{(2)} - \Omega^{(1)} = \epsilon^{(2)} - \epsilon^{(1)},$$

and, in general,

$$\delta^{(n)} = \Omega^{(n)} - \Omega^{(n-1)} = \epsilon^{(n)} - \epsilon^{(n-1)}$$

Then

$$\Omega^{(1)} + \delta^{(2)} + \delta^{(3)} + \cdots + \delta^{(n)}$$
$$= \Omega + \epsilon^{(1)} + [\epsilon^{(2)} - \epsilon^{(1)}] + [\epsilon^{(3)} - \epsilon^{(2)}] + \cdots + [\epsilon^{(n)} - \epsilon^{(n-1)}]$$
$$= \Omega + \epsilon^{(n)} = \Omega^{(n)}.$$

In the torsion problem under discussion, the differences $\delta^{(1)}$ can be found from the values of Ω calculated on the last two traverses (Fig. 55). In Fig. 56, these values of Ω are shown in the left-hand columns, while in the columns on the right are entered the differences $\delta^{(1)}, \delta^{(2)}, \delta^{(3)}, \delta^{(4)}$ obtained by applying (73.6) to the differences in the order $(\tfrac{1}{2}, \tfrac{1}{2}), (\tfrac{1}{2}, 0), (0, 0)$. Two advantages of working with the difference function δ rather than with the original function Ω are immediately seen. Both the zero boundary values and the smaller number of significant figures in the differences δ make for easier computation. A third and more powerful advantage is that use of the difference function δ makes possible the estimation of the effect of infinitely many traverses of the network. That is, inspection of the successive differences $\delta^{(2)}$, $\delta^{(3)}$, $\delta^{(4)}$ suggests that these and succeeding values may form a geometric progression of ratio one-half.

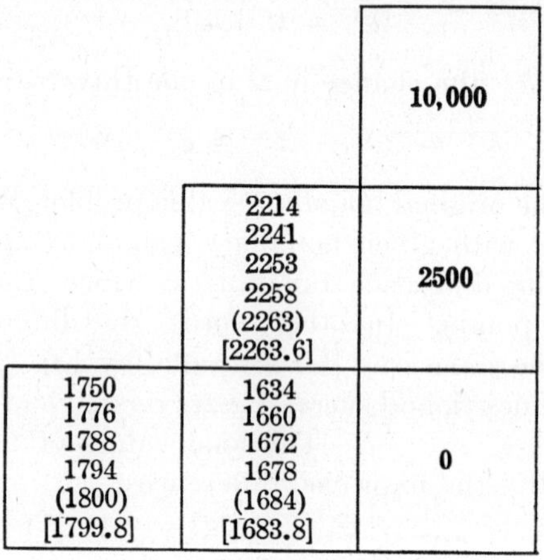

		10,000
	2214 2241 2253 2258 (2263) [2263.6]	2500
1750 1776 1788 1794 (1800) [1799.8]	1634 1660 1672 1678 (1684) [1683.8]	0

Fig. 56.

We hazard a guess that this is indeed the case and sum each infinite series of differences, getting, in this case,

$$\delta^{(5)} + \delta^{(6)} + \delta^{(7)} + \cdots = 2\delta^{(5)} = \delta^{(4)}.$$

The sum of the differences $\sum_{2}^{\infty} \delta^{(i)}$ is then added to $\Omega^{(1)}$ at each net point to obtain an estimate of $\Omega^{(\infty)}$ (see the third entry in each column on the left in Fig. 56). The assumption that the successive differences $\delta^{(n)}$ form a geometric progression can now be tested by using (73.6) to improve the values of Ω just obtained (see the last entry in the left-hand columns of Fig. 56).

Shortley, Weller, and Fried, in an investigation of the convergence of the method of finite differences,[1] have shown that

---

[1] G. H. Shortley, R. Weller, and B. Fried, "Numerical Solution of Laplace's and Poisson's Equations," *Ohio State University Studies*, Engineering Ser., (1940), p. 18.

§73   VARIATIONAL METHODS   343

this extrapolation to the limiting net value by summing the infinite geometric series of differences is possible in general.

The final net values in Fig. 56 satisfy the difference equation (to within one unit) but not the differential equation. This is shown by comparison with the exact ordinates given above, and the disagreement arises from the fact that in setting up the difference equation, terms of higher order in the net mesh were neglected [see (73.3)]. We proceed, therefore, to decrease the mesh of the net to one-half its original size.

In the process of interpolation leading to the values of Ω at the new net points, the difference equations (73.4) and (73.5)

				10,000
			4316	5625
		2263	*2672*	2500
	1857	*1853*	*1611*	625
1799	*1799*	1683	*1226*	0

Fig. 57.

are used to ensure that the interpolated values satisfy, at least approximately, the differential equation $\nabla^2 \Omega = 0$. The mean of the diagonal neighbors [Eq. (73.5)] furnishes the values of $\Omega(\frac{3}{4}, \frac{3}{4})$, $\Omega(\frac{3}{4}, \frac{1}{4})$, and $\Omega(\frac{1}{4}, \frac{1}{4})$. The immediate neighbors are used [Eq. (73.4)] to get $\Omega(\frac{1}{4}, 0)$ $\Omega(\frac{3}{4}, 0)$, $\Omega(\frac{1}{2}, \frac{1}{4})$, and $\Omega(\frac{3}{4}, \frac{1}{2})$. The resulting values are shown in Fig. 57. Without any further improvement, these values give $\Omega_1 = 1226$, $\Omega_2 = 1683$ [see Eq. (73.9)] and

$$\frac{1}{2\mu\alpha} \tau_{\max} = 0.661A,$$

which is 2 per cent below the exact value $0.675A$. The integral of Ω is found by Simpson's rule to be approximately 9292, and the approximate torsional rigidity is found to be $D/(\mu a^4) = 0.1414$, a value 0.6 per cent above the exact result of 0.1406.

Instead of traversing the lattice points in a fixed order and extrapolating to the limiting net value, as above, one can correct the lattice values in any way at all. Indeed, all that is required is that one arrive at values Ω for which the difference equation is satisfied—or, alternatively, for which the differences δ are zero. In this way, the experience and physical intuition of the computer can be used to good advantage.

Another variation in the finite difference method consists of replacing the lattice with square mesh, used above, by a lattice formed of regular polygons.[1]

The torsion problem of a beam of square cross section, considered above, is a particularly simple one in that the square cross section imposes no special complications at the boundary. When the boundary is curved, the derivatives are replaced by finite difference expressions involving unequal intervals.[2]

While the finite difference method has been illustrated by its application to the problem of Dirichlet,

$$\nabla^2 \Omega = 0 \quad \text{in } R,$$
$$\Omega \text{ given} \quad \text{on } C,$$

it can obviously be extended to a wide variety of problems in engineering and mathematical physics. We mention as examples[3] the plasticity problem of torsion of a shaft strained beyond the elastic limit and the problem of a two-dimensional magnetic field containing a triangular prism of iron.

---

[1] See, for example, D. G. Christopherson and R. V. Southwell, "Relaxation Methods Applied to Engineering Problems. III. Problems Involving Two Independent Variables," *Proceedings of the Royal Society* (London), ser. A, vol. 168 (1938), pp. 317–350.

[2] For this and other details, both theoretical and practical, relating to finite difference methods, see G. H. Shortley, R. Weller, and B. Fried, "Numerical Solutions of Laplace's and Poisson's Equations," *Ohio State University Studies*, Engineering Ser. (1940).

[3] D. G. Christopherson and R. V. Southwell, "Relaxation Methods Applied to Engineering Problems. III. Problems Involving Two Independent Variables," *Proceedings of the Royal Society* (London), ser. A, vol. 168 (1938), pp. 317–350.

§73 VARIATIONAL METHODS

Tables given by D. Moskovitz[1] can be used to obtain the exact solution of the *difference* equation corresponding to the Poisson equation

$$\frac{\partial^2 V}{\partial x^2} + \frac{\partial^2 V}{\partial y^2} = F(x, y),$$

when the region R is assumed to be rectangular.

An extensive bibliography of approximate mathematical methods of solving the problem of torsion has been given by T. J. Higgins.[2]

---

[1] "The Numerical Solution of Laplace's and Poisson's Equations," *Quarterly of Applied Mathematics*, vol. 2 (1944), pp. 148–163.

[2] "The Approximate Mathematical Methods of Applied Physics as Exemplified by Application to Saint Venant's Torsion Problem," *Journal of Applied Physics*, vol. 14 (1943), pp. 469–480.

APPENDIX

SUMMARY OF FORMULAS

The more important formulas of the text are included in this summary. The numbers on the left of the equations are equation numbers as they occur in the text. The numbers on the right are the numbers of the pages on which these equations first appear.

CHAPTER I

ANALYSIS OF STRAIN

The strains e_{ij} and the rotations ω_{ij} are connected with the displacements u_i by

(7.5) $\qquad e_{ij} = \frac{1}{2}(u_{i,j} + u_{j,i}), \qquad \omega_{ij} = \frac{1}{2}(u_{i,j} - u_{j,i}). \qquad$ (p. 19)

The dilatation ϑ is

$$\vartheta = e_{11} + e_{22} + e_{33}$$
$$= \frac{\partial u_1}{\partial x_1} + \frac{\partial u_2}{\partial x_2} + \frac{\partial u_3}{\partial x_3} = u_{i,i} \qquad \text{(p. 19)}$$

If we set $x_1 = x$, $x_2 = y$, $x_3 = z$, $e_{11} = e_{xx}$, $e_{12} = e_{xy}$, etc., and denote the components of the displacement vector (u_1, u_2, u_3) by (u, v, w), the components of the strain tensor become

$$e_{xx} = \frac{\partial u}{\partial x}, \qquad e_{yy} = \frac{\partial v}{\partial y}, \qquad e_{zz} = \frac{\partial w}{\partial z},$$

$$e_{zy} = \frac{1}{2}\left(\frac{\partial w}{\partial y} + \frac{\partial v}{\partial z}\right), \quad e_{xz} = \frac{1}{2}\left(\frac{\partial u}{\partial z} + \frac{\partial w}{\partial x}\right), \quad e_{yx} = \frac{1}{2}\left(\frac{\partial v}{\partial x} + \frac{\partial y}{\partial u}\right),$$

so that the dilatation ϑ is

$$\vartheta = e_{xx} + e_{yy} + e_{zz} = \frac{\partial u}{\partial x} + \frac{\partial v}{\partial y} + \frac{\partial w}{\partial z}. \qquad \text{(p. 20)}$$

The equations of compatibility are

(10.9) $\qquad e_{ij,kl} + e_{kl,ij} - e_{ik,jl} - e_{jl,ik} = 0, \qquad$ (p. 27) or, in unabridged notation,

$$(10.10) \begin{cases} \dfrac{\partial^2 e_{xx}}{\partial y\, \partial z} = \dfrac{\partial}{\partial x}\left(-\dfrac{\partial e_{yz}}{\partial x} + \dfrac{\partial e_{zx}}{\partial y} + \dfrac{\partial e_{xy}}{\partial z}\right), \\ \dfrac{\partial^2 e_{yy}}{\partial z\, \partial x} = \dfrac{\partial}{\partial y}\left(-\dfrac{\partial e_{zx}}{\partial y} + \dfrac{\partial e_{xy}}{\partial z} + \dfrac{\partial e_{yz}}{\partial x}\right), \\ \dfrac{\partial^2 e_{zz}}{\partial x\, \partial y} = \dfrac{\partial}{\partial z}\left(-\dfrac{\partial e_{xy}}{\partial z} + \dfrac{\partial e_{yz}}{\partial x} + \dfrac{\partial e_{zx}}{\partial y}\right), \\ 2\dfrac{\partial^2 e_{xy}}{\partial x\, \partial y} = \dfrac{\partial^2 e_{xx}}{\partial y^2} + \dfrac{\partial^2 e_{yy}}{\partial x^2}, \\ 2\dfrac{\partial^2 e_{yz}}{\partial y\, \partial z} = \dfrac{\partial^2 e_{yy}}{\partial z^2} + \dfrac{\partial^2 e_{zz}}{\partial y^2}, \\ 2\dfrac{\partial^2 e_{zx}}{\partial z\, \partial x} = \dfrac{\partial^2 e_{zz}}{\partial x^2} + \dfrac{\partial^2 e_{xx}}{\partial z^2}. \end{cases} \quad \text{(p. 27)}$$

If the Eulerian components of finite strains are denoted by η_{jk} and the Lagrangian components by ϵ_{jk}, then

(11.4) $\qquad 2\eta_{jk} = u_{j,k} + u_{k,j} - u_{i,j} u_{i,k},$ \hfill (p. 30)

and

(11.6) $\qquad 2\epsilon_{jk} = u_{j,k} + u_{k,j} + u_{i,j} u_{i,k}.$ \hfill (p. 30)

In unabridged notation these become

$$\eta_{xx} = \frac{\partial u}{\partial x} - \frac{1}{2}\left[\left(\frac{\partial u}{\partial x}\right)^2 + \left(\frac{\partial v}{\partial x}\right)^2 + \left(\frac{\partial w}{\partial x}\right)^2\right],$$

$$\epsilon_{aa} = \frac{\partial u}{\partial a} + \frac{1}{2}\left[\left(\frac{\partial u}{\partial a}\right)^2 + \left(\frac{\partial v}{\partial a}\right)^2 + \left(\frac{\partial w}{\partial a}\right)^2\right],$$

$$2\eta_{xy} = \frac{\partial u}{\partial y} + \frac{\partial v}{\partial x} - \left(\frac{\partial u}{\partial x}\frac{\partial u}{\partial y} + \frac{\partial v}{\partial x}\frac{\partial v}{\partial y} + \frac{\partial w}{\partial x}\frac{\partial w}{\partial y}\right),$$

$$2\epsilon_{ab} = \frac{\partial u}{\partial b} + \frac{\partial v}{\partial a} + \left(\frac{\partial u}{\partial a}\frac{\partial u}{\partial b} + \frac{\partial v}{\partial a}\frac{\partial v}{\partial b} + \frac{\partial w}{\partial a}\frac{\partial w}{\partial b}\right). \quad \text{(p. 31)}$$

CHAPTER II

ANALYSIS OF STRESS

The components of body force are denoted by F_i, the components of moments are M_i, the components of stress tensor are τ_{ij}, the stress vector is $\overset{i}{T}$, and the components of exterior unit normal are ν_i.

APPENDIX

The boundary conditions are

(13.3) $$\overset{\nu}{T}_i = \tau_{ji}\nu_j,$$ (p. 39)

or

$$\overset{\nu}{T}_x = \tau_{xx} \cos(x,\nu) + \tau_{yx}\cos(y,\nu) + \tau_{zx}\cos(z,\nu)$$
$$\overset{\nu}{T}_y = \tau_{xy}\cos(x,\nu) + \tau_{yy}\cos(y,\nu) + \tau_{zy}\cos(z,\nu)$$
$$\overset{\nu}{T}_z = \tau_{xz}\cos(x,\nu) + \tau_{yz}\cos(y,\nu) + \tau_{zz}\cos(z,\nu).$$ (p. 40)

Equations of equilibrium are

(15.3) $$\tau_{ji,j} = -F_i,$$ (p. 41)

or, when written out in full in the notation explained in Sec. 14,

$$\frac{\partial \tau_{xx}}{\partial x} + \frac{\partial \tau_{yx}}{\partial y} + \frac{\partial \tau_{zx}}{\partial z} = -F_x,$$
$$\frac{\partial \tau_{xy}}{\partial x} + \frac{\partial \tau_{yy}}{\partial y} + \frac{\partial \tau_{zy}}{\partial z} = -F_y,$$
$$\frac{\partial \tau_{xz}}{\partial x} + \frac{\partial \tau_{yz}}{\partial y} + \frac{\partial \tau_{zz}}{\partial z} = -F_z.$$ (p. 42)

In these equations,

(15.6) $$\tau_{ij} = \tau_{ji};$$ (p. 43)

that is, the stress tensor is symmetric.

If the principal stresses are denoted by τ_1, τ_2, τ_3, then the invariants of the stress tensor are

(17.11)
$$\begin{cases} \Theta_1 = \tau_1 + \tau_2 + \tau_3 = \tau_{11} + \tau_{22} + \tau_{33} \equiv \Theta, \\ \Theta_2 = \tau_1\tau_2 + \tau_2\tau_3 + \tau_3\tau_1 \\ = \begin{vmatrix} \tau_{22} & \tau_{23} \\ \tau_{23} & \tau_{33} \end{vmatrix} + \begin{vmatrix} \tau_{11} & \tau_{31} \\ \tau_{31} & \tau_{33} \end{vmatrix} + \begin{vmatrix} \tau_{11} & \tau_{12} \\ \tau_{12} & \tau_{22} \end{vmatrix}, \\ \Theta_3 = \tau_1\tau_2\tau_3 \\ = \begin{vmatrix} \tau_{11} & \tau_{12} & \tau_{13} \\ \tau_{21} & \tau_{22} & \tau_{23} \\ \tau_{31} & \tau_{32} & \tau_{33} \end{vmatrix}. \end{cases}$$ (p. 50)

The extremal shearing stresses are

(18.4) $\tau = \pm\tfrac{1}{2}(\tau_2 - \tau_3),\quad \tau = \pm\tfrac{1}{2}(\tau_1 - \tau_3),$
$$\tau = \pm\tfrac{1}{2}(\tau_1 - \tau_2),$$ (p. 54)

and these act, respectively, on planes whose normals have the direction cosines

$$(18.5) \quad \begin{cases} \nu_1 = 0, & \nu_2 = \pm \dfrac{1}{\sqrt{2}}, & \nu_3 = \pm \dfrac{1}{\sqrt{2}}, \\ \nu_2 = 0, & \nu_1 = \pm \dfrac{1}{\sqrt{2}}, & \nu_3 = \pm \dfrac{1}{\sqrt{2}}, \\ \nu_3 = 0, & \nu_1 = \pm \dfrac{1}{\sqrt{2}}, & \nu_2 = \pm \dfrac{1}{\sqrt{2}}. \end{cases} \quad \text{(p. 54)}$$

CHAPTER III

STRESS-STRAIN RELATIONS

Hooke's law for a homogeneous isotropic body is

$$(22.3) \qquad \tau_{ij} = \lambda \delta_{ij} \vartheta + 2\mu e_{ij}, \quad (i, j = 1, 2, 3). \qquad \text{(p. 66)}$$

Equation (22.3) yields a simple relation connecting the invariants $\vartheta = e_{ii}$ and $\Theta = \tau_{ii}$. Putting $j = i$ in (22.3) and noting that $\delta_{ii} = \delta_{11} + \delta_{22} + \delta_{33} = 3$, one finds that

$$\Theta \equiv \tau_{ii} = 3\lambda \vartheta + 2\mu e_{ii}$$

or

$$(22.4) \qquad \Theta = (3\lambda + 2\mu)\vartheta. \qquad \text{(p. 66)}$$

If one solves (22.3) for strains, one obtains

$$(22.5) \qquad e_{ij} = \frac{-\lambda \delta_{ij}}{2\mu(3\lambda + 2\mu)} \Theta + \frac{1}{2\mu} \tau_{ij}. \qquad \text{(p. 66)}$$

The constants of Lamé λ and μ are related to Young's modulus E and Poisson's ratio σ by

$$(23.3) \qquad \sigma \equiv \frac{\lambda}{2(\lambda + \mu)}, \qquad E \equiv \frac{\mu(3\lambda + 2\mu)}{\lambda + \mu}, \qquad \text{(p. 68)}$$

$$(23.5) \qquad \lambda = \frac{E\sigma}{(1 + \sigma)(1 - 2\sigma)}, \qquad \mu = \frac{E}{2(1 + \sigma)}. \qquad \text{(p. 68)}$$

The stress-strain relations (22.5), when written by making the substitutions from (23.5), assume the simple form

$$(23.10) \qquad e_{ij} = \frac{1 + \sigma}{E} \tau_{ij} - \frac{\sigma}{E} \delta_{ij} \Theta, \qquad \text{(p. 70)}$$

where $\Theta = \tau_{ii}$. If we recall the notation of Sec. 14, these relations can also be given in the following form:

$$(23.11) \begin{cases} e_{xx} = \dfrac{1}{E}[\tau_{xx} - \sigma(\tau_{yy} + \tau_{zz})], \\[4pt] e_{yy} = \dfrac{1}{E}[\tau_{yy} - \sigma(\tau_{zz} + \tau_{xx})], \\[4pt] e_{zz} = \dfrac{1}{E}[\tau_{zz} - \sigma(\tau_{xx} + \tau_{yy})], \\[4pt] e_{yz} = \dfrac{1+\sigma}{E}\tau_{yz}, \quad e_{zx} = \dfrac{1+\sigma}{E}\tau_{zx}, \\[4pt] \hspace{8em} e_{xy} = \dfrac{1+\sigma}{E}\tau_{xy}. \quad \text{(p. 70)} \end{cases}$$

The equilibrium equations of Navier are

$$(24.7) \qquad \mu \nabla^2 u_i + (\lambda + \mu)\frac{\partial \vartheta}{\partial x_i} + F_i = 0, \qquad \text{(p. 74)}$$

where

$$\vartheta = e_{ii} = u_{i,i} = \operatorname{div} \mathbf{u}.$$

The Beltrami-Michell compatibility equations are

$$(24.14) \quad \nabla^2 \tau_{ij} + \frac{1}{1+\sigma}\Theta_{,ij} = -\frac{\sigma}{1-\sigma}\delta_{ij}\operatorname{div}\mathbf{F} - (F_{i,j} + F_{j,i}).$$
$$\text{(p. 77)}$$

Equations (24.14), when written out in unabridged notation, yield the following six equations of compatibility:

$$(24.15) \begin{cases} \nabla^2 \tau_{xx} + \dfrac{1}{1+\sigma}\dfrac{\partial^2 \Theta}{\partial x^2} = -\dfrac{\sigma}{1-\sigma}\operatorname{div}\mathbf{F} - 2\dfrac{\partial F_x}{\partial x}, \\[6pt] \nabla^2 \tau_{yy} + \dfrac{1}{1+\sigma}\dfrac{\partial^2 \Theta}{\partial y^2} = -\dfrac{\sigma}{1-\sigma}\operatorname{div}\mathbf{F} - 2\dfrac{\partial F_y}{\partial y}, \\[6pt] \nabla^2 \tau_{zz} + \dfrac{1}{1+\sigma}\dfrac{\partial^2 \Theta}{\partial z^2} = -\dfrac{\sigma}{1-\sigma}\operatorname{div}\mathbf{F} - 2\dfrac{\partial F_z}{\partial z}, \\[6pt] \nabla^2 \tau_{yz} + \dfrac{1}{1+\sigma}\dfrac{\partial^2 \Theta}{\partial y\,\partial z} = -\left(\dfrac{\partial F_y}{\partial z} + \dfrac{\partial F_z}{\partial y}\right), \\[6pt] \nabla^2 \tau_{zx} + \dfrac{1}{1+\sigma}\dfrac{\partial^2 \Theta}{\partial z\,\partial x} = -\left(\dfrac{\partial F_z}{\partial x} + \dfrac{\partial F_x}{\partial z}\right), \\[6pt] \nabla^2 \tau_{xy} + \dfrac{1}{1+\sigma}\dfrac{\partial^2 \Theta}{\partial x\,\partial y} = -\left(\dfrac{\partial F_x}{\partial y} + \dfrac{\partial F_y}{\partial x}\right), \quad \text{(p. 77)} \end{cases}$$

If the field of body force $\mathbf{F}$ is conservative, so that

$$\mathbf{F} = \nabla \varphi$$

or

$$F_j = \varphi_{,j},$$

then
$$\text{div } \mathbf{F} \equiv F_{j,j} = \varphi_{,jj} \equiv \nabla^2\varphi,$$
and
$$F_{i,j} = \varphi_{,ij}, \qquad F_{j,i} = \varphi_{,ji} = \varphi_{,ij},$$
so that (24.14) can be written as

(24.16) $\quad \nabla^2\tau_{ij} + \dfrac{1}{1+\sigma}\Theta_{,ij} = -\dfrac{\sigma}{1-\sigma}\delta_{ij}\nabla^2\varphi - 2\varphi_{,ij}.$ (p. 78)

If $\mathbf{F}$ is constant, then φ is a linear function. In this case the right-hand member of (24.16) vanishes, and we obtain the equations of Beltrami,

(24.17) $\qquad\qquad \nabla^2\tau_{ij} + \dfrac{1}{1+\sigma}\Theta_{,ij} = 0.$ (p. 78)

From (24.13) it follows that in this case
$$\nabla^2\Theta = 0,$$
so that $\Theta = \tau_{ii}$ is a harmonic function. Equation (22.4) shows that the strain invariant $\vartheta = e_{ii}$ is also harmonic; that is,
$$\nabla^2\vartheta = 0$$
whenever Θ is harmonic. From (24.17) it is seen that if the τ_{ij} are of class $C^{(4)}$, the components of stress satisfy the *biharmonic* equation
$$\nabla^2\nabla^2\tau_{ij} \equiv \nabla^4\tau_{ij} = 0,$$
and since the strain components e_{ij} are linear functions of the τ_{ij}, we have
$$\nabla^4 e_{ij} = 0. \qquad\qquad \text{(p. 79)}$$

Dynamical equations for an isotropic elastic solid are

(25.2) $\qquad\qquad \mu\nabla^2 u_i + (\lambda + \mu)\dfrac{\partial \vartheta}{\partial x_i} + F_i = \rho \ddot{u}_i.$ (p. 82)

If the strain-energy density function is denoted by W, then

(26.7) $\begin{cases} W = \tfrac{1}{2}c_{ij}e_i e_j = \tfrac{1}{2}\tau_i e_i, & (i, j = 1, 2, \cdots, 6), \\ \quad = \tfrac{1}{2}\tau_{ij}e_{ij}, & (i, j = 1, 2, 3), \\ \quad = \tfrac{1}{2}(\tau_{11}e_{11} + \tau_{22}e_{22} + \tau_{33}e_{33} + 2\tau_{23}e_{23} + 2\tau_{31}e_{31} \\ \qquad\qquad\qquad\qquad\qquad\qquad\qquad\qquad + 2\tau_{12}e_{12}), \\ \tau_i = \dfrac{\partial W}{\partial e_i} = c_{ij}e_j, & (i, j = 1, 2, \cdots, 6), \\ c_{ij} = c_{ji}, & (i, j = 1, 2, 3), \quad \text{(p. 88)} \end{cases}$

(26.9) $\quad W = \tfrac{1}{2}\lambda(e_{11} + e_{22} + e_{33})^2$
$\qquad\qquad + \mu(e_{11}^2 + e_{22}^2 + e_{33}^2 + 2e_{23}^2 + 2e_{31}^2 + 2e_{12}^2),\quad$ (p. 89)

and

(26.10) $\quad W = \dfrac{-\sigma}{2E}(\tau_{11} + \tau_{22} + \tau_{33})^2 + \dfrac{1+\sigma}{2E}(\tau_{11}^2 + \tau_{22}^2 + \tau_{33}^2)$
$\qquad\qquad + \dfrac{1+\sigma}{E}(\tau_{12}^2 + \tau_{23}^2 + \tau_{31}^2).\quad$ (p. 89)

If a body is in equilibrium, then

(26.12) $\qquad \displaystyle\int_\tau F_i u_i\, d\tau + \int_\sigma \overset{\nu}{T}_i u_i\, d\sigma = 2\int_\tau W\, d\tau.\qquad$ (p. 90)

CHAPTER IV

EXTENSION, TORSION, AND FLEXURE OF HOMOGENEOUS BEAMS

PROBLEMS OF SAINT-VENANT

The complete problem of equilibrium of an elastic beam can be formulated in the following way. Determine the components of stress τ_{ij} and the displacements u_i that, in the region τ occupied by the beam, satisfy the systems of equations

(29.1) $\quad\begin{cases} \dfrac{\partial \tau_{xx}}{\partial x} + \dfrac{\partial \tau_{xy}}{\partial y} + \dfrac{\partial \tau_{xz}}{\partial z} = -F_x, \\[4pt] \dfrac{\partial \tau_{yx}}{\partial x} + \dfrac{\partial \tau_{yy}}{\partial y} + \dfrac{\partial \tau_{yz}}{\partial z} = -F_y, \\[4pt] \dfrac{\partial \tau_{zx}}{\partial x} + \dfrac{\partial \tau_{zy}}{\partial y} + \dfrac{\partial \tau_{zz}}{\partial z} = -F_z, \end{cases}\quad$ (p. 98)

(29.2) $\quad\begin{cases} \dfrac{\partial u}{\partial x} = \dfrac{1}{E}[\tau_{xx} - \sigma(\tau_{yy} + \tau_{zz})], \\[4pt] \dfrac{\partial v}{\partial y} = \dfrac{1}{E}[\tau_{yy} - \sigma(\tau_{zz} + \tau_{xx})], \\[4pt] \dfrac{\partial w}{\partial z} = \dfrac{1}{E}[\tau_{zz} - \sigma(\tau_{xx} + \tau_{yy})], \\[4pt] \dfrac{\partial v}{\partial x} + \dfrac{\partial u}{\partial y} = \dfrac{2(1+\sigma)}{E}\tau_{xy}, \\[4pt] \dfrac{\partial w}{\partial y} + \dfrac{\partial v}{\partial z} = \dfrac{2(1+\sigma)}{E}\tau_{yz}, \\[4pt] \dfrac{\partial u}{\partial z} + \dfrac{\partial w}{\partial x} = \dfrac{2(1+\sigma)}{E}\tau_{zx}, \end{cases}\quad$ (p. 98) and the boundary conditions

(29.3) $\quad \tau_{zx}, \tau_{zy}, \tau_{zz},\quad$ prescribed functions of x and y on the bases $z = 0$, $z = l$, (p. 98)

(29.4) $\quad\begin{cases} \tau_{xx}\nu_x + \tau_{xy}\nu_y = 0, \\ \tau_{yx}\nu_x + \tau_{yy}\nu_y = 0, \\ \tau_{zx}\nu_x + \tau_{zy}\nu_y = 0, \end{cases}$ on the lateral surface of the cylinder. (p. 98)

The functions τ_{ij}, naturally, must satisfy the Beltrami-Michell compatibility equations (24.15).

EXTENSION OF BEAMS BY LONGITUDINAL FORCES

If we assume

$$\tau_{zz} = p, \qquad \tau_{xx} = \tau_{yy} = \tau_{xy} = \tau_{yz} = \tau_{zx} = 0,$$

throughout the cylinder, then

$$u = -\frac{\sigma p}{E}x, \qquad v = -\frac{\sigma p}{E}y, \qquad w = \frac{p}{E}z. \quad \text{(p. 102)}$$

If the beam is extended by its own weight, then

$$\tau_{zz} = \rho g z, \qquad \tau_{xx} = \tau_{yy} = \tau_{xy} = \tau_{yz} = \tau_{zx} = 0, \quad \text{(p. 104)}$$

and

(31.5) $\quad u = -\dfrac{\sigma \rho g}{E} zx, \qquad v = -\dfrac{\sigma \rho g}{E} zy,$

$$w = \frac{\rho g}{2E}(z^2 + \sigma x^2 + \sigma y^2 - l^2). \quad \text{(p. 106)}$$

BENDING OF BEAMS BY TERMINAL COUPLE M

The stresses and displacements are given by

(32.2) $\quad \tau_{zz} = -\dfrac{M}{I}x,$

$$\tau_{xx} = \tau_{yy} = \tau_{xy} = \tau_{yz} = \tau_{zx} = 0, \quad \text{(p. 112)}$$

(32.10) $\quad \begin{cases} u = \dfrac{M}{2EI}(z^2 + \sigma x^2 - \sigma y^2), \\ v = \dfrac{M}{EI}\sigma xy, \\ w = -\dfrac{M}{EI}xz. \end{cases}$ (p. 114)

The curvature of a beam is related to the bending moment M by the Bernoulli-Euler law:

(32.1) $$R = \frac{EI}{M}.$$ (p. 110)

TORSION OF CYLINDERS

If the shaft is circular and is twisted so that α is the twist per unit length of the shaft, then

(33.2) $\tau_{zy} = \mu\alpha x, \quad \tau_{zx} = -\mu\alpha y,$
$$\tau_{xx} = \tau_{yy} = \tau_{zz} = \tau_{xy} = 0, \quad \text{(p. 120)}$$

and

(33.1) $$u = -\alpha zy, \quad v = \alpha zx, \quad w = 0.$$ (p. 120)

If the shaft is cylindrical and its cross section is bounded by a curve C, then

(34.3) $$u = -\alpha zy, \quad v = \alpha zx, \quad w = \alpha\varphi(x, y),$$ (p. 122)

(34.4) $$\begin{cases} \tau_{yz} = \mu\alpha\left(\frac{\partial\varphi}{\partial y} + x\right), \quad \tau_{zx} = \mu\alpha\left(\frac{\partial\varphi}{\partial x} - y\right), \\ \tau_{xy} = \tau_{xx} = \tau_{yy} = \tau_{zz} = 0, \end{cases}$$ (p. 122)

where $\varphi(x, y)$ satisfies the equation

(34.5) $$\nabla^2\varphi \equiv \frac{\partial^2\varphi}{\partial x^2} + \frac{\partial^2\varphi}{\partial y^2} = 0$$ (p. 122)

throughout the cross section of the cylinder, and

(34.6) $$\frac{d\varphi}{d\nu} = y\cos(x, \nu) - x\cos(y, \nu) \quad \text{on } C.$$ (p. 123)

The torsional rigidity D is given by

(34.10) $$D = \mu \int\int_R \left(x^2 + y^2 + x\frac{\partial\varphi}{\partial y} - y\frac{\partial\varphi}{\partial x}\right) dx\, dy,$$ (p. 125)

and the twisting moment M is

(34.11) $$M = D\alpha.$$ (p. 125)

The conjugate harmonic torsion function ψ is related to the torsion function $\varphi(x, y)$ by the Cauchy-Riemann equations, namely,

(35.1) $$\frac{\partial \varphi}{\partial x} = \frac{\partial \psi}{\partial y}, \quad \frac{\partial \varphi}{\partial y} = -\frac{\partial \psi}{\partial x},$$ (p. 127)

and satisfies the boundary condition

(35.3) $$\psi = \tfrac{1}{2}(x^2 + y^2) + \text{const.} \quad \text{on } C.$$ (p. 128)

The stress function Ψ is defined by

(35.5) $$\Psi = \psi(x, y) - \tfrac{1}{2}(x^2 + y^2).$$ (p. 129)

It satisfies the Poisson equation

(35.7) $$\nabla^2 \Psi \equiv \frac{\partial^2 \Psi}{\partial x^2} + \frac{\partial^2 \Psi}{\partial y^2} = -2 \quad \text{in } R,$$ (p. 129)

and on the boundary C of the region R assumes the value

$$\Psi = \text{const.}$$

The stresses can be calculated from

(35.6) $$\tau_{zx} = \mu\alpha \frac{\partial \Psi}{\partial y}, \quad \tau_{zy} = -\mu\alpha \frac{\partial \Psi}{\partial x}.$$ (p. 129)

The torsional rigidity D can be computed from

(35.10) $$D = 2\mu \iint_R \Psi \, dx \, dy.$$ (p. 132)

(*For particular cross sections see Secs. 36–38 and Secs. 45–47.*)

COMPLEX FORM OF FOURIER SERIES

(39.3) $$f(\theta) = c_0 + \sum_{k=1}^{\infty} c_k e^{ik\theta} + \sum_{k=1}^{\infty} c_{-k} e^{-ik\theta}$$

$$= \sum_{k=-\infty}^{\infty} c_k e^{ik\theta},$$ (p. 152)

where

(39.4) $$c_n = \frac{1}{2\pi} \int_0^{2\pi} f(t) e^{-int} \, dt, \quad (n = 0, \pm 1, \pm 2, \cdots).$$ (p. 152)

THE FORMULA OF SCHWARZ

(42.6) $$F(\zeta) = \frac{1}{2\pi i} \int_\gamma f(\theta) \frac{\sigma + \zeta}{\sigma - \zeta} \frac{d\sigma}{\sigma} + ib_0,$$ (p. 165) and the integral of Poisson is given by

$$(42.7) \quad \Re F(\zeta) \equiv u(\xi, \eta) = \frac{1}{2\pi} \int_0^{2\pi} \frac{(1 - \rho^2) f(\theta) \, d\theta}{1 - 2\rho \cos(\theta - \psi) + \rho^2}.$$
(p. 165)

THE GENERAL SOLUTION OF TORSION PROBLEM

The general solution of the torsion problem can be obtained by evaluating

$$(44.5) \quad f(\zeta) = \varphi + i\psi = \frac{1}{2\pi} \int_\gamma \frac{\omega(\sigma) \bar{\omega}(1/\sigma)}{\sigma - \zeta} \, d\sigma,$$
(p. 171)

where $\mathfrak{z} = \omega(\zeta)$ is a function that maps the region R on a unit circle γ.

CURVILINEAR COORDINATES

The expressions for the components of strain in general orthogonal curvilinear coordinates α_i are

$$(48.7) \quad e_{ii} = \frac{\partial}{\partial \alpha_i}\left(\frac{u_i}{\sqrt{g_{ii}}}\right) + \frac{1}{2g_{ii}} \sum_{k=1}^{3} \frac{\partial g_{ii}}{\partial \alpha_k} \frac{u_k}{\sqrt{g_{kk}}},$$
(p. 199)

$$(48.9) \quad e_{ij} = \frac{1}{2\sqrt{g_{ii}g_{jj}}}\left[g_{ii}\frac{\partial}{\partial \alpha_j}\left(\frac{u_i}{\sqrt{g_{ii}}}\right) + g_{jj}\frac{\partial}{\partial \alpha_i}\left(\frac{u_j}{\sqrt{g_{jj}}}\right)\right], \text{ if } i \neq j.$$
(p. 200)

The stress-strain relations are

$$(48.10) \quad \begin{cases} \tau_{ii} = \lambda \vartheta + 2\mu e_{ii} \quad \text{or} \quad \tau_{ii} = \dfrac{E\sigma}{(1+\sigma)(1-2\sigma)} \vartheta + \dfrac{E}{1+\sigma} e_{ii} \\ \tau_{ij} = 2\mu e_{ij} \quad \text{or} \quad \tau_{ij} = \dfrac{E}{1+\sigma} e_{ij}, \quad \text{if } i \neq j \end{cases}$$

where the invariant $\vartheta \equiv e_{11} + e_{22} + e_{33}$. Solving the system (48.10) for the components of strain yields

$$(48.11) \quad e_{ij} = \frac{1+\sigma}{E} \tau_{ij} - \frac{\sigma}{E} \delta_{ij} \Theta,$$
(p. 200)

where the invariant $\Theta = \tau_{11} + \tau_{22} + \tau_{33}$.

The equations of equilibrium are

$$(48.12) \quad \frac{\partial(g\tau_{ii})}{\partial \alpha_i} - \frac{1}{2}\sum_{j=1}^{3} \frac{g\tau_{jj}}{g_{jj}} \frac{\partial g_{jj}}{\partial \alpha_i} + \sum_{j \neq i} \frac{\partial}{\partial \alpha_j}\left(\frac{gg_{ii}\tau_{ij}}{\sqrt{g_{ii}g_{jj}}}\right)$$
$$+ F_i g \sqrt{g_{ii}} = 0, \quad (i = 1, 2, 3), \quad \text{(p. 201)}$$

where $g \equiv \sqrt{g_{11}g_{22}g_{33}}$, and the F_i are the components, in the directions of the coordinate axes, of the body force $\mathbf{F}$.

Inserting in Eqs. (48.7)–(48.12), the proper values of metric coefficients g_{ij} one gets the basic equations in

a. Polar Coordinates

$$(48.13) \quad \begin{cases} e_{rr} = \dfrac{\partial u_r}{\partial r}, \\ e_{\theta\theta} = \dfrac{1}{r}\dfrac{\partial u_\theta}{\partial \theta} + \dfrac{u_r}{r}, \\ e_{r\theta} = \dfrac{1}{2r}\left(\dfrac{\partial u_r}{\partial \theta} - u_\theta + r\dfrac{\partial u_\theta}{\partial r}\right), \end{cases}$$

while the equations of equilibrium (48.12) become

$$(48.14) \quad \begin{cases} \dfrac{\partial \tau_{rr}}{\partial r} + \dfrac{1}{r}\dfrac{\partial \tau_{r\theta}}{\partial \theta} + \dfrac{\tau_{rr} - \tau_{\theta\theta}}{r} + F_r = 0, \\ \dfrac{\partial \tau_{r\theta}}{\partial r} + \dfrac{1}{r}\dfrac{\partial \tau_{\theta\theta}}{\partial \theta} + \dfrac{2}{r}\tau_{r\theta} + F_\theta = 0. \end{cases}$$

b. Cylindrical Coordinates

$$(48.15) \quad \begin{cases} e_{rr} = \dfrac{\partial u_r}{\partial r}, \\ e_{\theta\theta} = \dfrac{1}{r}\dfrac{\partial u_\theta}{\partial \theta} + \dfrac{u_r}{r}, \\ e_{zz} = \dfrac{\partial u_z}{\partial z}, \\ e_{r\theta} = \dfrac{1}{2}\left(\dfrac{1}{r}\dfrac{\partial u_r}{\partial \theta} + \dfrac{\partial u_\theta}{\partial r} - \dfrac{u_\theta}{r}\right), \\ e_{rz} = \dfrac{1}{2}\left(\dfrac{\partial u_z}{\partial r} + \dfrac{\partial u_r}{\partial z}\right), \\ e_{\theta z} = \dfrac{1}{2}\left(\dfrac{\partial u_\theta}{\partial z} + \dfrac{1}{r}\dfrac{\partial u_z}{\partial \theta}\right), \end{cases} \quad \text{(p. 202)}$$

and the equations of equilibrium

(48.16)
$$\begin{cases} \dfrac{\partial \tau_{rr}}{\partial r} + \dfrac{1}{r}\dfrac{\partial \tau_{r\theta}}{\partial \theta} + \dfrac{\partial \tau_{rz}}{\partial z} + \dfrac{\tau_{rr} - \tau_{\theta\theta}}{r} + F_r = 0, \\ \dfrac{\partial \tau_{r\theta}}{\partial r} + \dfrac{1}{r}\dfrac{\partial \tau_{\theta\theta}}{\partial \theta} + \dfrac{\partial \tau_{\theta z}}{\partial z} + \dfrac{2}{r}\tau_{r\theta} + F_\theta = 0, \\ \dfrac{\partial \tau_{rz}}{\partial r} + \dfrac{1}{r}\dfrac{\partial \tau_{\theta z}}{\partial \theta} + \dfrac{\partial \tau_{zz}}{\partial z} + \dfrac{1}{r}\tau_{rz} + F_z = 0. \end{cases}$$
(p. 202)

c. Spherical Coordinates

(48.17)
$$\begin{cases} e_{rr} = \dfrac{\partial u_r}{\partial r}, \\ e_{\theta\theta} = \dfrac{1}{r}\dfrac{\partial u_\theta}{\partial \theta} + \dfrac{u_r}{r}, \\ e_{\alpha\alpha} = \dfrac{1}{r \sin\theta}\dfrac{\partial u_\alpha}{\partial \alpha} + \dfrac{u_r}{r} + u_\theta \dfrac{\cot\theta}{r} \\ e_{r\alpha} = \dfrac{1}{2}\left(\dfrac{1}{r\sin\theta}\dfrac{\partial u_r}{\partial \alpha} - \dfrac{u_\alpha}{r} + \dfrac{\partial u_\alpha}{\partial r}\right), \\ e_{r\theta} = \dfrac{1}{2}\left(\dfrac{1}{r}\dfrac{\partial u_r}{\partial \theta} - \dfrac{u_\theta}{r} + \dfrac{\partial u_\theta}{\partial r}\right), \\ e_{\alpha\theta} = \dfrac{1}{2}\left(\dfrac{1}{r}\dfrac{\partial u_\alpha}{\partial \theta} - \dfrac{u_\alpha \cot\theta}{r} + \dfrac{1}{r\sin\theta}\dfrac{\partial u_\theta}{\partial \alpha}\right), \end{cases}$$
(p. 203)

and the equations of equilibrium are

(48.18)
$$\begin{cases} \dfrac{\partial \tau_{rr}}{\partial r} + \dfrac{1}{r\sin\theta}\dfrac{\partial \tau_{r\alpha}}{\partial \alpha} + \dfrac{1}{r}\dfrac{\partial \tau_{r\theta}}{\partial \theta} \\ \qquad + \dfrac{2\tau_{rr} - \tau_{\alpha\alpha} - \tau_{\theta\theta} + \tau_{r\theta}\cot\theta}{r} + F_r = 0, \\ \dfrac{\partial \tau_{r\alpha}}{\partial r} + \dfrac{1}{r\sin\theta}\dfrac{\partial \tau_{\alpha\alpha}}{\partial \alpha} + \dfrac{1}{r}\dfrac{\partial \tau_{\alpha\theta}}{\partial \theta} + \dfrac{3\tau_{r\alpha} + 2\tau_{\alpha\theta}\cot\theta}{r} \\ \qquad\qquad\qquad\qquad\qquad\qquad\qquad\qquad + F_\alpha = 0. \\ \dfrac{\partial \tau_{r\theta}}{\partial r} + \dfrac{1}{r\sin\theta}\dfrac{\partial \tau_{\alpha\theta}}{\partial \alpha} + \dfrac{1}{r}\dfrac{\partial \tau_{\theta\theta}}{\partial \theta} \\ \qquad + \dfrac{3\tau_{r\theta} + (\tau_{\theta\theta} - \tau_{\alpha\alpha})\cot\theta}{r} + F_\theta = 0. \end{cases}$$
(p. 203)

FLEXURE OF BEAMS

Bending by a Load W_x along a Principal Axis

In terms of the harmonic function $\Phi(x, y)$, the stresses can be written as

$$(53.2)\begin{cases} \tau_{xx} = \tau_{xy} = \tau_{yy} = 0, \\ \tau_{zz} = -\dfrac{W_x}{I_y}(l-z)x, \\ \tau_{zx} = \mu\alpha\left(\dfrac{\partial\varphi}{\partial x} - y\right) \\ \qquad - \dfrac{W_x}{2(1+\sigma)I_y}\left[\dfrac{\partial\Phi}{\partial x} + \dfrac{1}{2}\sigma x^2 + \left(1 - \dfrac{1}{2}\sigma\right)y^2\right], \\ \tau_{zy} = \mu\alpha\left(\dfrac{\partial\varphi}{\partial y} + x\right) \\ \qquad - \dfrac{W_x}{2(1+\sigma)I_y}\left[\dfrac{\partial\Phi}{\partial y} + (2+\sigma)xy\right]. \end{cases} \quad \text{(p. 229)}$$

The harmonic function Φ must satisfy the condition

$$(53.3) \quad \dfrac{d\Phi}{d\nu} = -\left[\dfrac{1}{2}\sigma x^2 + \left(1 - \dfrac{1}{2}\sigma\right)y^2\right]\cos(x,\nu) \\ - (2+\sigma)xy\cos(y,\nu) \text{ on } C. \quad \text{(p. 229)}$$

The displacements are given by

$$(54.2)\begin{cases} u = -\alpha yz + \dfrac{W_x}{EI_y}\left[\dfrac{1}{2}\sigma(l-z)(x^2-y^2) - \dfrac{1}{6}z^3 + \dfrac{1}{2}lz^2\right], \\ v = \alpha xz + \dfrac{W_x}{EI_y}\sigma(l-z)xy, \\ w = \alpha\varphi(x,y) \\ \qquad - \dfrac{W_x}{EI_y}\left[\Phi(x,y) + xy^2 + \left(lz - \dfrac{1}{2}z^2\right)x\right]. \end{cases} \quad \text{(p. 231)}$$

If the beam is circular,

$$(55.2)\begin{cases} \tau_{zx} = \dfrac{(3+2\sigma)W}{2\pi a^4(1+\sigma)}\left(a^2 - x^2 - \dfrac{1-2\sigma}{3+2\sigma}y^2\right), \\ \tau_{zy} = -\dfrac{(1+2\sigma)W}{\pi a^4(1+\sigma)}xy, \\ \tau_{zz} = -\dfrac{4W}{\pi a^4}(l-z)x. \end{cases} \quad \text{(p. 235)}$$

(For other cross sections see Secs. 56 to 60.)

APPENDIX

CHAPTER V
VARIATIONAL METHODS

POTENTIAL ENERGY

The potential energy V of the system is defined as

$$(64.4) \qquad V = U - \int_\Sigma \overset{\nu}{T}_i u_i \, d\sigma - \int_\tau F_i u_i \, d\tau, \qquad \text{(p. 280)}$$

where it is understood that the integral $\int_\Sigma \overset{\nu}{T}_i u_i \, d\sigma$ is to be extended only over that portion of the surface Σ on which the surface forces $\overset{\nu}{T}_i$ are prescribed, and

$$U = \int_\tau W \, d\tau, \qquad \text{(p. 86)}$$

where W is the strain-energy density.

Of all displacements satisfying given boundary conditions, those that satisfy the equilibrium equations make the potential energy a minimum.

VARIATIONAL PROBLEMS

The problem of determining the function $u(x, y)$ minimizing the integral

$$(65.1) \qquad I = \int\int_R F\left(x, y, u, \frac{\partial u}{\partial x}, \frac{\partial u}{\partial y}\right) dx \, dy, \qquad \text{(p. 287)}$$

where F is a known function of the arguments $x, y, u, \dfrac{\partial u}{\partial x}$, and $\dfrac{\partial u}{\partial y}$, possessing continuous first and second partial derivatives with respect to these arguments can be reduced to the solution of Euler's equation,

$$(65.4) \qquad F_u - \frac{\partial F_{u_x}}{\partial x} - \frac{\partial F_{u_y}}{\partial y} = 0. \qquad \text{(p. 288)}$$

The Euler equation for the minimal problem characterized by the integral

$$I = \int\int_R [(\nabla u)^2 + 2ku] \, dx \, dy = \text{minimum} \qquad \text{(p. 289)}$$

is found to be

$$\nabla^2 u = k.$$

If k is set equal to zero, one obtains the Laplace equation

$$\nabla^2 u = 0,$$

whose solution minimizes the integral

(65.5) $$I = \iint_R (\nabla u)^2 \, dx \, dy$$
$$= \iint_R \left[\left(\frac{\partial u}{\partial x}\right)^2 + \left(\frac{\partial u}{\partial y}\right)^2\right] dx \, dy \quad \text{(p. 290)}$$

The Euler equation for the minimal problem defined by

$$I = \int_{x_0}^{x_1} F\left(x, y, \frac{dy}{dx}\right) dx = \text{minimum}$$

is

(65.6) $$F_y - \frac{d}{dx} F_{y'} = 0. \quad \text{(p. 290)}$$

If two independent variables enter into the integrand, as in

$$I = \int_{x_0}^{x_1} F(x, y, z, y', z') \, dx = \text{minimum},$$

then the Euler equation must hold for each variable:

$$F_y - \frac{d}{dx} F_{y'} = 0,$$
$$F_z - \frac{d}{dx} F_{z'} = 0.$$

If the minimal problem involves higher derivatives, as in

$$I = \int_{x_0}^{x_1} F(x, y, y', y'', \cdots, y^{(n)}) \, dx = \text{minimum},$$

then the Euler equation is

(65.7) $$F_y - \frac{d}{dx} F_{y'} + \frac{d^2}{dx^2} F_{y''} - \cdots$$
$$+ (-1)^n \frac{d^n}{dx^n} F_{y^{(n)}} = 0. \quad \text{(p. 290)}$$

The Euler equation for the minimal problem

$$I = \iint_R F(x, y, u, u_x, u_y, u_{xx}, u_{xy}, u_{yy}) \, dx \, dy = \text{minimum}$$

is

(65.8) $$F_u - \frac{\partial}{\partial x} F_{u_x} - \frac{\partial}{\partial y} F_{u_y} + \frac{\partial^2}{\partial x^2} F_{u_{xx}}$$
$$+ 2 \frac{\partial^2}{\partial x \, \partial y} F_{u_{xy}} + \frac{\partial^2}{\partial y^2} F_{u_{yy}} = 0. \quad \text{(p. 290)}$$

APPENDIX

Theorems of Work and Reciprocity

Reciprocal Theorem of Betti and Rayleigh:

(67.2) $\int_\Sigma \overset{\nu}{T}_i u'_i \, d\sigma + \int_\tau F_i u'_i \, d\tau$

$$= \int_\Sigma \overset{\nu}{T}'_i u_i \, d\sigma + \int_\tau F'_i u_i \, d\tau. \quad \text{(p. 297)}$$

(67.3) $\quad \int_\Sigma \overset{\nu}{T}_i u'_i \, d\sigma + \int_\tau F_i u'_i \, d\tau = \int_\tau \tau_{ij} e'_{ij} \, d\tau \quad \text{(p. 298)}$

with

$$\int_\tau \tau_{ij} e'_{ij} \, d\tau = \int_\tau \tau'_{ij} e_{ij} \, d\tau.$$

Equation (67.3) is an alternative form of the Reciprocal Theorem,

Theorem of Castigliano:

(67.7) $\qquad\qquad \dfrac{dU}{dT_i} = u_i. \qquad\qquad$ (p. 301)

Galerkin's Method

The problem of minimizing the integral

$$I = \int\int_R F\left(x, y, u, \frac{\partial u}{\partial x}, \frac{\partial y}{\partial y}\right) dx \, dy$$

leads to the equation

(69.1) $\qquad \int\int_R L(u)\eta(x, y) \, dx \, dy = 0,$

where

(69.2) $\qquad L(u) \equiv F_u - \dfrac{\partial F_{u_x}}{\partial x} - \dfrac{\partial F_{u_y}}{\partial y} = 0$

is the Euler equation associated with the minimal problem and $\eta(x, y)$ is an arbitrary function that vanishes on the boundary C on the region R.

An approximate solution u_N may be obtained by choosing

(69.3) $\qquad u_N(x, y) = \sum_{i=1}^{N} c_i f_i(x, y), \qquad$ (p. 313)

where the functions $f_i(x, y)$ satisfy the boundary conditions imposed on $u(x, y)$. The constants c_i are determined from

(69.4) $\quad \int\int_R L(u_N) f_i \, dx \, dy = 0, \quad (i = 1, 2, \cdots, N).$ (p. 314)

AUTHOR INDEX

A

Arutinyan, N. C., 316
Avazashvili, D. Z., 253

B

Baes, L., 190
Basu, N. M., 328
Bautz, W., 191, 209
Beltrami, E., 78
Bergmann, S., 186
Bernstein, B., 186
Bieberbach, L., 170
Biezeno, C. B., 191, 195, 300, 301, 321, 323
Biot, M. A., 33
Blinchikov, T. N., 204
Brillouin, L., 91, 204

C

Caratheodory, C., 168, 170
Carslaw, H. S., 154
Cesaro, E., 27
Christoffel, E. B., 169
Christopherson, D. G., 344
Churchill, R. V., 154
Coulomb, C. A., 121
Courant, R., 288, 311, 313, 322, 328, 334
Cranz, H., 191

D

Deimel, R. F., 212
Den Hartog, J. P., 190
Deutler, H., 190
Duncan, W. J., 312, 315

E

Engelmann, F., 195
Evans, G. C., 165

F

Forsyth, A. R., 170
Föppl, A., 80, 81, 118, 212
Föppl, L., 190
Fried, B., 337, 342, 344
Friedrichs, K., 328

G

Galerkin, B. G., 149, 150, 314, 323
Geckeler, J. W., 142
Golusin, G., 169
Goodier, J. N., 96, 101
Goursat, E., 163
Grammel, R., 191, 195, 300, 301, 320
Graustein, W. C., 136
Green, S. L., 170
Greenhill, A. G., 182
Griffith, A. A., 187, 262

H

Hatamura, M., 190
Higgins, T. J., 184, 275, 345
Hilbert, D., 288, 313
Hooke, R., 57

J

Jackson, D., 154
Jacobsen, L. S., 191, 209
Janelidze, G., 274

K

Kantorovich, L., 169, 315
Kappus, R., 33
Kellogg, O. D., 126, 129, 131, 133, 165
Kirchhoff, G., 40
Knopp, K., 154
Koch, J. J., 321, 323

Kolossoff, G., 149
Kopf, E., 190
Korn, A., 92
Kryloff, V., 169
Kufarev, P. F., 186

L

Lamé, G., 67
Lechnitzky, S. G., 276
Lecornu, L., 33, 64
Leibenson, L., 139
Lichtenstein, L., 92
Lourie, A. I., 274
Love, A. E. H., 17, 23, 24, 28, 33, 41, 45, 54, 59, 64, 71, 83, 91, 95, 96, 107, 116, 127, 133, 142, 150, 182, 191, 201, 203, 212, 217, 231, 274

M

McConnell, A. J., 203
McGivern, J. G., 190
MacMillan, W. D., 133
MacRobert, J. M., 210
March, H. W., 187, 191, 195, 217
Matuyama, T., 190
Melentieff, P., 169
Michell, J. H., 78
Mooratoff, M., 169
Morris, R. M., 174, 184, 228, 243
Moskovitz, D., 345
Murnaghan, F. D., 33, 89
Muschelišvili, N. I., 97, 160, 171
Mushtari, N. M., 252, 253

N

Navier, C. L. M. H., 121
Neményi, P., 195, 263
Neuber, H., 209

O

Odqvist, F., 203, 204
Ollendorff, F., 170
Osgood, W. F., 136, 163, 170

P

Palerino, D., 276
Panov, D. Y., 33, 275
Pearson, K., 40
Picard, E., 159, 163
Piccard, A., 190
Pippard, A. J. S., 302
Platrier, C. F. F., 276
Pohlhausen, K., 170
Poincaré, H., 64
Pöschl, T., 208, 275, 302
Prandtl, L., 131
Privaloff, I. I., 131

R

Radó, T., 131
Reichenbächer, H., 190
Reissner, E., 43
Ritz, W., 304, 323
Riz, P. M., 33, 274, 275
Rothe, R., 170
Ruchadze, A. K., 275, 276

S

Saint-Venant, B., 27, 95
Salet, G., 191, 209
Schaefer, C., 91, 95
Schwarz, H. A., 169
Seth, B. R., 33
Shepherd, W. M., 243, 252
Shermann, D. L., 92
Shortley, G. H., 337, 342, 344
Smirnov, V. I., 169
Sokolnikoff, E. S. (see I. S.)
Sokolnikoff, I. S., 127, 154, 180, 184, 186, 210
Sonntag, R., 212
Southwell, R. V., 24, 28, 41, 45, 54, 96, 117, 127, 191, 195, 196, 208, 344
Specht, R. D., 184
Stenin, N., 169
Stevenson, A. C., 228, 243
Sunatani, C., 190
Synge, J. L., 273

AUTHOR INDEX

T

Taylor, G. I., 187, 262
Thiel, A., 190
Thirring, H., 204
Thum, A., 191, 209
Timoshenko, S., 17, 23, 24, 28, 41, 54, 71, 83, 91, 95, 96, 107, 111, 116, 118, 127, 142, 150, 187, 189, 191, 195, 196, 208, 212, 242, 261, 262, 272, 309
Titchmarsh, E. C., 163
Trayer, G. W., 187, 191, 195, 217
Trefftz, E., 12, 23, 41, 45, 54, 71, 83, 91, 95, 117, 127, 186, 201, 203, 305, 328

V

Vantorin, V. D., 276
Vekoua, I., 276

Voigt, W., 64, 101
Volterra, V., 27, 204
von Mises, R., 96

W

Walker, M., 170
Watson, G. N., 154, 210
Weber, C., 141
Weber, E., 190
Webster, A. G., 17, 24, 89, 91, 127
Weller, R., 337, 342
Westergaard, H. M., 81
Whittaker, E. T., 154
Williams, D., 296, 302
Wolf, K., 212

Z

Zanaboni, O., 96
Zvolinsky, N. V., 33, 212, 274

SUBJECT INDEX

A

Affine transformation, 2, 5, 18
Analysis, of strain, 1–34
  of stress, 35–36
Anisotropic media, 61, 212, 217, 274, 276

B

Beams, anisotropic, 212, 217, 276
  bent by couples, 106
  bent by distributed loads, 274
  bent by terminal loads, 217, 228
  cantilever, 269, 282, 291, 296, 302, 312, 323
  compound, 276
  curved, 275
  extension, torsion, and flexure of, 97–276
  flexure of, cardioid, 242
    circular, 235, 253
    elliptical, 237, 258
    rectangular, 240, 272
    triangular, 252, 255
  flexural rigidity of, 111
  initially twisted, 274
  technical theory of, 263
  torsion of, anisotropic, 212, 274
    cardioid, 245
    circular, 119
    compound, 276
    conical, 208
    curved, 275
    elliptical, 134, 196
    hollow, 191, 194, 196
    initially twisted, 274
    polygonal, 186, 187, 275, 316
    rectangular, 143, 308, 317, 318, 320, 333, 338
    semicircular, 181

Beams, torsion of, triangular, 139, 143
  torsional rigidity of, 125, 141, 175, 189
  of variable cross section, 275
  (*See also* Extension; Flexure; Torsion)
Beltrami-Michell equations, 78
Bending of beams (*see* Beams; Flexure)
Bending-moment diagram, 268
Bernoulli-Euler law, 110, 115, 233, 264, 274
Betti and Rayleigh Theorem, 297
Biharmonic function, 79
Boundary-value problems in elasticity, 73
  dynamical, 83
  existence of solution of, 92
  mixed, 93
  uniqueness of solution of, 92

C

Cantilever beam, 269, 282, 291, 296, 302, 312, 323
Cartesian tensors, 13
Castigliano's Theorem, 301
Cauchy, integral formula of, 156
  Integral Theorem of, 155
Cauchy's quadric, 11, 46
Cauchy-Riemann equations, 127
Center of flexure, 224, 227
Central line of a beam, 108
  curvature of, 110, 233
Circular pipe, bending of, 253
  torsion of, 196
Collocation method, 317
  application, to torsion problem, 317
  estimates of errors in, 323

Compatibility, equations of, 24, 27, 34, 77
Complete systems of functions, 313
Complex variable theory, 154
Conformal mapping, 165
  applications of, 176
  Riemann's theorem on, 168
Continuity of class $C^{(n)}$, 18
Curvilinear coordinates, 197
  cylindrical, 202
  plane polar, 201
  spherical, 203

D

D'Alembert, Principle of, 82
Deformation, 1
  Eulerian components of, 29
  finite, 28
  homogeneous, 4, 8
  infinitesimal, 5, 18, 28
  Lagrangian components of, 29
  nonhomogeneous, 19
  plane, 22
  pure, 2, 8, 19
  quadric of, 11
Dilatation, 17, 19, 20, 32, 117
  strain energy of, 118
  uniform, 21
Dirichlet, conditions of, 151
  problem of, 128
    for a circle, 133, 163
    method of finite differences for, 335
    Poisson's formula for, 165
    Schwartz's formula for, 165
Displacement, components of, 18
Displacements, continuity of, 27
  neighboring, 281
Divergence, 20
Dynamical equations, 82

E

Elastic coefficients, 60, 65, 67
  symmetry of, 63
Elastic limit, 58
Elastic moduli, 67

Elastic moduli, interpretation of, 68
  relations among, 71, 72
Energy, complementary, 286
  of deformation, 90, 278
  minimum principle of, 278, 313
  potential, 90, 280, 282, 306
  of strain, 83, 87, 284
Equations, of Beltrami-Michell, 78
  of Cauchy-Riemann, 127
  of compatibility, 24, 27, 34, 77
  of elasticity, in curvilinear coordinates, 201
  of equilibrium, 41, 72, 201
Equilibrium equations, 41, 72
  in curvilinear coordinates, 201
Error functions, 314, 316, 323
Eulerian strain components, 30, 63
Euler's equation, 287, 289
Extension, 8, 12, 31
  of beams, 101, 104

F

Finite deformations and strains, 28, 89
Finite differences, application of, to torsion, 338
  method of, 334
  use of, in Poisson's equation, 345
Flexural rigidity, 111
Flexure, of beams, anisotropic, 274
  cardioid, 242
  circular, 235
  compound, 276
  curved, 275
  elliptical, 237, 258
  hollow, 253
  initially twisted, 274
  rectangular, 240, 272
  (*See also* Beams)
  center of, 224, 227, 230, 258
  function of, 221, 225, 229, 231, 263
Forces, body, 35, 73
  surface, 35
Fourier series, 150

SUBJECT INDEX

G

Galerkin's method, 313, 322
   application of, to torsion, 315
   estimates of errors in, 323
   modification of, 315
Gauss' Theorem, 20
Generalized Hooke's law, 59, 88
Generalized Kronecker delta, 16
Gradient, 20
Green-Gauss Theorem, 20
Grooves, effect of, 139, 191

H

Harmonic function, 79
Harnack's Theorem, 160
Hölder's condition, 164
Homogeneous media, 60, 65
Homogeneous strain, 4, 8, 33
Hooke's law, 57
   connection of, with energy, 83
   in curvilinear coordinates, 200
   generalized, 59, 88
   for isotropic bodies, 70, 200
   symmetry of coefficients in, 88

I

Infinitesimal deformation, 5, 18, 28
Influence coefficients, 300, 302
Invariants, 13, 17, 50, 90, 117
   equations satisfied by, 79
Isotropic media, 60, 64, 65

K

Kronecker delta, 2, 16

L

Lagrangian strain components, 29, 30, 63
Lamé's constants, 67
Laurent's series, 157
Least-squares method, 318
   application of, to torsion, 318
Lipschitz condition, 164

Load, point of, 217
   plane of, 232
Local effects, 209, 273

M

Maxwell's influence coefficients, 300
Membrane analogy, 131, 132, 187, 255, 257, 261, 314, 318
Membrane deflection, 188
Membrane function, 262
   connection of, with flexure and torsion, 263
Minimum potential energy, theorem on, 281, 304
   application of, 291, 312
Minimum principle of stress, 286
Moduli of elasticity, 67
   bulk, 70
   compression, 70
   determination of, 111
   Poisson's, 68
   relations among, 71, 72
   rigidity, 69
   shear, 69
   values of, 71
   Young's, 68

N

Navier's equations, 74
Neumann's problem, 123, 163
Neutral plane, 111, 115
Nonhomogeneous deformation, 19
Normal stresses, 52

O

Orthogonal transformation, 12
Orthotropic medium, 62, 64

P

Plane, of bending, 115, 232
   of load, 232
   neutral, 111, 115, 232
Plane deformation, 22
Plane strain, 22

Plane stress, 56
Poisson, integral of, 165
Poisson's ratio, 68
  measurement of, 111
Potential energy, 90, 280
  minimum principle of, 278, 313
  of twisted shaft, 282
Principal strain, 13, 15
  axes of, 67
Principle, of D'Alembert, 82
  of minimum energy, 278, 312
  of minimum stress, 286
  of Saint-Venant, 95
  of superposition, 5, 28, 79
Proportional limit, 58
Pure deformation, 2, 18, 19
Pure shear, 11

Q

Quadric, of deformation, 11
  of stress, 46

R

Rayleigh-Betti Theorem, 297
Rayleigh-Ritz method, 304
  applications of, 305, 308
  estimates of error in, 324
Reciprocal theorems, 297, 302, 303
Relaxation of boundary conditions, 329
Residues, 157
Riemann's theorem, 168
Rigid body motion, 2, 5, 19
  conditions for, 6
Rigidity, flexural, 111
  modulus of, 69
  torsional, 125
Ritz's method, 304
  applications of, 305, 308
  estimates of errors in, 324
Rotation, components of, 7, 19, 20, 21, 23

S

Saint-Venant, principle of, 95, 99, 209
  semi-inverse method of, 100

Schwarz-Christoffel transformation, 169
Schwarz's formula, 163, 165
Section modulus, 268
Shear, 10, 38
  maximum, 52, 54
  modulus of, 69
Shearing, force, 266
  lines of, 130
  strain, 22
  stress, 10, 38, 52, 54
Simply connected regions, 25
Skew symmetry, 6
Statically determinate systems, 270, 273
Strain, analysis of, 1–34
  components of, 20, 29, 30
  description of, 29
  deviations in, 117
  finite, 28, 89
  homogeneous, 4, 8, 33
  infinitesimal, 28
  invariants of, 17
  notations for, 23
  plane, 22
  principal, 13
  quadric of, 11
  shearing, 21, 200
  tensor, 8, 19, 20, 31, 199
Strain energy, 83, 87, 278, 284
  in a bent beam, 118, 292
  of deviation, 117
  of dilatation, 118
  of distortion, 118
  of twisted beam, 282
Stress, actual, 57
  analysis of, 35–56
  concentration of, 189, 191
  deviations in, 117
  examples of, 54
  function in torsion, 127, 129
  invariants of, 45, 50
  maximum, 52, 54
  nominal, 57
  normal, 38, 52, 55
  notations for, 40
  plane, 56
  principal, 49

Stress, quadric of, 46
  shearing, 38, 55
  tensor, 37, 50
    symmetry of, 43
    transformation of, 44
  ultimate, 58
  units for, 49
Stress-strain relations, 57–96
  for isotropic bodies, 70, 72, 191
Subharmonic functions, 130
Superposition, 5, 28, 79

T

Tension, 55
Tensors, alternating, 36
  cartesian, 13
  strain, 8, 31
  stress, 37, 50
  transformation of, 44, 45
Torsion, of anisotropic beams, 274
  of beams with variable cross section, 205, 212, 275
  of cardioid, 176, 245
  of circular beams, 119
    with grooves, 141
  of compound beams, 276
  of conical shafts, 208
  of curved beams, 275
  of cylinders, 121, 305, 312, 328
  of elliptical beams, 134, 196
  of hollow beams, 191, 194, 196
  of initially twisted beams, 274
  of inverse of an ellipse, 183
  of lemniscate, 178
  of nonisotropic beams, 212, 217, 274
  of polygonal beams, 186, 187, 275, 316
  of rectangular beams, 143, 308, 317, 318, 320, 333, 338
  of regions bounded by circular axes, 177, 180, 181
  of semicircular beam, 181
  of triangular beams, 139, 143

Torsion analogies, 131, 132, 187, 191, 195
Torsion problem, solution by, collocation method, 317
  conformal mapping, 170
  electrical analogies, 191
  Galerkin's method, 308, 315
  hydrodynamic analogies, 191
  least-squares method, 318
  membrane analogies, 131, 132, 187, 195
  Ritz's method, 305
  Trefftz's method, 333
  (*See also* Beams)
Torsional rigidity, 125, 141, 175, 189
Trefftz's method, 329
  application of, to torsion problem, 333
Twist, angle of, 119
  local, 220, 257

U

Ultimate stress, 58
Uniqueness of solution, 79, 92

V

Variational methods, 277–345
  of Galerkin, 313
  of Kantorovitch, 315
  of Rayleigh, 304
  of Ritz, 304
  of Trefftz, 329
Virtual work, theorems on, 284, 314

W

Work, rate of doing, 83, 90
  virtual, 284, 314
Work and reciprocity theorems, 297

Y

Yield point, 58
Young's modulus, 68, 111